NRAES–171

Animal Production Systems for Pasture-Based Livestock Production

Edited by:
Edward B. Rayburn, Extension Forage Agronomist
West Virginia University

Written by:

William J. Bamka
Larry E. Chase
John W. Comerford
Darrell L. Emmicʰ
Scott P. Greiner
John B. Hall
Harold W. Harpst
Daniel Kluchinski
Jean-Marie Luginbuhl
Bill R. McKinnon

Lawrence D. Muller
Carl E. Polan
Edward C. Prigge
Frederick D. Provenza
Edward B. Rayburn
William L. Shockey
William P. Shulaw
Jeremy W. Singer
Kenneth E. Turner
Mark L. Wahlberg

Steven P. Washburn

Plant and Life Sciences Publishing (PALS)
34 Plant Science Building • Ithaca, New York 14853

NRAES–171

December 2008

© 2007 by NRAES (Natural Resource, Agriculture,
and Engineering Service). All rights reserved. Inquiries invited.

ISBN: 978-1-933395-04-3

Library of Congress Cataloging-in-Publication Data

Animal production systems for pasture-based livestock production / edited by Edward B. Rayburn ; written by William J. Bamka
.. [et al.].
    p. cm. -- (NRAES ; 171)
  December 2008.
  Includes bibliographical references.
  ISBN 978-1-933395-04-3 (pbk.)
  1. Animal nutrition. 2. Forage. 3. Livestock productivity. I. Rayburn, Edward B. II. Bamka, William J. III. Natural Resource,
Agriculture, and Engineering Service. Cooperative Extension. IV. Series: NRAES (Series) ; 171.
  SF95.A637 2008
  636.08'5--dc22

2007009951

Requests to reprint parts of this publication should be sent to PALS.
In your request, please state which parts of the publication you would like to reprint and
describe how you intend to use the material. Contact PALS if you have any questions.

Plant and Life Sciences Publishing (PALS)
34 Plant Science
Ithaca, New York 14853
Phone: (607) 255-7654 • Fax: (607) 254-8770
E-mail: palspublishing@cornell.edu • Web site: palspublishing.com
Marty Sailus, PALS Director

Reprinted March 2012

Reprinted August 2015

Table of Contents

CHAPTER 2: BASIC ANIMAL NUTRITION *(continued)*

CHAPTER 5: SHEEP NUTRITION AND MANAGEMENT *(continued)*

CHAPTER 6: GOAT NUTRITION AND MANAGEMENT **153**

CHAPTER 7: HORSE NUTRITION AND MANAGEMENT *(continued)*

**CHAPTER 8: PARASITE CONTROL: BASIC BIOLOGY AND
CONTROL STRATEGIES FOR PASTURE-BASED SYSTEMS 209**

About the Authors

CHAPTER 1: ANIMAL ECOLOGY AND FORAGING BEHAVIOR

Darrell L. Emmick
State Grazing Land Management Specialist, U.S. Department of Agriculture, Natural Resources Conservation Service, New York

Frederick D. Provenza
Professor, Department of Wildland Resources
Utah State University

CHAPTER 2: BASIC ANIMAL NUTRITION

Kenneth E. Turner
Research Animal Scientist
U.S. Department of Agriculture, Agricultural Research Service, West Virginia

Harold W. Harpster
Associate Professor of Animal Sciences
Department of Dairy and Animal Science
Pennsylvania State University

William L. Shockey
Extension Agent and Extension Associate Professor
West Virginia University Extension Service

CHAPTER 3: BEEF NUTRITION AND MANAGEMENT

Edward C. Prigge
Professor Emeritus
Department of Animal and Veterinary Sciences
West Virginia University

John B. Hall
Associate Professor
Extension, Beef Nutrition and Reproduction
Department of Animal and Poultry Sciences
Virginia Polytechnic Institute and State University

John W. Comerford
Associate Professor of Dairy and Animal Science
Department of Dairy and Animal Science
Pennsylvania State University

CHAPTER 4: DAIRY NUTRITION AND MANAGEMENT

Lawrence D. Muller
Professor of Dairy Science
Department of Dairy and Animal Science
Pennsylvania State University

Carl E. Polan
Professor Emeritus, Nutrition
Department Emeritus of Dairy Science
Virginia Polytechnic Institute and State University

Steven P. Washburn
Professor and Extension Specialist
Department of Animal Science
North Carolina State University

Larry E. Chase
Professor and Extension Specialist in Dairy Nutrition
Department of Animal Science
Cornell University

CHAPTER 5: SHEEP NUTRITION AND MANAGEMENT

Scott P. Greiner
Associate Professor
Extension Animal Scientist, Beef/Sheep
Department of Animal and Poultry Sciences
Virginia Polytechnic Institute and State University

Mark L. Wahlberg
Associate Professor
Extension, 4-H Livestock
Department of Animal and Poultry Sciences
Virginia Polytechnic Institute and State University

Bill R. McKinnon
Executive Secretary
Virginia Cattlemen's Association

CHAPTER 6: GOAT NUTRITION AND MANAGEMENT

Jean-Marie Luginbuhl
Associate Professor
Crop Science and Animal Science
North Carolina State University

Edward B. Rayburn
Extension Forage Agronomist
West Virginia University

CHAPTER 7: HORSE NUTRITION AND MANAGEMENT

William J. Bamka
Associate Professor and County Agricultural Agent
Department of Agricultural and Resource Management Agents
Rutgers Cooperative Extension

Daniel Kluchinski
County Agent I (Professor) and Chair
Department of Agricultural and Resource Management Agents
Rutgers Cooperative Extension

Jeremy W. Singer
Research Agronomist
Agricultural Land and Watershed Management Research
U.S. Department of Agriculture, Agricultural Research Service, Iowa

CHAPTER 8: PARASITE CONTROL: BASIC BIOLOGY AND CONTROL STRATEGIES FOR PASTURE-BASED SYSTEMS

William P. Shulaw
Extension Veterinarian, Cattle/Sheep
Department of Veterinary Preventive Medicine
Ohio State University Extension

Acknowledgments

The authors wish to thank the following peer reviewers for offering comments to improve the quality and accuracy of the text:

Gary J. Bergmann
Vice President and General Manager
Stonegate Standardbred Farms, Inc.
Glen Gardner, NJ

Jacqueline Bird
Associate Professor of Biology
Biology Department
Northern Michigan University

Chad Broyles
Farm Owner and Manager
Chestnut Springs Farm
Bluefield, West Virginia

Keith A. Bryan
Formerly with Department of Dairy and Animal Sciences
The Pennsylvania State University

Mike Carpenter
Virginia Department of Agriculture

G. L. Monty Chappell†
Extension Professor
University of Kentucky

Corey Childs
Director, Loudoun County Extension Office
Virginia Cooperative Extension

Daryl Clark
Extension Agent, Agriculture and Natural Resources
Ohio State University Extension

E. Ann Clark
Associate Professor
Department of Plant Agriculture
University of Guelph

Ben H. Cooper
Conservation Planner
Maryland Department of Agriculture
Allegany Soil Conservation District

Thomas M. Craig
Professor
Department of Veterinary Pathology
Texas A & M University

Sam Dixon
Dairy Manager
Shelburne Farms

Bill Epperson
Extension Veterinarian
South Dakota State University

John Fike
Assistant Professor
Forage-Livestock Research
Virginia Polytechnic Institute and State University

John Freeborn
Extension Agent (Former)
West Virginia University

Louis Gasbarre
Bovine Functional Genomics Research Leader
U.S. Department of Agriculture, Agricultural Research Service

David L. Greene
Principal Agent Emeritus
University of Maryland, College of Agricultural and Natural Resources

Betsy Greene
Extension Equine Specialist
University of Vermont

George F. W. Haenlein
Professor Emeritus, Department of Animal and Food Science
University of Delaware

Harold Harpster
Associate Professor of Animal Sciences
Department of Dairy and Animal Science
Pennsylvania State University

D. W. Hartman
Extension Agent
Penn State Cooperative Extension

Gary W. Hornbaker
Extension Agent
Virginia Cooperative Extension

John Thomas Johns
Extension Professor, Beef Nutrition and
Management
Department of Animal Sciences
University of Kentucky

Richard Kersbergen
Extension Educator
University of Maine Cooperative Extension

Cleon Kimberling
Professor and State Extension Veterinarian
Veterinary Teaching Hospital, Colorado State
University

R. Clif Little
Assistant Professor
Ohio State University Extension

Larry Lohr
Dairy Farmer
Cold Ridge Farms

Mark Matheny
Farmer
Morgantown, WV

Michael McCormick
Director, Southeast Research Station
Louisiana State University

Rory Miller
Sheep Producer

James Neel
Research Animal Scientist
U.S. Department of Agriculture, Agricultural
Research Service, West Virginia

Amy Ordakowski Burk
Assistant Professor/Horse Extension Specialist
Department of Animal and Avian Sciences
University of Maryland

Paul R. Peterson
Assistant Professor and Extension Agronomist
University of Minnesota

Gail Pratt
Owner/Manager
Saddlebrook Ridge Equestrian Center

Craig Reinemeyer, DVM
President, East Tennessee Clinical Research, Inc.

Susan Schoenian
Area Agent, Sheep and Goats
Maryland Cooperative Extension

Michael E. Scott
Division of Waste Management
North Carolina Department of Environment and
Natural Resources

William L. Shockey
Extension Agent and Extension Associate
Professor
West Virginia University Extension Service

Kathy Soder
Animal Scientist
U.S. Department of Agriculture, Agricultural
Research Service, Pennsylvania

Roger William Stich
Associate Professor, Department of Veterinary
Preventive Medicine
Ohio State University

Richard Swartzentruber
Farmer Representative for Delaware

Joe Tritschler
Extension Animal Scientist, Small Ruminants
Virginia Cooperative Extension

J. Craig Williams
Extension Educator
Penn State Cooperative Extension

David R. Wolfgang
Field Studies Director/Extension Veterinarian
Pennsylvania State University

Charles M. Young
Agricultural Extension Agent
North Carolina State University Cooperative
Extension

† Deceased

CHAPTER 1
Animal Ecology and Foraging Behavior

Darrell L. Emmick and Frederick D. Provenza

INTRODUCTION

To the casual observer, the foraging behavior of livestock may appear to be merely the random meanderings of animals in search of something to eat, a drink of water, or a place to rest. However, on closer inspection, one would see that herbivores have evolved a very sophisticated strategy for survival in a world that is ever changing, tremendously complex, and inherently unpredictable (21).

Unlike livestock that are kept in confinement and fed prepared rations with little choice but to eat what is put in front of them, grazing animals face many challenges in selecting what, where, and when to eat (2, 20). In this chapter we explore some of the adaptations and mechanisms that allow animals to make foraging decisions, cope with change, and survive in an ever changing world. We illustrate how this knowledge can improve the profitability and efficiency of pasture-based livestock production systems.

THE ECOLOGICAL CONTEXT

Ecology is the branch of biology that identifies and studies the mutual relationships among organisms and between organisms and their environment. The interrelationships between grazing animals and the plants they consume are ecological in nature. "The grazing animal is a part of the plant's environment and the plant a part of the animal's. So long as the two live together, the welfare of each is dependent upon the other" (31).

When placed in an ecological context, the "living together" of plants and herbivores provides an excellent example of coevolution—the joint evolution of two populations interacting interdependently in which selection pressures are reciprocal (30). Parasite and host, predator and prey, and pollinator and plant are other commonly observed examples of coevolved relationships. For coevolving species to continue their living arrangement, each must continually adapt (30). Taken to the extremes, in the plant-herbivore dynamic, failure to adapt can result in either plants being eaten to extinction or animals overingesting toxins and dying.

Plants and herbivores have coexisted for millions of years. Although each depends upon the other for continued existence, the relationship is not as amicable as might be imagined. Beyond the facade of tranquility, a contest of strategy and counterstrategy and adaptation and counteradaptation is continually underway. The adaptive changes brought about through this process involve ongoing interactions between the genome and the environment. The process is not unlike the interaction that occurs between two teams involved in an athletic contest (3, 30, 34).

To win an ice hockey game for example,—or to continue the relationship between coevolved species—the teams in the contest must continually adapt and counteradapt to changing conditions. As one team develops a more efficient means of scoring goals, capturing prey, or consuming forages, the team being scored on, preyed on, or consumed must, in turn, develop more efficient ways of scoring goals of their own, or avoid being scored on, preyed on, or consumed. Each team is attempting to exploit the weaknesses of the other to gain an advantage. When an advantage is gained, unless the disadvantaged team adjusts or adapts, it will be at risk of losing the contest. Conversely, if the team scoring the first goal finds that the opposition has effectively countered the goal scoring strategy, to score another goal, they must now adapt or lose the game.

If an herbivore attains a higher level of proficiency in harvesting a plant than the plant has ability to tolerate or defend against, the herbivore wins and the plant loses. Conversely, if a plant develops means to rapidly regrow following herbivory or if a plant develops defenses such as toxins, thorns, or a change in growth form making it more difficult for an herbivore to consume, the plant stays in the game. However, unlike athletic contests in which one team wins and one team loses, in coevolved relationships, the contest generally continues but does so as an ongoing kind of biological arms race (24).

HERBIVORE ADAPTATION AND FORAGING HABITS

Millions of years of coevolutionary pressures have resulted in genetic change. The subsequent emergence of behavioral, morphological, and physiological characteristics has allowed herbivores to develop a diverse array of anatomical and physiological adaptations to exploit various food sources in different environments. Herbivores vary in body size, dentition, mouth size and structure, digestive tract specialization, size of digestive system in relation to body weight, and other anatomical, morphological, and physiological features (5, 28).

As a result, herbivores often are broadly classified by their primary diet choices into *graz-*

ers or bulk and roughage feeders, b*rowsers* or concentrate selectors, or *intermediate* or mixed feeders. Generally, grazers' diets are grass-dominated and contain less than 25% browse (8). North American grazers include cattle, bison, horses, elk, bighorn sheep, mountain goats, and musk oxen (9). Browsers are those animals that select diets containing at least 75% woody plant foliage, shrub and forb stems and leaves, and the fruits of various plants (8). Domestic goats, moose, pronghorn, mule deer, and white-tailed deer are common examples of North American browsers (9). Intermediate feeders are those animals that have the capability to adjust their diets to the available food supply. They can consume both grasses and browse species (8). Examples of North American intermediate feeders include domestic sheep, burros, and caribou (9).

The aforementioned categories are general in nature, and individuals within any species can—as a result of history, necessity, and chance—end up eating any of a variety of different plant species (20). Nonetheless, this broad classification reflects the fact that there are fundamental differences (table 1-1)—anatomically and physiologically—that enable different kinds of herbivores to use different sources of food and to exploit different habitats. These differences extend beyond the choice of what an animal prefers to eat to what an animal can most efficiently harvest and extract nutrients from based on its own specialized anatomical and physiological characteristics (28).

The differences between browsers and grazers simply demonstrate that each is adapted to do something a little different—secure different types of food, digest different kinds of plant materials, and live in different habitats. For example, grazers tend to have wider muzzles than browsers, smaller mouth openings, stiffer lips, and lower incisors of similar size that project forward and to the sides of the mouth in a spatulate fashion (12). These adaptations allow

Table 1-1. A relative comparison of digestive anatomy between grazers and browsers.

Characteristic	Grazers	Browsers
Foregut	Large	Small
	Subdivided	Simple
	Smaller opening between reticulum & omasum	Larger opening between reticulum & omasum
	Sparser, more uneven papillae	Denser, more even papillae
True stomach	Smaller	Larger
Hindgut	Smaller cecum and intestines	Larger cecum and intestines
Salivary glands	Smaller parotid salivary glands	Larger parotid salivary glands
Liver	Smaller	Larger
Mouth	Wider muzzle and incisor row	Smaller muzzle and incisor row
	Lower incisors of similar size	Central incisors broader than outside ones
	Incisors project forward	Incisors more upright
	Smaller mouth opening and stiffer lips	Larger mouth opening with longer lips tongue
Teeth	Higher crowns in some species	Lower crowns in some species

Sources: Based on Hoeck, H. N. 1975. Differential feeding behavior of the sympatric hyrax Procavia johnstoni *and* Heterohyrax brucei. *Oecologia 22: 15–47; Hofman, R. R. 1989. Evolutionary steps of ecophysical adaptation and diversification of ruminants: A comparative view of their digestive system. Oecologia 78: 443–457. 25. Robbins, C. T., D. E. Spalinger, and W. Van Hoven. 1995. Adaptation of ruminants to browse and grass diets: Are anatomical-based browser-grazer interpretations valid? Oecologia 103: 208–213. As presented in Shipley, L. A. 1999. Grazers and browsers: How digestive morphology affects diet selection. pp. 20–27, In: K. L. Launchbaugh, J. C. Mosley, and K. D. Saunders (ed.).* Grazing Behavior of Livestock and Wildlife. *Idaho Forest, Wildlife and Range Experiment Station. Moscow, ID. Used by permission.*

grazers to take very large bites, thus maximizing the harvest rate from grasslands exhibiting a fairly uniform continuous plant cover (12). The down side to these adaptations is that they make it more difficult for grazers to select the most nutritious diet from grasslands that exhibit wide variability in plant species and cover (12). There is also a tendency for grazers to have a larger, subdivided, and more muscular rumen/reticulum with smaller passageways between the reticulum and the omasum than do browsers (28). This adaptation may slow the passage of digesta to the lower tract, thus providing a longer fermentation time, which, in turn, would allow grazers to more completely utilize the cellulose in grasses, enabling them to extract the greatest amount of energy per unit of feed ingested (28).

In contrast, browsers generally have more narrow muzzles than grazers, larger mouth openings extending back towards the jaw, more flexible lips, and lower incisors occurring in a more upright position. Also, the central incisors are broader than those more laterally situated (12). These adaptations allow browsers to be much more selective in their choice of diet, strip leaves more easily from shrubs and forbs, and evade structural defenses on browse, such as thorns (12). The rumen/reticulum in browsers tends to be smaller, simpler, and have a larger opening between the reticulum and omasum as compared with grazers (28). This adaptation allows for a very rapid passage of highly nutritious digesta through the animal and reflects browsers' tendency to eat foods high in concentrates. Considering that most browse species contain a high percentage of lignin, which is indigestible, the fast rate of passage allows the indigestible food particles to quickly pass through the animal, which in turn promotes a higher overall intake (28). To accommodate the highly nutritious forages and to compensate for the low retention time of the digesta, the rumen of browsers tends to have an extensive network of very dense papillae. These papillae enlarge the surface area of the rumen by 22 times, thus allowing for an efficient absorption of volatile fatty acids, even with a high rate of passage (28).

PLANT CHARACTERISTICS THAT INFLUENCE SELECTION AND INTAKE

Herbivores make choices from an array of plants and plant parts that vary in kinds and concentrations of nutrients and toxins (17). They often prefer some plants or parts of plants and limit or avoid intake of others. This selection process is in response to a combination of stimuli (15). The senses of site, smell, touch, taste, and postingestive feedback mechanisms all influence which foods are eaten and which are avoided (1, 13, 17).

Physical Signals

The physical characteristics of plants vary considerably. Each species looks a little different, has a slightly different color, and perhaps a different texture. Species also vary in tensile strength, shear strength, and water content (15). These physical attributes all play important roles in defining what an animal eats.

Although it is generally accepted that most herbivores lack color vision, this does not appear to compromise their ability to discern one plant from another (15). One need only spend time in a pasture watching animals graze to observe them "looking" for specific plants. Some plants are taller, some are shorter, some are darker shades. Others vary by having thorns, hairy leaves or stems, or serrated leaf edges. At any given point in time, some plants may be in the leafy vegetative stage while others may be in the reproductive stage, complete with seed heads. All of these attributes serve as visual clues to herbivores as they select or avoid different plants (15).

Once an herbivore observes a plant and identifies it as potentially acceptable forage, other factors—tensile strength, shear strength, and moisture content—come into play. Although few studies have related these factors to diet selection, it seems likely that the harder it is for an animal to tear or separate leaves and buds away from any particular plant, the less likely the plant will be selected over another, provided the nutrient content is similar. Studies on the amount of water contained in plants have demonstrated no clear influence on selectivity. However, it is hypothesized that if water is not limiting in the diet, herbivores will select plants that are easier to harvest or have higher nutritional characteristics regardless of the water content. However, if water is limiting in the diet, herbivores may select plants that are higher in moisture content until their water requirements are satisfied (15).

Chemical Signals

Not only do plants differ in how they look, they vary also in their chemistry. Plants smell different, taste different, and possess a high degree of variability in nutrient content (15).

Odors or smells have the ability to elicit any number of different behavioral responses. The odor of a skunk puts us on alert. The smell of pine trees on a warm spring day relaxes us. The smell of our favorite food cooking makes us hungry. One of the primary ways in which humans evaluate the foods we eat is to give them the "sniff test." If the food item smells good, we give it the "taste test." If the food tastes good, we eat it. Conversely, if a food does not smell good, we generally don't bother tasting it. This is especially true if the food item being evaluated is unknown or new to us and different from what we have experienced, in which case the food will typically be perceived as smelling bad. The volatile compounds released by plants have similar influences on herbivores.

In a preference trial involving eight cultivars of tall fescue (*Festuca arundinacea*), researchers observed that upon entering the pastures, cattle (*Bos taurus* L.) walked through the various cultivars with their muzzles in the forage canopy, only occasionally taking a bite. However, within one hour, one cultivar in each of the replicates was grazed more heavily than the others (29). The researchers reported that it did not appear necessary for the animals to taste the forage to determine preference. They simply passed their muzzles over the forage canopy and decided what to eat (29). This behavior appears to indicate that the animals used smell to detect and evaluate the various volatile compounds emanating from each of the cultivars, and touch to evaluate the texture of the forage canopy (15).

Such rapid responses to seemingly unfamiliar foods likely occur because animals generalize preferences, based on previous experiences with odors of familiar foods, from familiar to unfamiliar foods (23). This is consistent with observations that low preference tall fescue is made more acceptable to cattle by spraying it with the juice of highly preferred Italian ryegrass *(Lolium multiflorum)* (15, 26, 27). It is also consistent with the observation that the acceptability of Italian ryegrass is decreased when it is sprayed with the juice of tall fescue. These studies demonstrate that aroma plays an important role in determining what an animal will or will not eat, and that animals generalize from familiar to unfamiliar foods based on familiar aromas (33).

Flavor is the combination of taste and odor (15). Whether or not a particular flavor is acceptable rests with the species of animal and the preference of the individual within the species. Studies have also shown that different species exhibit varying abilities to discern specific flavors (4, 15) (see table 1-2, p. 6).

The data in table 1-2 represent the order based on the lowest concentration level of the flavor that the animals can discern. However, as seen in table 1-3 (p. 6), this order changes when a different threshold limit is used.

The data presented in tables 1-2 and 1-3 demonstrate that each kind of animal is unique in response to flavor concentrations. Although some kinds of animals are sensitive to specific flavors at low concentrations, other kinds of animals are sensitive at high concentrations.

Table 1-2. Sensitivity to chemical solutions based on the lowest concentration discriminated.

Flavor	Livestock rank
Sweet	Cattle > Normal goats > Pygmy goats > Sheep
Salty	Cattle > Pygmy goats > Normal goats > Sheep
Sour	Cattle > Pygmy goats = Sheep > Normal goats
Bitter	Pygmy goats = Normal goats > Sheep > Cattle

Source: Based on Goatcher, W. D., and D. C. Church. 1970. Taste responses in ruminants. IV. Reactions of pygmy goats, normal goats, sheep and cattle to acetic acid and quinine hydrochloride. J. Anim. Sci. 31: 373–382. *As presented in Mayland, H. F., and G. E. Shewmaker. 1999. Plant attributes that affect livestock selection and intake. pp. 70–74, In: K. L. Launchbaugh, J. C. Mosley, and K. D. Saunders (ed.).* Grazing Behavior of Livestock and Wildlife. *Idaho Forest, Wildlife and Range Experiment Station. Moscow, ID. Used by permission.*

Table 1-3. Concentration at which solutions are rejected (< 40% total fluid intake).

Flavor	Livestock rank
Sweet	No rejection thresholds were found
Salty	Cattle > Sheep > Normal goats > Pygmy goats
Sour	Cattle > Sheep > Normal goats = Pygmy goats
Bitter	Sheep = Cattle > Normal goats = Pygmy goats

Source: Based on Goatcher, W. D., and D. C. Church. 1970. Taste responses in ruminants. IV. Reactions of pygmy goats, normal goats, sheep and cattle to acetic acid and quinine hydrochloride. J. Anim. Sci. 31: 373–382. *As presented in Mayland, H. F., and G. E. Shewmaker. 1999. Plant attributes that affect livestock selection and intake. pp. 70–74, In: K. L. Launchbaugh, J. C. Mosley, and K. D. Saunders (ed.).* Grazing Behavior of Livestock and Wildlife. *Idaho Forest, Wildlife and Range Experiment Station. Moscow, ID. Used by permission.*

It is important to note that the data in these tables reflect not only responses to flavor, but responses to the postingestive effects of the solutions as well, as discussed below (6).

The nutrient density and toxic properties of forages also play important roles in diet selection and intake. Van Soest suggested that "there are two fundamental aspects of plant survival and evolution relevant to the nutritive quality of forage: storage of nutrients and defense against the environment" (32). Through photosynthetic activity, plants convert light energy from the sun to energy to maintain plant life processes and to provide structural and defensive compounds (31). As a result, plants exist as a complex of chemical compounds, such as carbohydrates, protein, minerals, vitamins, amino acids, fatty acids, and fat, which are used by herbivores as a source of food (30). Conversely, other compounds, such as alkaloids, terpenes, and phenols, synthesized by plants deter or prevent ingestion (32).

The nutritive value of a plant can be expressed as the sum of its positive chemical and physical attributes minus the sum of its negative chemical and physical attributes. What may chemically and physically help prolong the survival of a plant may hinder the survival of the herbivore. The well-being and continued survival of herbivores depend on their ability to evaluate the nutritive value and toxic properties of foods, to select those that generally meet their requirements, and to avoid those that are nutritionally excessive, deficient, or toxic (17).

Each food encountered by an herbivore represents a unique combination of nutrients and toxins. Although animals use the senses of site, smell, and taste to initially discriminate among foods, it is the uniqueness of each food's chemical composition and subsequent effect on a particular animal postingestion that has the final say on whether or not a food is acceptable. Each food has a different impact on the chemical, osmotic, and mechanical receptors within a particular animal (17). Simplistically, if a plant is eaten and shortly afterward, the animal experiences illness, the plant will generally be selected against. Conversely, if plant consumption leads to satiation, the plant will likely be selected. Hence, postingestive feedbacks from the nutrients and toxins contained within foods serve as a primary influence in diet selection (17).

ORIGINS OF DIET SELECTION

It has been long thought that diet selection is a simple matter of herbivores eating what they like and avoiding what they don't like (14). However, research suggests that there is much more to the process than this naïve view would indicate (18). The real challenge is understanding how herbivores know what and what not to eat. Is it something they are born with? Is it something they learn along the way? Or is it both?

Nature Versus Nurture

Much has been written concerning how animals select their diets. One school of thought suggests that animals are born knowing what is good for them and what is not, and thus possess some genetically innate knowledge about food sources (i.e., *nature* dominates). Another school of thought suggests that animals learn from their mothers, peers, and through their own trial and error experiences, thus suggesting a learned behavior through social interaction and trial and error (i.e., *nurture* dominates) (14). However, neither explanation is entirely accurate given the interconnectedness of both learned and inherited behaviors (14) and observations that animals can turn on to or away from foods for reasons not explained by either nature or nurture alone. Perhaps the most plausible explanation is a basis partly on genetics (morphological and physiological adaptation), partly on learned behaviors (social interaction), and partly on the postingestive consequences of ingesting various foods relative to the requirements of the animal.

Behavior is a function of its consequences: positive consequences increase, and aversive consequences decrease, the likelihood of a behavior recurring. Because animals satiate, behavior-consequence relationships are dynamic and transitory. *Consequences* are a function of how animals process sensory information neurologically, morphologically, and physiologically: the nervous system integrates morphological and physiological environments with social and physical environments. The *genome* contains information with the potential to develop in various ways—neurologically, morphologically, physiologically—depending on context. *Context* reflects cellular and abiotic/biotic environments: although experiences occurring on each of these levels during development in utero and early life are critical in neurological, morphological, and physiological growth and development, genome-environment interactions continue through life. The temporal scales of behavior-consequence-genome relationships vary. Although behavior and consequences interact with the genome on a short-term basis, the genome itself typically changes over a longer time period. Changes in context, for example, the availability of alternative foods in confinement, on pastures, or on rangelands, alter the expression of the behavior-consequence-genome relationship. Because contexts change continually as systems evolve, so too do behavior-consequence-genome relationships (35).

Palatability, Preference, and Postingestive Consequences

Palatability can be defined as the relative attractiveness of plants to animals as feed. *Preference* refers to the selection of these plants by animals (2). In the plant-herbivore dynamic, palatability and preference are considered distinct by definition but linked through functional association. Although most of the common definitions for palatability and preference relate intake to a food's flavor, chemical composition, or physical characteristics, none of them integrate these factors (17, 19, 23).

Palatability may best be described as the relationship between a food's recognizable flavor—distinctive taste, texture, and odor—and its postingestive effects, which are the result of nutrients and toxin concentrations in foods. Palatability depends on the nutritional state of a particular animal at a particular time and place (17). Nutrient requirements vary with the age and physiological condition of an animal and with environmental factors (17). Postingestive feedbacks calibrate the senses—palatability or liking for the flavor of a food—in accord with a food's utility to the body (17). Ultimately, it is the collective interaction between palatability and preference that determines what, when, and how much an animal will actually eat (2).

Palatability and preference exist as codependent variables. For either to have relevance, each must be considered in light of the other. For example,

> "Palatability and preference interact simultaneously along the lines of a continuum in a functional relationship that can be described in much the same manner as the phrase *beauty is in the eye of the beholder*. As the eye (preference) of the beholder undergoes change, so does the beholder's perception of beauty (palatability). Conversely, as that which is perceived as beauty undergoes change, so must the eye of the beholder continually re-define its perception of beauty" (2).

Because plants are alive and actively growing, they are continually changing in nutrient and toxin concentrations, how they taste, and how they look over the course of a growing season, between seasons, and across landscapes. Livestock nutritional requirements also vary considerably over time. As a result of these continuing transformations, the palatability of any given plant, at any given point in time or space, can range from high to low, and thus be preferred or not (2).

For example, despite minor differences in their overall quality, most grasses, when in the leafy vegetative stage, are nutritious, palatable, and preferred by cattle. The same grasses, when allowed to grow to the reproductive stage, become increasingly less nutritious and increasingly unpalatable, and as a result, are much less preferred.

A similar phenomenon can be observed with plants growing in different environments. A plant species grown under marginal environmental conditions—less than adequate moisture, fertility, pH, or temperature—will generally be less preferred than the same plant grown under ideal conditions. Although plants may be highly nutritious, palatable, and thus preferred in one location, they may be extremely unpalatable and avoided in another. The "packaging" of the plants is so dissimilar that to the herbivore, they are essentially two different foods, one preferred and one not (2).

Another situation that can occur to change the palatability and preference rankings of a particular plant is, fundamentally, the reverse of the previous example. The packaging, instead of being dissimilar, is so similar that it becomes monotonous. The same food eaten bite after bite, day after day, and week after week tends to become less palatable and thus less preferred (19). This phenomenon, known as a conditioned taste aversion, causes transient aversions to foods eaten too often or in excessive amounts.

Feedback Mechanisms and Behavior

Grazing and browsing animals have a remarkable ability to select diets that are higher in nutrients and lower in toxins than the average available in the foraging environment. This process is not just the "luck of the bite," but rather, the functioning of two interrelated systems: affective and cognitive (11). These two systems, working together, are expressed as a deliberate course of action (behavior) by the animal (2).

The affective or involuntary system represents a subconscious connection between an animal's brain and gut that links the taste of a particular food with its unique postingestive feedback (11, 17, 23). When a food is ingested, its nutritional and toxicological properties are sensed by an animal's chemical, osmotic, and mechanical receptors, and this information feeds back to the brain. Although this process functions with no conscious effort on the part of the animal, it represents a fundamental means by which foods are evaluated, and based on the outcome, preferences and intake adjustments are made (11, 17, 23). For example, if an animal consumes a food and becomes ill, the taste of the food item will become aversive and the animal will subsequently avoid the food. The strength of the aversion will depend on the severity of the discomfort experienced. Conversely, if a food is ingested and an animal becomes satiated, the food will be selected (11, 17, 23).

The other system at work is the cognitive or voluntary system. This system integrates the senses of taste, sight, and smell with information received from mother, other members of the herd or flock, and trial and error experiences (previous postingestive consequences) to allow animals to differentiate between and make conscious choices concerning what food items to select or avoid (11). In other words, if an animal encounters a familiar food item, past experiences will determine whether or not the food is consumed again. If an animal has previously consumed a plant and experienced illness, the taste of the particular plant will become objectionable and the animal will use its senses of sight and smell to avoid consuming the plant again (22). The converse of this would occur if the initial encounter were positive.

The amount of a familiar food consumed is influenced by the most recent postingestive feedback. Plants and other food items can change in nutrient density and toxicity over time. Just because a plant was palatable and highly preferred at the first encounter does not mean that it will exhibit the same characteristics at the next. If the most recent postingestive feedback is negative, the plant will be less palatable and, as a result, will be less preferred (2).

The affective and cognitive systems function as two separate systems, but they are integrated through the senses of taste, sight, smell, and postingestive feedback (11). The affective system evaluates the postingestive consequences of ingesting particular foods. The cognitive system then modifies the animal's foraging behavior according to whether the postingestive feedback was negative or positive (11, 16, 17, 22, 23). Through this information exchange, herbivores continually monitor the foods they eat and alter their diets in relationship to their own requirements and changes in the foraging environment (11).

Conditioned Taste Responses

Conditioned taste responses operate through the affective system and can be either aversive or positive. If a food tastes good to an animal and the postingestive feedback is positive, that is, it produces no toxic effects and is nutritionally adequate, most likely the animal will form

a preference for this food and will continue to seek it out. This is known as a conditioned taste preference (11, 16, 22). Conversely, if a food item is ingested and the subsequent postingestive feedback is negative, that is, it is nutritionally inadequate or high in toxin content, most likely the animal will form an aversion to this particular food and avoid or limit intake of it. This is known as a conditioned taste aversion (11, 16, 22). It is believed that herbivores evolved conditioned taste aversions as a survival mechanism to avoid overingesting foods that may taste good but are nutritionally inadequate or are nutritionally adequate but contain toxins (11, 16, 22). Keep in mind that the concept of nutritional inadequacy involves both excessive nutrient concentrations as well as deficits.

Dietary Social Facilitation and the Influence of Mom

Dietary social facilitation is the influence of one animal on the diet selection of another (16, 18, 22). Because most domestic herbivores are social animals, they frequently have the opportunity to watch each other and modify their diet selection based on what their companions are eating (16, 18, 22). An additional and perhaps more important influence is the mother. While still in the womb, flavors of various foods can be passed to the fetus in the amniotic fluid. As a result, animals may be born knowing something about diet selection by experiencing the influence of mom's dietary choices (16, 22). Mom's milk may also pass these same flavors to the nursing young (16, 22). Mom also teaches by example. Researchers and producers have observed many times that if mom eats it and baby watches, baby will generally eat what mom eats (18).

Learning from mom and social interaction with peers, especially early in life, help young herbivores to select nutritious diets and good locations to forage. There is much variability across the landscape in the kinds and amounts of foods available to herbivores, and certainly these foods vary in nutrient as well as toxin concentrations. Learning from mom and peers not only what to eat but where decreases the time spent acquiring foraging skill and reduces the likelihood that young animals will overingest toxic foods (18, 22).

Familiar Versus Unfamiliar Foods and Foraging Locations

The difference between familiar and unfamiliar foods and foraging locations is generally a matter of upbringing. When young animals are raised with their mothers foraging in a particular location, they learn what to eat, when to eat, where certain foods can be found, and where water and shelter are located (22). Familiar foods in known foraging locations increase foraging efficiency and afford animals a sense of comfort and safety. They know where they are going and know what to eat when they get there.

Conversely, when animals are introduced to a new location with unfamiliar or novel foods, they do not have a clue as to what or where to eat, where to find a drink of water, a place to get out of the sun, wind, or rain, or hide from predators. As a result, forcing animals to forage in unfamiliar locations that contain unfamiliar foods elicits fear, discomfort, and stress. Under these conditions, animals generally walk farther distances, spend a greater amount of time foraging but consume less, are more likely to suffer from malnutrition and predation, and ingest a greater amount of toxic plants (22).

Learning efficiency and, thus, survivability is greatly enhanced when young herbivores have the opportunity early in life to forage with their mothers, family members, and other members of their social groups. They learn through social interaction where to locate food and what to consume when they get there, and through

postingestive feedback, how much to consume (11, 17). The nutrient densities and toxic properties of plants can change in a matter of minutes or hours depending on environmental conditions and previous herbivory (11). This applies to familiar plants in familiar locations as well as unfamiliar plants in unfamiliar environments. As a result, animals may frequently change what they are eating. In unfamiliar environments and with unfamiliar plants, the challenges of foraging are vastly more difficult. Thus, "animals prefer familiar foods to novel foods and familiar foraging locations to unfamiliar locations" (21).

MANAGEMENT IMPLICATIONS

The foraging behavior of herbivores represents coevolutionary adaptations that enable animals to seek foraging environments that best meet their needs and to disregard those that do not. As managers, it is important to recognize that animals are not machines. They are living, breathing, feeling, social creatures with likes and dislikes, and they feel pain, stress, and discomfort. They prefer familiar foods to novel foods, they prefer familiar environments to unfamiliar environments, and they prefer to be with companions rather than strangers (21). The more we can accommodate, rather than dictate, their needs, the more contented and productive they will be.

Never Keep Animals Guessing

Animals are creatures of habit, and old habits die hard. Once animals become familiar with a foraging location or particular forage types, they will exert a great deal of effort to stay in a particular area and spend a lot of time searching for familiar food items. Animals that are born and raised on a farm or ranch generally perform better than animals that have been trucked in. Raising your own replacements keeps animals in familiar surroundings, reduces stress, and improves performance.

For an animal to readily consume a particular plant, the animal must recognize the plant as familiar and possessing some desirable attribute. Animals readily seek out known plants containing desirable nutritional qualities and ignore unknown plants or plants with undesirable qualities. They view familiar foods as safe and unfamiliar foods as potentially dangerous (17, 11).

Adult dairy cows put out to graze after a lifetime of confinement feeding are at a distinct disadvantage in learning how to graze efficiently. They have had no social model, that is, mom, to teach them how to graze or what to eat, and they certainly don't view pasture as a familiar habitat. Although most dairy cows can, and do, make the transition from barn to pasture, it is often a long process for both the animal and the producer. In the interim, milk production generally drops and frustration levels rise.

To avoid this situation, it is often best to start by conditioning the animals to the outside environment first. Feed them and put them out to a high quality pasture for a few hours each day. At first, they may not recognize pasture as food and will eat very little. Once they get used to the routine of going outside after milking, and you have observed that they are grazing some, you can start cutting back feed in the barn. Do not be surprised if cows strongly resist cutting back feed. Remember that cows are creatures of habit, and they are conditioned to having their feed provided to them by you. When they have not been fed their customary barn ration, they will often stand at the pasture gate and bellow their indignation rather than increase their grazing time. However, by keeping the cows on high quality pasture and not giving in to their bellowing, in time they will begin to graze more and complain less. The secret is to not give in to their demands to feed them. Giving in and providing them feed only delays their learning to forage for themselves.

Generally, young animals adapt to change more quickly than adults. Thus, when possible, it is best to start grazing heifers at an early age. By the time they are part of the milking herd, they will already recognize pasture as a source of feed and will know how to graze. You can also precondition animals to graze unfamiliar forages by feeding these forages as hay or green chop prior to going out on pasture. Although hay and green chop are not exactly the same as fresh forage in a pasture, animals can generalize based on smell and taste.

Maintaining high within-pasture plant species diversity allows animals to select from a variety of plants and parts of plants diets that most closely complement their particular nutritional requirements. However, having too many pastures seeded to distinctly different combinations of plant species causes animals to spend more time evaluating plants and less time eating. When dissimilar plant complexes, such as perennial ryegrass (*Lolium perenne* L.) and white clover (*Trifolium repens* L.) as compared with reed canarygrass (*Phalaris arundinacea* L.) and birdsfoot trefoil (*Lotus corniculatus* L.), are required to accommodate differences in soil characteristics or to provide a more uniform seasonal distribution of forage, seed enough of each different mixture to allow animals time to familiarize themselves with the new species, evaluate their nutrient and toxic properties, and adjust intakes accordingly. Generally, if dissimilar plant complexes are used, provide enough acres of each type to accommodate 10–14 days of grazing before the livestock are moved to a distinctively different complex (2). Because animals do learn, and remember for long time periods, this is more important early in the transition to pasture and in the first encounter with a new forage species than it is after the animals have become conditioned to grazing or in subsequent encounters with the new forage species.

Avoid monocultures, or single species pastures, if possible. Although animals will generally eat what is available—including monocultures—this does not mean they like it. When provided with a choice, livestock, like people, include a wide variety of food items in their diets. Presumably, this occurs because no single food contains all of the necessary nutrients required throughout an animal's lifetime. Thus, to ensure that animals have the opportunity to balance their diets, make a variety of foods available. Diverse pasture mixes allow each individual animal a greater opportunity to select foods that most closely meet its nutritional requirements over the long term and to maximize dry matter intake on a sustained basis (19). You can use monocultures of cereal forages, brassicas, annual ryegrass, and the like to fill in shortfalls in forage production or to extend the length of the grazing season. However, for the reasons explained above, their use should be minimized rather than maximized.

Never Work Animals Too Hard

The "Law of Least Effort" influences all creatures. In the long term, no animal can afford to expend a greater amount of energy in the acquisition of its food than it will obtain from consuming the food (2). This is readily observed in the natural world in predator-prey relationships. It makes little sense energetically for predators to chase after food items that they stand little chance of catching. Generally, the weak, the young, the old, the injured, and the slow are preyed upon while the strong, the agile, and the fast are usually ignored (2). You can observe this same phenomenon in the foraging activities of grazing animals.

From a behavioral perspective, the feed intake of herbivores equals the product of the time spent grazing or browsing, the rate of biting during grazing or browsing, and the amount of herbage taken in with each bite (10). Anything that managers do that interferes with this process will lower or limit intake and, as a result, decrease or limit animal performance.

To ensure that grazing animals are not working harder than necessary, managers must make certain that there is an ample quality and quantity of forage available. In some instances, pastures may need complete renovation. In other situations, improving grazing management on existing pasture, using higher yielding land, controlling weeds, or liming and fertilizing may be required. Providing 2 acres of low plant density, low-yielding pasture is not an equal substitute for 1 acre of high plant density, high-yielding pasture. When plant densities and yields are low, animals have to work harder and longer. They are forced to cover more ground, spend a greater amount of time foraging, and generally consume less feed (2).

Dairy cows that have to travel long distances down laneways to pasture expend energy in walking and concurrently consume no food. To minimize production losses, keep travel distances to a minimum and maintain laneways for easy travel. This is particularly important for high-producing dairy cows. The farther animals must walk and the longer they are away from feed, the lower the milk production. For dairy cows, travel distances from the barn should be kept to less than 1 mile.

Topographical features of the landscape are also important management considerations. The larger the body size of an animal, the lower will be its ability to negotiate steep slopes or rough terrain. Lactating dairy cows should graze on flat to slightly rolling land. Beef cattle, horses, dairy heifers, and dry dairy cows can readily graze flat to moderately steep land. Sheep, goats, llamas, and deer have the best ability to utilize the steepest land or land with large rocks or rock outcrops. Beef cattle and dry dairy cows can also graze wetter land types.

Always Provide Water

Although the actual water requirements of livestock vary with the kind of stock, weather conditions, and the nature of the forage (31), all animals require clean water. Provide water in ample quality and quantity within 300 feet of lactating dairy cows and within 1,000 feet for all other livestock. The closer water sources are to where animals are foraging, the less disruptive it is to the herd or flock when an animal goes for a drink. When water is supplied to dairy cows in the paddock they are grazing, it is not uncommon to observe one or two cows at a time going to the trough while the remainder of the herd continues to graze.

Maintain Soil Fertility

Maintaining adequate pH and soil nutrients is generally recommended as a means to improve forage yields and to ensure high plant densities. However, there is another reason for concern about soil fertility. The nutritive value and preference of any given plant depends on the environmental conditions under which it grows. As previously mentioned, plants growing in stressed or marginal environments—less than adequate moisture, fertility, pH, or temperature—are less preferred than the same plants grown under ideal conditions. This is the result of variation in the chemical composition of the plants, that is, nutrients and toxic mineral or organic compounds, which is expressed as a change in palatability (32). To reduce this variability, maintain pasture fertility levels based on soil test results. The pH levels should be 6.0–6.5 or slightly higher, because plant nutrient availability and soil microorganism activity are near optimum in this range.

SUMMARY

As coevolved organisms in the plant-herbivore dynamic, plants and animals are locked in a game of adaptation and counteradaptation with the survival of each dependent on the other. Various species of herbivores are adapted a little bit differently through a combination of genetics, social interaction, and postingestive feedback to consume different foods and to function in different ways within the ecosystem. Although these processes and interactions are knowable, they are inherently dynamic and do not necessarily follow any predictable order (21, 23). However, by understanding the dynamics of these processes, we are all in a better position to manage both plants and animals in pasture-based livestock production systems.

CHAPTER 2

Basic Animal Nutrition

Kenneth E. Turner, Harold W. Harpster, and William L. Shockey

INTRODUCTION TO RUMINANT DIGESTION

A major advantage of ruminants over simple-stomached animals, such as pigs, is that ruminants have the ability to convert plant fiber into energy. Ruminants utilize high forage diets by fermenting forage in their rumen. They chew their cud (ruminate) to aid in the extraction of nutrients from plants. The cell walls of plants are composed of fiber (cellulose and hemicellulose) that cannot be broken down by mammalian digestive enzymes. Only the cellulase and hemicellulase enzymes produced by the billions of microorganisms living in the rumen can break down the fiberous fraction of forages. To fully utilize high forage diets, digestion by the ruminant involves their multicompartmented stomach; each compartment has a specialized function. Cattle *(Bos taurus and B. indicus)*, sheep *(Ovis aries)*, and goats *(Capra hircus)* have a four-compartmented stomach that includes (i) a rumen, (ii) a reticulum, (iii) an omasum, and (iv) an abomasum.

The rumen and reticulum, often referred to as the rumino-reticulum or reticulo-rumen, function together as a large fermentation vat containing many billions of specialized microorganisms. The cud-chewing action of ruminants damages plant cell walls and grinds feedstuff, releasing cell contents. Microorganisms in the rumen utilize cell contents or the more readily soluble carbohydrates, proteins, fats, vitamins, and minerals released from the plant cells to further break down the cell walls of forages using specialized cellulase enzymes to release additional carbohydrates, fats, proteins, vitamins, and minerals, as well as to synthesize some nutrients and volatile fatty acids.

Volatile fatty acids are short-chain fatty acids such as acetic, propionic, butyric, valeric, and iso-valeric acids. These fatty acids are absorbed from the rumen and used by the liver and other tissues to support metabolic functions. Propionic acid is used by the liver to synthesize glucose; acetic acid is used to synthesize fat.

The reticulum or honeycomb aids the ruminant in regurgitation and cud chewing. The omasum or many plies acts to remove most of the water from the digesta before it enters into the abomasum or true stomach, where secretion of hydrochloric acid aids in feed/forage breakdown and secretion of pepsin aids in protein breakdown, similar to the monogastric (single compartment) stomach of horses and swine.

Digesta then flows into the small intestine, where the nutrient absorption process is similar to that in monogastrics. At the junction of the small and large intestines, ruminants have a functional caecum, which is a site of secondary fermentation for forage residues prior to moving into the large intestine, where much of the water is removed prior to excretion. Herbivorous monogastrics also have a caecum, which is the main reason horses and swine can utilize forages. The secondary fermentation process in the caecum is not as efficient as that in the rumen, as there is limited time for absorption of nutrients from the large intestine.

Nutritional management, or more importantly the balance of nutrients, on offer to the grazing animal becomes important in optimizing growth, milk, or wool production. The nutritional needs of production livestock are as variable as are species of animals, but the requirement for basic nutrients (energy [from carbohydrates, fats, and protein], minerals, vitamins, and water) is common to all animals.

NUTRITIONAL REQUIREMENTS

Energy

Forages provide many of the nutrients required by livestock, and chief among these is energy. Energy is defined as the potential to do work. In the United States the calorie is the basic unit of measuring energy; it is defined as the amount of heat required to raise the temperature of 1 gram of water 1° C. When it comes to livestock requirements and feed values, the calorie is a very small amount of energy, so we more commonly talk about kilocalories (1 kcal = 1,000 calories) and megacalories (1 Mcal = 1,000 kcal = 1 million calories).

The flow of energy through an animal following ingestion of a feedstuff is outlined in figure 2-1. This figure and the description of terms that follows are adapted from the National Research Council (23). Refer to figure 2-1 as you consider the following terms.

Intake of Food Energy (IE) is the gross energy content from carbohydrates, fats, and protein in the food consumed. IE is the weight of food consumed multiplied by the gross energy of a unit weight of food.

Fecal Energy (FE) is the gross energy in the feces. FE is the weight of feces multiplied by

the gross energy of a unit weight of feces. FE can be partitioned into energy from undigested food (F_iE) and energy from compounds of metabolic origin (F_mE).

Digestible (Apparent) Energy (DE) is energy in food consumed less energy in feces: DE = IE – FE.

Gaseous Products of Digestion (GE) includes combustible gases produced in the digestive tract incident to fermentation of food by microorganisms. Methane makes up the major proportion of combustible gas normally produced in both ruminant and nonruminant species. Hydrogen, carbon monoxide, acetone, ethane, and hydrogen sulfide are produced in trace amounts and can reach high levels under certain dietary conditions. Present knowledge indicates that energy lost as methane in ruminants and nonruminant herbivores is quantitatively the most significant GE loss.

Waste Energy from Urine (UE) is the total gross energy in urine. It includes energy from nonutilized absorbed compounds from the food (U_iE), end products of metabolic processes (U_mE), and end products of endogenous origin (U_eE).

Metabolizable Energy (ME) is the energy in the food less energy lost in feces, urine, and combustible gas: ME = IE – (FE + UE + GE).

Total Heat Production (HE) is the energy lost from an animal system in a form other than as a combustible compound. Heat production may be measured by either direct or indirect calorimetry.

Basal Metabolism (H_eE) reflects the need to sustain the life processes of an animal in the fasting and resting state. This energy is used to maintain vital cellular activity, respiration,

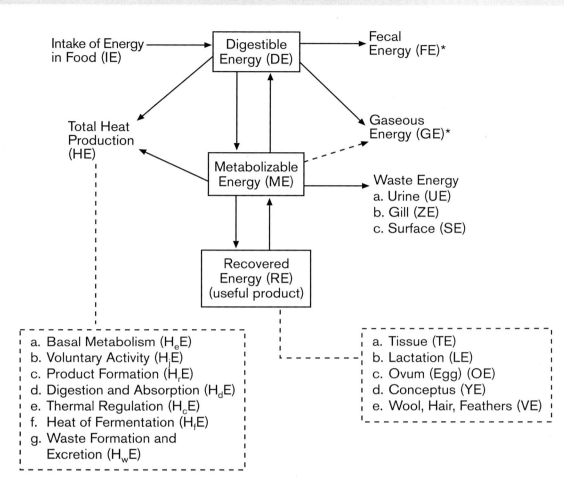

Figure 2-1. The idealized flow of energy through an animal.

*Source: Reprinted with permission from the National Academies Press,
Copyright 1981, National Academy of Sciences.*

and blood circulation and is referred to as the basal metabolic rate (BMR).

Voluntary Activity (H_jE) is the heat production resulting from muscular activity required in, for example, getting up, standing, moving about to obtain food, grazing, drinking, and lying down.

Digestion and Absorption (H_dE) is the heat produced as a result of the action of digestive enzymes on the food within the digestive tract and the heat produced by the digestive tract in moving digesta through the tract as well as in moving absorbed nutrients through the wall of the digestive tract.

Fermentation (H_fE) is the heat produced in the digestive tract as a result of microbial action. In ruminants, H_fE is a major component often included in the heat of digestion (H_dE).

Product Formation (H_rE) is the heat produced in association with the metabolic processes of product formation from absorbed metabolites. In its simplest form, H_rE is the heat produced by a biosynthetic pathway.

Thermal Regulation (H_cE) is the additional heat needed to maintain body temperature when environmental temperature drops below the zone of thermal neutrality, or it is the additional heat produced as the result of an animal's efforts to maintain body temperature when environmental temperature goes above the zone of thermal neutrality.

Waste Formation and Excretion (H_wE) is the additional heat production associated with the synthesis and excretion of waste products.

Recovered Energy (RE), commonly called *energy balance*, is that portion of the feed energy retained as part of the body or voided as a useful product. In animals raised for meat, RE = tissue energy (TE). In a lactating animal, RE is the sum of TE, lactation energy (LE), and energy in products of conception (YE): RE = TE + LE + YE.

Understanding the flow of energy through an animal is useful to the understanding of applied aspects of pasture management. Forage species, maturity, fertility, and environment as well as animal species, productivity level, activity, and countless other factors are all important determinants of the net productivity of animals grazing forage plants.

Caloric Density and Rate and Extent of Digestion

We know that in general forages are bulky—they occupy space in the gut of animals consuming them. Thus the concept of calories per unit weight (such as Mcal per pound) is central to the amount of daily energy a given animal can consume of a given forage. Why is animal performance usually lower on high forage versus high grain diets? The simple answer is that the digestible caloric density of forages is usually much less than that of concentrate.

Related to the concept of forage caloric density are *rate* and *extent* of digestion. In general forage fiber digests rather slowly in the rumen and to a lesser extent than concentrates. It is clear then that "physical fill" limits the productivity of animals consuming high forage diets. This explains much of the reduced animal performance on mature forages with high **neutral detergent fiber** (NDF) and **acid detergent fiber** (ADF) levels. The forages are digested so slowly and to such a limited extent that total daily energy intake and thus animal performance is compromised.

Bottom Line

The emphasis continually placed on "forage quality" should now be obvious. Providing a dense, vegetative, relatively low fiber sward provides the best opportunity for that animal to maximize daily energy intake and performance. Forage quality is often considered in terms of the intake of DE from forages (20).

Protein

Deficiency of protein in diets of growing, mature, and lactating animals can result in poor growth, delayed onset of puberty, loss of weight, impaired fertility, low birthing percentage, and reduced milk production. Protein metabolism is very complex in the ruminant, especially because protein can leave the rumen as ammonia and not through absorption (7).

True proteins in forages and feedstuffs are long chains of amino acids composed of carbon, hydrogen, oxygen, and nitrogen (N); sulfur (S) is also contained in three important amino acids: methionine, cystine, and cysteine. Many of the individual amino acids cannot be synthesized by the ruminants or cannot be synthesized in sufficient quantities either by the animal or microbes to satisfy nutrient demands. These amino acids are referred to as essential amino acids (EAA), meaning that they must be contained in the diet offered to the animal. The EAA requirements have not been quantified sufficiently for the ruminant. Poppi and McLennan (32) demonstrated that methionine, lysine, histidine, arginine, threonine, and cysteine were the six limiting amino acids for growing ruminants grazing cool-season grass and legume pastures. In ruminants, some EAA can be supplied by ruminal microorganisms. Once the microorganisms in the digesta flow into the abomasum, these microbes are broken down by mammalian enzymes to supply amino acids, which are absorbed from the small intestines. Forages and feedstuffs supply both protein and nonprotein nitrogen (NPN). Urea and biuret (NPN sources) can be used by rumen microorganisms to synthesize protein. The ability to use NPN is useful when the diet contains NPN sources or low levels of **crude protein** (CP).

Crude Protein Determination

In determining CP, total percent N in forages and feedstuffs can be determined via the Kjeldahl wet-chemistry procedure. Another faster method uses near infrared reflectance spectroscopy to bounce a light beam off the forage or feed sample and measure the light wavelength reflected by the N molecules in the sample to determine total percent N. Both procedures estimate CP by multiplying total percent N by the factor 6.25, because true protein, on average, contains 16% N. Included in the total N fraction is N contained in amino acids, glycoproteins, and NPN—thus the term "crude" protein. The CP values are used to help classify feeds and forages and are used in ration formulation.

Protein from Forages and Browse

The overall CP concentration in forages depends on the plant species, plant part, stage of maturity, season, and preservation method (26).

Grasses are usually grouped into cool-season or warm-season grasses based on photosynthetic (carbon dioxide fixation) pathways. Cool-season grasses such as orchardgrass (*Dactylis glomerata* L.), tall fescue (*Festuca arundinacea* Schreb.), Kentucky bluegrass (*Poa pratensis* L.), and others use the C3 pathway to fix carbon dioxide. Warm-season grasses such as big bluestem (*Andropogon gerardii* Vitman) and others capture carbon dioxide via the C4 pathway. Cool-season grasses typically yield 70% of their total dry matter (DM) production by June. Warm-season grasses have a different growth pattern and yield 65–75% of their total DM during midsummer. The CP content of cool- and warm-season grasses depends highly upon available soil N levels. Warm-season grasses usually have a higher proportion of stems in herbage. Leaves contain greater CP than stems; thus, overall protein levels may be lower in warm-season grasses (more stems) than cool-season grasses, especially late in the growing season.

Legumes include plants that have the ability to fix atmospheric N into nodules on roots. Legumes such as alfalfa (*Medicago sativa* L.), white clover (*Trifolium repens* L.), red clover (*Trifolium pratense* L.), lespedezas (*Kummerowia* spp. and *Lespedeza* spp.), peanut (*Arachis* spp.), sanfoin (*Onobrychis* spp.), trefoil (*Lotus* spp.), and others have higher CP than cool- and warm-season grasses.

Nontraditional plants for forages include forbes and browse. Forbs are herbaceous broadleaf plants (weeds), and browse is plants other than grasses with woody stems (shrubs, briars, and vines). Forbs and browse become important components of goat diets because these animals are considered browsers instead of grazers like cattle and sheep. Forbs can contain high CP levels (16), but only for a very short period of time. Browse such as multiflora rose (*Rosa multiflora* Thunb.), autumn olive (*Elaeagnus umbellate* Thunb.), and honeysuckle (*Lonicera morowii* Gray) often contains high levels of CP (36).

Chicory is considered a weed in U.S. cropping situations (12), but plant breeding efforts in New Zealand led to an improved cultivar specifically designed for pasture and livestock production. Chicory could be a valuable summer forage. Turner et al. (35) suggested that high ammonium nitrate fertilization to chicory swards may have had a negative impact on forage digestibility, selectivity by grazing lambs, and total weight gain by growing lambs.

Plant Part

In general, leaves of plants contain about twice the amount of CP as stems. Young, vegetative leaves contain higher crude protein than older, mature stems. Forage managers should strive to maintain a vegetative sward that has high amounts of leaf area in relation to stem.

Plant Maturity

Advancing forage maturity decreases plant protein degradation (14). Protein and overall quality of grasses usually decline faster than do legumes. Plant maturity very much depends upon the growing season and sward management.

Season

Throughout the growing season, the CP of plants varies, but overall the protein concentration declines with advancing season. The amount of soluble protein is highest in the spring and fall and lowest in midsummer (1). In another study, spring-harvested grass had lower CP and higher carbohydrate than autumn-harvested grass (15). Grazing or harvesting frequency can be used to maintain swards in a more vegetative stage throughout the growing season. Maintaining vegetative stage helps to maintain high CP levels. Forbs and browse, such as multiflora rose, autumn olive, and honeysuckle, contain high but variable levels of CP (37) throughout the growing season with a general decline in total protein as the season advances. Varying CP and soluble fractions over the growing season need to be considered when developing protein supplementation strategies for lactating and young growing ruminants grazing pastures.

Forage Preservation

The goals of any forage preservation program are to conserve herbage energy, CP, and fiber and to maintain protein in a form that can be used effectively and efficiently by ruminants (30). The overall nutritive value of conserved feeds is directly related to the nutritive value of the herbage at time of harvest. In other words, high quality hay or silage will be obtained only if the herbage is harvested at an early growth stage or one considered optimal for mixed swards. For most grasses this is in the boot to early seedhead stage, and for legumes this is in the bud or early flower stage.

Care must also be taken in the hay- or silage-making process because nutritive value of the forage can be further decreased if significant leaf loss occurs. The main difference between silage and hay is the level of water contained in the product for preservation. To produce good fermentation products and maintain protein in silages, chopped forage should contain 50–70% water. To produce good storage characteristics, hays should contain less than 20% water. These limited water concentrations help to control the heat up during the initial stages of storage. Excess heat produced from excess water will result in chemical reactions between proteins and carbohydrates, termed the Maillard reaction or browning reaction. This reaction reduces the overall availability and digestibility of protein in the hay or silage. Ensiled forages have greater concentrations of degraded plant protein than hays (14) and have higher levels of NPN.

Protein Supplementation

Deficiency of CP in ruminant diets can be corrected by supplementary feeding of protein or NPN sources, fertilizing grass pastures with N, interseeding legumes into grass pastures, feeding to increase rumen-undegradable protein levels, or **ionophore** feeding (see page 22) (19).

Protein supplements are probably the most expensive feeds to purchase. A general rule of thumb is that when forage CP is less than 7%,

supplemental protein is necessary because insufficient ammonia is present in the rumen to support optimal microbial activity. Simply offering high CP legume hays, soybean meal, or dehydrated alfalfa meal will usually improve animal performance.

Nitrogen fertilization can increase plant DM production and quality, livestock carrying capacity, and livestock production. Nitrogen can also be supplied by interseeding legumes, which have N-fixing capability, into grass-dominated swards. In general, grass-legume mixtures are more productive than N-fertilized grass (17). Legume addition to grass pasture increased steer average daily gain over that of N-fertilized grass pasture (8). Management to maintain legumes in pastures must be adapted to the specific grass/legume mixture. A specific botanical composition does not represent a stable density of legumes in all pastures, especially because grazing ruminants are highly selective for the legume component of the sward. Soil, light, temperature, rainfall, frequency, and height of defoliation of a canopy and selective grazing pressure by livestock can interact to influence legume persistence in mixed swards.

Much of the herbage protein when livestock graze pastures is highly soluble, is degraded rapidly by rumen microorganisms, and results in high rumen concentrations of ammonia. Feeding protein sources with high rumen-undegradable value is recommended to improve N use in the grazing ruminant. This supplementation strategy may involve feed protein sources low in rumen-soluble protein such as that contained in dried distillers' grains, corn gluten meal, dried blood meal, and hydrolyzed feather meal. High forage diets with low protein solubility would involve feeding soluble protein sources (highly degradable in the rumen) using feedstuffs such as soybean meal, corn gluten feed, or alfalfa hay.

Ionophores were developed and approved for use in ruminant diets in the mid-1970s (33). These feed additives generally improve feed efficiency by decreasing feed intake. Ionophores have also been reported to inhibit the growth of ruminal bacteria responsible for breaking down protein (27), thereby reducing rumen ammonia concentrations and improving N use in the rumen.

In any supplementation strategy, the resulting CP:energy ratio in the overall diet becomes an important consideration to optimize protein use and forage utilization. The CP of well managed grass and legume pastures is seldom deficient relative to the digestible energy. However, addition of corn grain supplement (for energy; corn grain is low in CP when used as a supplement) can increase use of plant proteins in ruminants (13) by providing adequate fermentable energy for capture of excess N by microorganisms in the rumen and improving the CP:energy ratio when grazing alfalfa pastures. Corn supplementation is recommended not to exceed 0.5% body weight/head/day. In grass hay and corn silage, CP can be low relative to DE and protein supplementation is necessary.

The nutritive value of forages, the level of intake by livestock, and the overall efficiency of nutrient utilization help to characterize the forage quality. One measure that has been applied to forage is the quality index (21). This index is the ratio of the voluntary free-choice intake of forage DE divided by the animal's maintenance energy requirement. When the DE:CP ratio is less than 5, there is a balance of energy and protein. In extremely low ratios, there may be need for supplemental energy. When total digestible nutrients (TDN):CP is greater than 5, there is usually a need for supplemental protein in dairy cattle. A ratio of 8 is usually used for beef cows, which can survive at lower CP levels and per-

form adequately. Various supplementation and management strategies can help improve the protein use in grazing livestock.

Ruminant Protein Utilization

In livestock production situations using forages, N utilization by ruminants is inefficient (3). In grazed environments, N use is poor, with as much as 75–95% of N ingested by ruminants grazing temperate pastures lost via urine and feces (2, 40). The efficiency of utilization is lower for all forage diets compared to forage plus concentrate or all-concentrate diets. Efficiency of utilization also varies with harvest date of the forage; for ruminants grazing pasture, this results in a daily variation in the efficiency of utilization.

In ruminants, proteolytic microorganisms in the rumen break down ingested feed and plant proteins to amino acids and some can further break down amino acids into ammonia (an N-containing compound). When livestock consume more protein than is needed to meet requirements, the excess is excreted and represents a waste. As a result of rumen fermentation, ruminants have two sources of protein presented to the lower digestive tract: (i) microbial protein and (ii) undegraded dietary protein (5, 6).

Forages, browse, and forbs supply N in both protein and NPN forms for use by ruminal microorganisms. Excess protein consumed by ruminants causes proteolytic bacteria in the rumen to metabolize excess amino acids (28), resulting in high levels of ammonia in the rumen. Excess ammonia must be eliminated. The intermediate between breakdown of protein in the rumen by microorganisms to ammonia and excretion as urea via the kidney is blood urea nitrogen (BUN) synthesized by the liver. Preston et al. (31) and Pfander et al. (29) reported high positive correlation between dietary N (protein) level and BUN concentration. BUN concentrations have been monitored to determine N use and to determine level and feeding time of protein supplements for lambs (29) and cattle (10) to improve N use efficiency.

Minerals

Minerals serve three broad functions (39). These are as structural components of body organs and tissues; as electrolytes concerned with osmotic pressure and acid-base balance; and as catalysts in enzyme and hormone reactions. Examples include calcium (Ca) and phosphorus (P) in bone; sodium (Na), magnesium (Mg), potassium (K), and chloride (Cl) in the blood; and selenium (Se) in glutathione peroxidase.

For purposes of discussion, minerals are generally divided into two classes: macrominerals and microminerals. This division is based on the amount of mineral needed to perform its function. Macrominerals are usually measured in units of percent, pounds, or grams. Microminerals are usually measured in milligrams, micrograms, and parts per million. Macrominerals include Ca, P, K, Na, Cl, Mg, and S. Microminerals include iron (Fe), iodine (I), copper (Cu), molybdenum (Mb), cobalt (Co), nickel (Ni), manganese (Mn), zinc (Zn), Se, and fluorine (Fl). The list of microminerals can vary depending upon the animal species because requirements for any given micromineral may or may not have been demonstrated for any given animal species.

Calcium

The concentration of Ca in forages ranges from 0.3 to 0.7% in grass and weed species to 1.0–1.9% in legumes and improved pastures. For large ruminants consuming 15–20 pounds of forage DM per day, this equates to 20–173 grams (e.g., 15 lb $\times$ 0.003 Ca $\times$ 450 g/lb) of Ca from forage each day. Small ruminants that consume 1–5 pounds of forage DM per day will get 2–40 grams of Ca from forage per day.

Calcium is required for skeletal development, enzyme systems, and milk production. When more Ca is required than is present in the forage, supplemental Ca can be added to the diet. Supplementation needs depend on the physiological state and type of animal. For example, a pregnant brood cow in the first or second trimester can easily obtain her Ca requirement by consuming 15 pounds of grass hay containing 0.3% Ca. A lactating dairy cow producing 80 pounds of milk daily must receive 30 grams of supplemental Ca even though she is consuming 12 pounds of legume hay containing 1.9% Ca. The two most common sources of Ca are limestone (40% Ca) and dicalcium phosphate (26% Ca). Livestock that are most likely to require Ca supplementation beyond the amounts supplied by forages are those that are lactating, in the final trimester of pregnancy, or growing in early stages of life.

Phosphorus

Forage P concentrations range from 0.2 to 0.5% in grass and weed species to 0.2 to 0.4% in legumes and improved pastures. For large ruminants, consuming 15–20 pounds of forage DM per day, this equates to 14–45 grams of P from forage each day. Small ruminants that consume 1–5 pounds of forage DM per day will get 1–10 grams of P from forage per day. As for Ca, supplementation requirements of P depend on the physiological state and type of animal. Pregnant brood cows in the first or second trimester can easily obtain their daily P requirement by consuming 15 pounds of grass hay containing 0.2% P. A lactating dairy cow producing 80 pounds of milk daily must receive 55 grams of supplemental P even though she is consuming 12 pounds of legume hay containing 0.5% P.

Dietary deficiencies of P are manifested in abnormalities of the bones and teeth, slow growth rates, low DM intake, and reduced reproductive performance. When supplemented in mineral form, P is usually supplied as dicalcium phosphate (20% P), especially when supplemental Ca is also required, or sodium phosphate (24% P) when required alone. When balancing rations with supplemental protein, it is important to consider the P that comes from soybean meal (0.7% P). Livestock that are most likely to require P supplementation are those that are lactating or are ready to be bred. Many times these two stresses occur simultaneously. Growing animals may also require supplemental P.

Potassium, Sodium, and Chloride

Concentrations of K, Na, and Cl in grasses range from 0.5 to 3.5%, 0.01 to 0.2%, and 0.2 to 0.6%, respectively. In legumes, concentrations range from 0.5 to 2.5%, 0.05 to 0.2%, and 0.2 to 0.8%, respectively. Large ruminants consuming 15–20 pounds of forage DM per day will consume 34–320 grams of K, 1–20 grams of Na, and 14–73 grams of Cl from forages each day. Small ruminants that consume 1–5 pounds of forage DM per day will consume 2–80 grams of K, 0.05–5 grams of Na, and 1–18 grams of Cl from forages each day. Forage is usually the primary source of K in most diets.

When forage K concentrations are so low that the K concentration in the whole diet drops below 0.6–0.8%, supplemental potassium chloride (50% K) or potassium carbonate (55% K) should be provided. Salt, the most economical and common source of Na and Cl, should constitute 0.1–0.5% of the diet DM. Because of the low cost of salt supplementation and the low concentration of Na in most forages, salt should always be supplemented. Salt intakes up to 2.2 pounds per day or up to 9% of diet DM have been fed to large ruminants (22, 24) without

deleterious effects. When salt is supplemented at recommended levels, animal Cl status is adequate.

Potassium, Na, and Cl are broadly similar in both function and requirements in livestock (18, 39). All are involved in controlling water metabolism in body tissues. Deficiency signs of Na are usually first seen as a craving for salt. This observation should be confirmed with blood, urine, or saliva analysis. Supplementation of salt, by free choice or blended with grains, corrects the deficiency. The K deficiency signs are not well documented, likely because of the high concentration of K in most forages. Poor growth, muscle weakness, stiffness, and paralysis have been reported with K deficiency.

Salt is usually provided free choice to livestock. Salt, usually sodium chloride, can be provided alone, but mixtures of sodium chloride and potassium chloride also can be provided. The salt mixture can also be used to provide other important macro- and micronutrients. Trace mineralized salt is commonly used to provide important micronutrients.

Magnesium

Because of its association with the etiology of **grass tetany**, Mg is one of the most studied macrominerals. Forage Mg concentrations range from 0.1 to 0.5% in both grass and legume species. Each day large ruminants will consume 7–45 grams of Mg, while small ruminants will consume 0.5–11 grams. When the concentration of Mg in the diet is less than 0.2% of DM, supplemental Mg is necessary. Magnesium can be supplemented by adding magnesium oxide (56% Mg) or magnesium sulfate (Epsom salts, 10% Mg) to a supplemental grain mix. Using dolomitic limestone (50% calcium carbonate + 50% magnesium carbonate) instead of limestone (100% calcium carbonate) when liming may increase the forage Mg concentration.

Subclinical Mg deficiency is manifested in low feed intake in all animals. Acute Mg deficiency usually occurs in early lactation cattle as nervousness, unsteady gate, and uncontrolled movement of rear legs. Death will result unless Mg can be administered intravenously within a short time.

Sulfur

Forage S concentrations range from 0.1 to 0.3% for grasses and 0.1 to 0.4% for legumes. Sulfur is usually present in higher concentrations in high protein feedstuffs because of S-containing amino acids. Large ruminants will consume 7–36 grams of S each day, while small ruminants will consume 1–9 grams. The S requirements are not well defined. Diets containing 0.1% S can sustain beef cattle without adverse effects. High producing dairy cows require approximately 0.2% S in the diet for optimum performance. Diets containing more than 0.26% S can cause reduced feed intake and result in lower performance. Sources of S, such as water, must be considered when evaluating total dietary S. Supplementation of diets with S is seldom necessary because forage S levels and other high protein feedstuffs contain adequate amounts of S.

Microminerals

Exact requirements of microminerals, also known as trace elements, by ruminant livestock are not well understood. It is difficult to measure very small amounts of microminerals without variation. High costs of sample analysis make it impractical, on a routine basis, to monitor microminerals, so most nutritional consultants rely on published values. The concentrations of microminerals in forages range from 0.1 to 0.4 ppm for Co; 3 to 14 ppm for Cu; 0.15 to 0.25 ppm for I; 100 to 400 ppm for Fe; 30 to 130 ppm for Mn; 0.3 to 0.5 ppm for Se; and 15 to 30 ppm for Zn.

Because they are present in such small quantities, microminerals are susceptible to a number of environmental factors. One such factor is interactions of microminerals with other elements. Copper, Mb, and S can form insoluble complexes that lower the biological availability of all three elements. For example, the requirement for Cu by a dairy cow is 10 ppm. In the presence of Mb and S, however, 15–20 ppm Cu may be necessary to achieve a true biological requirement. Another factor is the regional differences in the amount of Se available in the soil for absorption by forages. Where Se is adequate, there is no need for supplemental Se; however, in regions with Se-deficient soils, it is necessary to supplement rations with the legal limit of 0.3 ppm SE to prevent white muscle disease and reproductive problems.

Such variation in trace element availability prompts many nutritional consultants to provide the National Research Council (NRC)-recommended requirement of trace minerals in supplemental form at all times. Even though trace minerals are expensive on a unit weight basis, they are required in such small amounts that ration costs are low.

Vitamins

Vitamins are relatively new in the field of nutritional study. In the early 1800s, William Prout recognized the dietary need for carbohydrate, fat, and protein (18). It was not until the early 1900s, when scientific instrumentation was sufficiently improved, that vitamins were discovered and their roles understood.

For purposes of discussion vitamins are divided into two main groups based upon their physical properties. Vitamins A, D, E, and K are soluble in fat, so they are called fat-soluble vitamins. The B-complex vitamins and vitamin C are soluble in water, so they are called water-soluble vitamins.

Fat-Soluble Vitamins

In outdoor management systems, where animals are regularly exposed to sunlight, only two fat-soluble vitamins require supplementation, A and E. Vitamin D is generated when sunlight (ultra-violet radiation) strikes the skin. Vitamin K is a common product of rumen microbial growth and is readily available to healthy ruminating livestock. In a confinement management system, vitamin D supplementation is recommended.

Forage vitamin A and E levels vary greatly depending on the growth stage of the plant. Early growth alfalfa, containing 22% crude protein, will contain approximately 250 ppm of both vitamins A and E and little supplementation would be required. Turner et al. (38) reported that alfalfa or perennial ryegrass pastures were supplying 10 times NRC recommendations for vitamin E (alpha-tocopherol) for growing lambs.

As the plant matures and the crude protein concentration drops to 15%, the concentration of vitamins A and E drops to about 80 ppm and supplementation would be recommended. Again, as with trace minerals, because laboratory analysis is expensive, most nutritional consultants do not analyze forages for vitamin levels, but supplement recommended levels at all times.

Water-Soluble Vitamins

Water-soluble vitamins include the B-complex vitamins and vitamin C. Rumen microorganisms produce all of these vitamins. Therefore, it is not necessary to supplement these vitamins to healthy, ruminating livestock. However, if animals are sick or are under other stresses that limit intake and rumination activity, supplementation or injection of B-complex vitamins is needed. In addition to rumen microorganisms, ruminant livestock can manufacture vitamin C

in the liver, thus negating the need for vitamin C supplementation, even under conditions of stress.

Water

Water is not normally thought of as a nutrient. Yet this essential dietary component can cause more direct effects on animal health and performance than any other nutrient or combination of nutrients.

Water makes up more than 95% of the body's composition. It dissipates heat in maintaining body temperature, transports nutrients through digesta and blood, facilitates chemical reactions, lubricates joints, provides structure to certain organs such as the eye, and facilitates waste removal via the kidney.

Water deprivation results in discomfort, reduced DM intake, illness, and death if severe deprivation occurs. Individual water requirements vary depending upon age, physiological state, dietary components, livestock species, and environment. Water turnover rate is faster in young and highly productive animals; the opposite is true in older and less productive animals.

Water in the animal's body comes from three sources: drinking water, water found in forages and feedstuffs, and water from nutrient metabolism within various organs in the body.

Drinking Water

Ruminants in general are efficient users of water but need access to a clean, potable source of water. Anything that restricts the amount of water to which livestock have access will reduce productive performance, usually through a loss in DM intake. The rate at which an animal consumes water is directly related to the environment. Livestock will obviously drink more water on a hot day than on a day during which temperatures are close to the comfort zone of the animal.

Ruminants usually replace about 15–20% of their body weight when they first drink (34). This factor is an important consideration for intensive grazing systems for livestock management. Each pasture or paddock should contain a source of drinking water. In some cases the design or layout of the pasture system allows for a centralized water source. When a centralized water source is used, care must be taken not to allow livestock to lounge there for extended periods of time because nutrients from manure will tend to be highest around watering sources and not distributed around the pasture or paddock.

On many farms, ponds, wells, and springs serve as sources of drinking water for livestock. Water samples should be collected periodically and tested to be sure dissolved mineral levels in water are within tolerable guidelines for livestock. Water samples should be tested for arsenic, cadmium, chromium, Cu, Fe, lead, mercury, nitrate, sulfate, Se, and Zn. In some cases, Ca, Mg, and Na concentrations should also be checked, especially in areas where hard water is a known concern.

It is also important to evaluate levels of bacteria, especially *E. coli* and other intestinal bacteria, in water sources for livestock because livestock and poultry manures are being used more frequently as fertilizers for pastures and this use can contaminate water sources.

Water Contained in Forages and Feeds

Forages and feedstuffs contain varying amounts of water. Fresh herbage can contain as much as 90% water; often this water is on the leaves of the plant as a result of dew formation or rain.

Often the term "as-fed basis" is used when referring to a particular nutrient contained in the forage or feedstuff as is. "Air-dry basis" is a subclass of as-fed basis, meaning that the forage or feedstuff has 10% water or 90% DM. When nutrients are reported on a DM basis, there is no water in the sample; it is 100% DM. Because there is such a wide variation in the amount of water contained in forages and feedstuffs used in ruminant diets, most forages and feedstuffs need to be converted to a common DM basis for comparison of weight or nutrients, especially when buying or selling feedstuffs.

Growing forages contain 75–80% water; in spring this number may be closer to 90%. Hays generally contain less than 15% water; silages contain greater than 50% water. Grazing animals need to consume less liquid water than livestock consuming the same forage preserved as hay. Because of the low ration cost, it is most feasible to provide free access to fresh, clean water.

Water from Nutrient Metabolism

Water can be supplied from basic biochemical reactions in the body. In protein synthesis, the joining of two amino acids to form a dipeptide releases a molecule of water.

FORAGE INTAKE

Intake by animals is very complex, but intake in grazing livestock is probably controlled by three major sets of stimuli: energy/nutrient demand, physical satiety, and behavioral inhibition (11). Selectivity for specific plants (usually legumes are preferred to grasses), plant parts (leaves preferred to stems), growth stage (vegetative preferred to mature), and total grazing time are important factors determining overall intake and ultimately performance by grazing ruminants. Grazing site and selectivity by ruminants are influenced by management, stage of cultivar growth, dung and urine patterns, ground slope, temperature, humidity, and grazing pressure (19). Sward structure and **phenological** stage affect biting rate and grazing time. Ultimately, herbage intake and digestibility are positively correlated with weight gain when forages provide the sole source of nutrients (36).

Environmental Effects on Livestock Influencing Intake

The thermoneutral zone or temperature comfort zone for livestock is 59–77° F; for cattle this range may be lower. Environmental temperature extremes will influence intake by livestock. During periods of low temperature stress, the animal needs more energy, thus intake may increase by 30% to supply additional energy. However, during heat stress, the desire to eat is low, so intake may decline 30%. Air temperature, wind speed, precipitation, and sunshine can all interact to influence intake by livestock.

Environmental Effects on Plants Influencing Intake

Environmental effects on plants also impact intake by livestock. Quantity of herbage consumed by grazing ruminants depends on herbage physical and chemical composition and overall availability. High temperatures and drought conditions tend to force plants to early maturation for self preservation. Thus, high temperatures and drought narrow the window of when forages are considered of high nutritive value. Animals eat to meet specific nutrient demands. Upon maturation, forage plants tend to have higher proportions of structural carbohydrates (fiber) and lower proportions of protein and cell solubles. Voluntary intake declines with advancing plant maturation. Increased fiber in forages limits the overall amount of feed that can be consumed as a result of slowed rates of digestion and passage in the rumen.

Changes in Nutrient Needs Over the Production Cycle

In general, forages supply needed energy, protein, vitamins, and minerals for many livestock species in varying growth and physiological states, and can compose 70–90% of the total diet. The seasonality of herbage growth patterns is a major limitation to optimized livestock production (9) because inadequate herbage supply may not meet the requirements of growing livestock grazing pastures. Livestock should be managed by grouping animals with similar nutritional demands in order to provide forages and feedstuffs to optimize performance. The reader is referred to other chapters in this book dealing with cattle, sheep, and goat production for specific information on changing nutrient needs with changing physiological condtions.

Animal Class

Calves, lambs, and kids have very high nutritional requirements. Next are weanling animals, followed by animals to be used in developing replacement animals, then young animals. Mature animals have the lowest nutritional requirements, but physiological state is also important.

Influence of Animal Physiological State on Requirements

Growing livestock have high nutritional requirements, but intake based on body size is less compared with mature, nonlactating animals, which have less stringent nutritional demands. The rapidly growing fetus in the last trimester of pregnancy greatly increases the need for nutrients compared to animals in the first trimester of pregnancy. Another factor that comes into play is the rapidly growing fetus's restriction on the capacity of the rumen, especially if the dam is carrying twins or triplets. Thus, pregnant animals in the last trimester need to be offered a supplemental, nutrient-dense feedstuff or forage.

Lactating animals have high nutritional demands related to the amount of milk produced daily. Growing animals bred to give birth at 1 year of age have a higher nutritional requirement than mature, pregnant animals. After giving birth young females need to be fed not only to satisfy their own growth requirements and that of milk production for the offspring, but also to prepare the animals for rebreeding.

SEASONALITY OF HERBAGE PRODUCTION

Over the growing season, plant growth progresses from the vegetative stage to boot stage to bud to early bloom to full bloom and finally to seedhead development. These developmental changes are associated with less leaf, more stem, lower protein, and higher fiber in plants, resulting in lower digestibility and intake of the forage by livestock. However, there is a greater amount of harvestable aboveground biomass as the plant matures. Thus, hay- or silage-making is a balance between optimal plant nutrients for livestock and quantity of forage for economical harvest. Pasture management is an important factor influencing nutritive value and ultimately livestock performance (36).

Grasses

Grasses differ in their optimal temperature for growth. Cool-season grasses grow well in the spring when temperatures are between 40 and 55° F; warm-season grasses prefer midsummer temperatures of 65+° F for optimal growth. Warm-season grasses have greater fiber concentration and lower digestibility compared to cool-season grasses.

Legumes

In general terms, legumes are considered to be of higher quality than grasses, mainly due to more rapid digestion than grasses at a similar maturity stage. Higher quality is probably due to legumes having higher protein and lower structural fiber concentrations, which results in a greater rate of passage from the rumen when compared to cool-season grasses.

Nontraditional Plants

Weeds (16) and browse (37) can provide nutrients for browsing animals such as goats. Health problems associated with nutritive value and mineral concentration in chicory would be minimal in grazing livestock (4). Phosphorus supplementation may be necessary when browsing goats consume multifora rose, autumn olive, or honeysuckle for long periods of time because the Ca:P ratio is unbalanced in mid- to late growing season (K. E. Turner, unpublished data).

RATION BALANCING

The total amount of forage or feed that an animal can consume is limited to the capacity of the digestive tract and the rate of passage through the various compartments of the digestive tract as defined by the physiological demands of that animal. Knowing how to combine feedstuffs or forages to supply needed nutrients based on these limitations is imperative to optimize nutrient use, animal performance, and economics.

The formulation of rations for animals has evolved from an "art" based primarily on experience and trial and error to a sophisticated science as our knowledge of nutrient requirements and feed technology has undergone continual refinement. The development of computer systems has greatly facilitated our abilities to simultaneously meet an incredible number of nutritional goals. Despite our reliance on computers, producers should master the basic mathematical calculations necessary to balance simple rations. These skills are necessary to truly understand applied nutrition and to adequately interpret and evaluate computer output.

Unfortunately, it is typically difficult to employ sophisticated methods of ration formulation for pastured animals. At this point there are simply no practical reliable methods of measuring feed intake in the grazing animal, and because forage intake is uncontrolled and largely unknown, we rely on estimates based on experience.

RATION BALANCING PROCESS

In theory, ration balancing is an uncomplicated process in which the goal is to offset the animal's requirements with an appropriate level of feed nutrients. Clearly there are economic and quite possibly animal welfare consequences when the ration is out of balance in either the surplus or deficient direction. Obviously the outcome of the overall process will only be as reliable as the information provided for each side of the balance (nutrient requirements and feed nutrient values). Regardless of the animal species or the type of diet being formulated, the producer should master several basic concepts and types of calculations.

Nutritional Goals and Sources of Information

Diet formulators must be keenly aware of the nutritional goal when devising feeding programs for animals. In many livestock situations the goal is maximum production, but other purposes arise. For example, maintenance of body weight and condition may be the goal for dry pregnant beef cows or horses with low activity levels. Diets promoting health and longevity with little attention to performance may be of primary importance when formulating rations for sedentary animals. Finally, the ability to sustain work or athletic performance may be the primary nutritional goal for work- and racehorses. Although we must always rely on the published nutrient requirement information to begin the formulation process, in many of these situations experience is important, including visual assess-

ment of the animal's response to the diet being fed.

The first step in diet formulation is to obtain a reliable source of nutrient requirements for the target animal (see other chapters in this book regarding the various livestock species). In the United States, the National Research Council is generally regarded as the authoritative source. The council publishes a "Nutrient Requirements of Domestic Animals" series on a wide variety of animal species. Experts in the field compile the available research for each animal species at periodic intervals, and the information is updated and disseminated. The growing popularity of the personal computer for diet formulation is evident as a number of the species publications now contain a CD with appropriate ration balancing software.

Other sources of information may include extension publications and the results of university research. In some situations this information may provide useful modifications to published requirements based on "local conditions." Special environmental, animal, or feed conditions prevalent in a given state or region may be addressed in these sources.

In addition to reliable animal nutrient requirement information, it is essential that accurate feed nutrient values be available for the diet formulation process. The NRC nutrient requirement publications typically contain a listing of common feedstuffs fed to that particular animal species. These "book values" simply list the average levels expected for common feedstuffs and as such will seldom match the actual nutrient content of the feed being used. The degree of ration formulation error introduced by using book values varies. For example, concentrate ingredients tend to have far less variation in nutrient content than forages, which are drastically affected by local harvest and storage conditions. Thus, it is always preferable to have the actual feedstuff to be fed analyzed by a feed testing service. As we will see in subsequent sections, this is becoming increasingly challenging, as a number of comprehensive diet formulation models now require an extensive array of information. Not only is content required for an increasing number of nutrient fractions, but also descriptive utilization data such as the rate of digestion are required.

Requirements

The nutrient requirements for various classes of animals are commonly described in two ways: nutrient concentration or density, and daily amount of nutrients required by an animal. Because the critical factor is amounts of a given nutrient consumed daily, why are requirements also expressed in terms of nutrient concentration? This relates to the way animals are fed in practice with the requirements expressed in two ways for the convenience of the person formulating the ration. When animals are fed a ration ad lib (they may consume all the feed they desire), we typically balance the ration to meet recommended nutrient concentrations. We

may not know exactly how much each animal consumes, but as long as consumption meets the minimum recommended in table 2-1, the minimal *amounts* of nutrients required daily will be met. It is most common to balance rations for market animals on a nutrient concentration basis because we want them to consume feed ad lib for maximum performance. In other cases (typically mature animals or female replacements) we may want to *limit* or restrict feed to prohibit excessive fatness. In this case it may be easiest to express the animal's requirements in terms of nutrients per day and then calculate the amount of a given feed or mixed ration necessary to supply that amount (table 2-2).

Using the National Research Council's nutrient requirements of sheep (25) as an example, note how the two types of tables are related.

From table 2-1, if the animal consumes 1.2 kg of ration DM containing 55% TDN and 9.4% total protein, 0.66 kg of TDN (1.2 × .55) and 113 grams (rounded) of total protein (1.2 × .094 × 1000) will be consumed. These correspond to the daily amounts recommended in table 2-2.

Table 2-1. Daily nutrient requirements of sheep (as-fed basis).				
	Body wt./animal	**Daily ration DM**	**TDN (%)**	**Total protein (%)**
Ewe maintenance	70 kg (154 lb)	1.2 kg (2.6 lb)	55	9.4

Table 2-2. Daily nutrient requirements of sheep (100% DM basis).				
	Body wt./animal (lb)	**Daily ration DM (lb)**	**TDN**	**Total protein**
Ewe maintenance	70 kg (154 lb)	1.2 kg (2.6 lb)	0.66 kg	113 g

Thus, both ways of expressing the requirements are related and are presented for ease in ration formulation for a given purpose. It is also clear that accurate estimates of expected feed intake are required.

Feed Composition Values

Because feed ingredients in rations may vary widely in DM content, we typically balance rations on a DM basis. The nutrient values for feedstuffs may be expressed on either an "as fed" (air-dry) basis or a DM basis. When using the latter type of feed composition table (table 2-2, p. 31) to calculate the amount of a nutrient in a feed, we must first calculate the amount of DM present.

Example:

20 lb of 32% DM corn silage
8% crude protein (DM basis)

20 x 0.32 = 6.4 lb DM x 0.08 = 0.51 lb of protein

The other approach would be to first change the crude protein content to an as-fed basis:

8% crude protein (DM basis) x
0.32 (DM content) =
2.56% crude protein (as fed)

20 lb corn silage x 0.0256 =
0.51 lb crude protein

Thus, *units must agree* when calculating contents of nutrients (as-fed amounts multiplied by nutrient concentration on an as-fed basis and DM amounts by nutrient concentration on a DM basis). Usually it is most convenient to first formulate rations on a DM basis and as a last step, convert those amounts to as fed.

Feed Formulas and Conversions

There are several basic types of data manipulation used in formulating a given ration:

A. Amounts of DM to amounts as fed:

10 lb of DM required and feed is 30% DM:

10 lb of DM ÷ 0.30 lb DM per lb as fed =
33.3 lb as fed

B. Amounts as fed to amounts of DM:

50 lb of corn silage as fed and
DM content is 35%:

50 lb x 0.35 = 17.5 lb DM

C. Mixed ration formulas – We know DM composition and want to convert to an as-fed formula:

(See next page)

D. Mixed rations – We know as-fed composition and want to convert to a DM formula:

(See next page)

E. Nutrients consumed from as-fed amounts of feed:

(See next page)

Simple Balancing

In many cases, only a computer has the capacity to simultaneously calculate all nutrients required at least cost. However, the following approaches may be used in various situations depending on the level of refinement needed in formulation of the ration.

Trial and Error

One must realize that hand-balancing a ration is often not an exact science. Probably one of the most often used techniques is simply trial and error. A producer might decide to feed "X" amount of forage to a given animal. He then calculates the amounts of nutrients supplied in that

C. Mixed ration formulas – We know DM composition and want to convert to an as-fed formula:

Feed	% of ration DM	DM of feed	lb as fed	Total lb as fed	% as fed
Corn silage	60[a]	0.35	171.4	225.4	76.1
High moisture shelled corn	30	0.70	42.9	225.4	19.0
Supplement	10	0.90	11.1	225.4	4.9
Totals	100		225.4		100

[a] Calculation sample: 60 ÷ 0.35 = 171.4 ; 171 ÷ 225.4 = 0.761 x 100 = 76.1
Note that the percent DM of the mixed ration would be 100 lb DM ÷ 225.4 lb as fed = 44.3% DM.

D. Mixed rations – We know as-fed composition and want to convert to a DM formula:

Feed	% of ration as fed	DM of feed	lb of DM	lb of ration as fed	% of ration DM
Corn silage	76.1[a]	0.35	26.64	44.35	60.1
High moisture shelled corn	19.0	0.70	13.30	44.35	30.0
Supplement	4.9	0.90	4.41	44.35	9.9
Totals	100		44.35		100

[a] Calculation sample: 76.1 x 0.35 = 26.64 ; 26.64 ÷ 44.35 = 0.601 x 100 = 60.1

E. Nutrients consumed from as-fed amounts of feed:

Feed	As fed (lb/d)	Feed DM (%)	lb as DM	% CP (DM basis)	CP (lb/d)
Corn silage	30[a]	0.35	10.5	0.08	0.84
High moisture shelled corn	10	0.70	7.0	0.10	0.70
Supplement	2	0.90	1.8	0.40	0.72
Totals			19.3		2.26

[a] Calculation sample: 30 x 0.35 = 10.5 ; 10.5 x 0.08 = 0.84
Note that the percent CP in the ration (DM basis) would be (2.26 ÷ 19.3) x 100 = 11.7%.

forage. He compares this to the requirement, and by trial and error (repeatedly checking against the deficiencies left from the forage), he formulates a grain/vitamin/mineral mix that will meet the animals' needs.

Pearson's Square

Pearson's square is a useful technique for balancing one nutrient using two feeds. One feed must have a nutrient concentration above and one feed below the desired nutrient level.

Example:

Balance a ration for 11.5% crude protein using alfalfa hay (17.1% CP) and ground ear corn (9.3% CP). The desired protein content is placed in the center of the square and the feed protein levels at the left corners. Then subtract the smaller number from the larger number diagonally across the square. Finally, convert the DM parts to percentage DM from each feed.

Feed protein level	Desired protein content (%)	DM parts
Alfalfa hay 17.1		2.2
	11.5	
Ear corn 9.3		5.6

$$17.1 - 11.5 = 5.6$$

$$11.5 - 9.3 = 2.2$$

$$2.2 + 5.6 = 7.8$$

$$(2.2 \div 7.8) \times 100 = 28.2\% \text{ DM}$$

$$(5.6 \div 7.8) \times 100 = 71.8\% \text{ DM}$$

Check:

$$28.2 (0.171) + 71.8 (0.093) = 11.5\% \text{ protein}$$

Algebraic Method

The algebraic method works similarly to Pearson's square and is perhaps easier for some producers.

From example above:

$$\text{Let } X = \text{DM needed from hay}$$

$$100 - X = \text{DM needed from ground corn}$$

Then:

$$0.171 (X) + 0.093 (100 - X) = 11.5$$

$$0.171X + 9.3 - 0.093X = 11.5$$

$$0.078X = 2.2$$

$$X = 28.2 \text{ (\% of DM from hay)}$$

$$100 - X = 71.8 \text{ (\% of DM from ground corn)}$$

Modified Pearson's Square

The square method can be expanded to include three or more feeds if the decision is made to "lock in" two or more of the feeds in a fixed proportion.

Example:

$$\text{Alfalfa hay—20\% CP}$$

$$\text{Corn grain—10\% CP}$$

$$\text{Wheat—12\% CP}$$

A decision is made to fix the two grains at 60% corn grain and 40% wheat on a DM basis. How should the alfalfa hay and grain mixture be combined to produce a total ration containing 15% CP?

Grain protein mix: $0.60 (10) + 0.40 (12) = 10.8\%$

Balance for 15% in total ration:

Feed protein level	Desired protein content (%)	DM parts
60:40 Grain mix 10.8		5
	15	
Alfalfa hay 20		4.2

$$15 - 10.8 = 4.2$$

$$20 - 15 = 5$$

$$5 + 4.2 = 9.2$$

$$(5 \div 9.2) \times 100 = 54.3\% \text{ DM}$$

$$(4.2 \div 9.2) \times 100 = 45.7\% \text{ DM}$$

Final ration:

Alfalfa hay		45.7%
Corn grain	$54.3 \times 0.60 =$	32.6
Wheat	$54.3 \times 0.40 =$	21.7
Total:	$45.7 + 32.6 + 21.7 =$	100%

Check:

$$0.457 \,(20) + 0.326 \,(10) + 0.217 \,(12) = 15\% \text{ CP}$$
$$\text{Hay} \qquad\quad \text{Corn} \qquad\quad \text{Wheat}$$

Modified Algebraic Equations

$$\text{Grain mix} \qquad \text{Hay}$$
$$0.108 \,(X) + 0.20 \,(100 - X) = 15$$
$$X = 54.3 \text{ grain mix}$$
$$100 - X = 45.7 \text{ alfalfa hay}$$

Final ration:

$$45.7\% \text{ alfalfa hay}$$
$$54.3 \times 0.60 = 32.6\% \text{ corn}$$
$$54.3 \times 0.40 = 21.7\% \text{ wheat}$$

Simultaneous Equations

Simultaneous equations can often be used to calculate the amounts of two feeds needed to meet *both* protein and TDN requirements (nutrients per day).

Example:

Formulate a ration to meet the daily total protein (TP) and TDN needs of a 50-kg finishing lamb.

Step 1

List daily nutrient requirements (protein and TDN) of animal in question.

Daily requirements for a 50-kg finishing lamb:

total protein	198	g/d
TDN	1,260	g/d

Step 2

      a. select two feedstuffs
      b. record TP and TDN values

	Shelled corn	Red clover hay
TP	10%	14.9%
TDN	89%	59%

Step 3

Let   X = wt. of shelled corn DM required
           Y = wt. of red clover hay DM required

Then:

Equation 1 (Protein equation)

$$0.10X + 0.149Y = 198 \text{ g TP}$$

Equation 2 (TDN equation)

$$0.89X + 0.590Y = 1260 \text{ g TDN}$$

Step 4

Solve the two simultaneous equations:

$$0.10X + 0.149Y = 198$$
$$0.89X + 0.59Y = 1260$$
$$0.59Y = 1260 - 0.89X$$
$$Y = 2135.6 - 1.508X$$
$$0.10x + 0.149 (2135.6 - 1.508X) = 198$$
$$X = 964 \text{ g shelled corn DM}$$
$$Y = 2135.6 - 1.508 (964)$$
$$Y = 682 \text{ g red clover hay DM}$$

Step 5

Check:

$$(0.10)(964) + (0.149)(682) =$$
$$(0.89)(964) + (0.59)(682) =$$
$$1,260 \text{ g TDN supplied}$$

The final step would be the conversion of the DM amounts to as-fed amounts.

Algebraic Method for Formulating a Grain/Mineral/Vitamin Mix

The method involves the computation of a grain ration for a certain percent of protein or percent of other nutrients such as Ca, P, etc., if desired. For this discussion we will consider protein. We basically formulate to 100 parts or 100% with the amounts of mineral and vitamins locked at the desired levels, the grains or energy sources being X, and the protein sources being the remaining "floating" portion minus the energy (grain sources).

Basic Formula (see below for explanation)

$$\underset{\text{desired}}{100(\% \text{ CP})} = \underset{\text{grain}}{X (\% \text{ CP})} + \underset{\text{protein}}{(97 - X) (\% \text{ CP})} +$$

$$\underset{\text{mineral}}{1.5(0)} + \underset{\text{salt}}{1.0 (0)} + \underset{\text{vitamin}}{.5 (0)}$$

In the above we are formulating for 100 parts with a specified percent CP on the left. Our ingredients on the right of the equal sign are:

(1) Locked ingredients: we lock in a mineral mix at 1.5%, salt at 1.0%, and vitamin premix at 0.5%, with each containing 0% protein.
(2) Grain – we let X equal the grain ingredient (can be several grains if in locked proportions)
(3) Protein source – Because three parts are locked above, then 97 parts are either grain or protein, thus (97 – X).

The method has some advantages, the main one of which is its flexibility in using several ingredients. Several ingredients can be locked in and several grain sources may be used. Perhaps a limitation is that we are only using one nutrient—in our example, *protein*.

To illustrate, we will assume the following, also locking in 5% molasses:

Formulate an 18% CP grain ration

Use corn (10% CP) and oats (13% CP) as the grains in a 2:1 ratio

Use soybean meal (SBM) (50% CP) as the protein source

Lock in 5% molasses (3% CP)

Lock in similar minerals and vitamins as above:

$$\underset{\text{desired}}{100 \times 18} = \underset{\text{corn}}{[2X (10)]} + \underset{\text{oats}}{[X (13)]} +$$

$$\underset{\text{SBM}}{[(92.0 - 3X) \times 50]} + \underset{\text{mineral}}{1.5(0)} + \underset{\text{TM salt}}{1.0(0)} + \underset{\text{vitamin}}{.5(0)} + \underset{\text{molasses}}{5(3)}$$

Solve for X (oats)

$$1800 = 20X + 13X + 4600 - 150X + 15$$
$$2815 = 117X$$

$$\begin{array}{lll} X & = & 24.0 \text{ oat} \\ 2X & = & 48.0 \text{ corn} \\ 92 - 3X & = & 20.0 \text{ SBM} \\ & & 5.0 \text{ molasses} \\ & & 1.5 \text{ mineral} \\ & & 1.0 \text{ TM salt} \\ & & 0.5 \text{ vitamins} \end{array}$$

Check the protein level of the final mix:

$$48(0.10) + 24(0.13) + 20(0.50) + 5(0.03) +$$

$\quad$ corn $\qquad$ oats $\qquad$ SBM $\qquad$ molasses

$$1.5(0) + 1.0(0) + 0.5(0) = 18.07\% \text{ CP}$$

$\quad$ mineral $\quad$ TM salt $\quad$ vitamins

This method can be expanded to include other locked ingredients such as urea, to include other grains in different proportions, etc. Perhaps the most confusing aspects are to remember the locked ingredients and subtract from 100 to get the amount of protein, and to total up the number of X's and get that number in the amount of protein (in the above example, $92 - 3X$).

Micro-Math

Many minerals, vitamins, and feed additives are included in the rations of animals in extremely small quantities. It is critical that this process be conducted accurately. In some cases an over-consumption of a micronutrient or a feed additive can negatively affect animal performance or even be life-threatening. In many instances, especially in the case of additives classified as drugs, only a specially licensed mill may handle the product, and producers must purchase the additive as a component of a premix or complete supplement. A good rule of thumb to follow with typical farm-level mixing equipment is to never add any ingredient to a ration at a level less than 20 pounds per ton. This is essential to ensure thorough mixing and dispersion of the ingredient. If the desired level of inclusion is less than

20 pounds per ton, it should first be premixed carefully with a carrier substance such as fine ground corn, soybean meal, corn gluten meal, or other feeds of relatively small particle size.

An example of incorporating required levels of minerals into a total ration is presented in a subsequent section. For our purposes here, the conversion factors in table 2-3 (p. 38) should be useful in calculating appropriate levels of micronutrients and feed additives to be added to rations. See pp. 228–230 for more information about unit conversions.

Pasture Allowance Example

As previously discussed, intake is largely unknown in free-grazing animals; thus ration balancing is an imprecise process at best. In practice we rely on the available forage, consumed to appetite, to meet most of the animal's nutrient needs. Typically a mineral/vitamin supplement is offered ad lib or possibly in limited amounts (for supplemental protein and/or energy source). Supplementation is discussed in chapter 4.

Therefore, the primary challenge in ration balancing for grazing animals is providing an adequate allowance of forage for the production level desired. With experience, reasonable estimates of forage intake can be made. If one has knowledge of the approximate nutritive value of the grazed forage, the nutritional adequacy of the diet can be estimated with acceptable accuracy.

Let's assume a group of 1,300-pound lactating beef cows are rotationally grazing a mixed red clover-orchardgrass sward in June. The forage has been maintained in a high quality vegetative state, thus our emphasis will be on providing an adequate amount of total DM.

The expected seasonal DM yield of our pasture under rotational grazing is 3.4 tons per acre (actual yield estimates could be used in place of

Table 2-3. Conversion factors for calculating appropriate levels of micronutrients and feed additives for rations.[a]

Convert from	To	By
g/t[b]	%	x 0.00011
g/t	lb/t	x 0.0022
g/t	g/lb	x 0.0005
g/t	mg/lb	x 0.5
%	g/t	÷ 0.00011
%	ppm	Move decimal four places to right
mg/kg	mg/lb	÷ 2.2
mg/lb	g/t	÷ 0.5
mg/lb	mg/kg	x 2.2
mg/lb	mcg/g	x 2.2
mg/lb	mg/g	÷ 454
mg/lb	ppm	x 2.2
mcg/g	mcg/lb	÷ 2.2
mcg/kg	mcg/lb	÷ 2.2
ppm	g/t	÷ 1.1
ppm	%	Move decimal four places to left
ppm	mg/lb	÷ 2.2

[a] See also pp. 228–230 for additional unit conversions.
[b] g = grams, t = tons, mg = milligrams, kg = kilograms, mcg = micrograms, ppm = parts per million.

this average value. Also, we note that the proportion of total yield available in June for this species mix (Group I) is 30%. Plugging these numbers into our worksheet (table 2-4) and completing the steps indicated below table 2-4, we can calculate **pasture carrying capacity**. We have a month-long DM demand of 1,560 pounds per head (1,300 x 0.04 x 30) and an available forage of 2,040 pounds per acre (3.4 x 2,000 x 0.30). In this example a reasonable allowance to meet the needs of our target animals is 1.3 animal units per acre (2,040 ÷ 1,560). Again, this assumes that the forage is of good quality in vegetative state.

Table 2-4. Pasture carrying capacity worksheet.

Month	Animal demand				Pasture supply				Carrying capacity
	1	2	3	4	5	6	7	8	9
	Body wt. (Avg. lb)	Intake factor	Days	(= 1x2x3) DM (Demand/hd)	Grazing method (C or R)	Seasonal DM yield (lb.)	Monthly available (% ÷ 100)	(= 6x7x2000) DM available (lb/ac)	(= 8 ÷ 4) Anim./ac
Jan									
Feb									
Mar									
Apr									
May									
Jun									
July									
Aug									
Sep									
Oct									
Nov									
Dec									

1 Estimate the average body weight of the animals on pasture.

2 Use your best estimate of DM intake expressed as a fraction of body weight. For example, you might use a liberal average DM intake of 4% of body weight. This allows 3% for actual intake, 0.5% trampling loss, and 0.5% safety factor.

3 Enter the days in a given month.

4 Multiply columns 1 × 2 × 3 to calculate the amount of DM needed per animal for the month.

5 Enter "C" for continuous or "R" for rotational grazing.

6 Obtain the seasonal DM yield from table 2-5 (p. 41) for a given forage species(s) based on grazing method used. If you have knowledge of actual yield, use your own data.

7 Enter the monthly availability of the seasonal yield from table 2-6 (p. 41).

8 Multiply column 6 × column 7 × 2000 to calculate the amount of DM available per acre for the month.

9 Divide column 8 by column 4 to estimate the carrying capacity per acre for the month.

Group number	Species	Seasonal hay	DM Yield CG[b]	T/Ac[a] RG[c]
I	Alfalfa	5.4	4.1	4.6
	Alfalfa — orchardgrass	5.5	4.1	4.6
	Alfalfa — smooth brome	5.4	4.1	4.6
	Alfalfa — timothy	5.4	4.1	4.6
	Alfalfa — perennial ryegrass	5.4	4.1	4.6
	Alfalfa — reed canarygrass	5.4	4.1	4.6
	Red clover	4.1	3.1	3.4
	Red clover — orchardgrass	4.1	3.1	3.4
	Red clover — smooth brome	4.1	3.1	3.4
	Red clover — timothy	4.1	3.1	3.4
	Red clover — perennial ryegrass	4.1	3.1	3.4
	Red clover — reed canarygrass	4.1	3.1	3.4
II	Tall fescue	4.5	3.4	3.9
	Red clover — tall fescue	4.1	3.1	3.4
	Birdsfoot — tall fescue	2.7	2.1	2.3
III	Birdsfoot — trefoil	2.7	2.1	2.3
	Birdsfoot — orchardgrass	2.7	2.1	2.3
	Birdsfoot — smooth brome	2.7	2.1	2.3
	Birdsfoot — timothy	2.7	2.1	2.3
	Birdsfoot — perennial ryegrass	2.7	2.1	2.3
	Birdsfoot — reed canarygrass	2.7	2.1	2.3
IV	Stockpiled forage:			
	Tall fescue	5.2	4.0	4.4
	Tall fescue — red clover	4.5	3.4	3.9
V	Kentucky bluegrass — white clover	2.5	1.9	2.4
	Orchardgrass	4.5	3.4	3.9
	Smooth brome	3.6	2.9	3.1
	Timothy	4.1	3.1	3.4
	Perennial ryegrass	3.2	2.3	2.7
	Reed canarygrass	4.5	3.4	3.9
VI	Sorghum x sudangrass	5.4	4.1	4.6
VII	Spring-seeded brassicas Turnips, rape, kale, swedes	4.5	3.4	3.9

Table 2-5. Forage species and yields for typical Northeast pastures.

Table 2-5 (continued). Forage species and yields for typical Northeast pastures.

Group number	Species	Seasonal hay	DM Yield CG[b]	T/Ac[a]RG[c]
VIII	Fall-seeded brassicas	3.6	2.7	3.1
IX	Warm-season grasses Big bluestem, switchgrass, indiangrass	4.1	3.1	3.4

[a] Total seasonal DM yield, tons per acre.
[b] Continuous grazing. Be aware that some species will not persist under continuous grazing management, especially poor legume stands.
[c] Rotational grazing.

Source: Adapted from The Penn State Agronomy Guide. 2002. *Eston Martz, Ed. College of Agricultural Sciences. The Pennsylvania State University, University Park, PA.* © *The Pennsylvania State University. Used with permission.*

Table 2-6. Seasonal distribution of yields for typical Northeast pastures.[a]

Group No.	Month											
	Jan	Feb	Mar	Apr	May	Jun	Jul	Aug	Sept	Oct	Nov	Dec
I		--	--	--	--	30	20	15	15	15	5	--
II		--	--	--	5	25	20	15	15	15	5	--
III	--	--	--	--	27	25	15	15	15	3	--	--
IV	--	--	--	5	30	20	13	--	--	15	12	5
V	--	--	--	5	38	20	10	10	12	5	--	--
VI	--	--	--	--	--	20	33	33	14	--	--	--
VII	--	--	--	--	--	--	40	40	20	--	--	--
VIII	--	--	--	--	--	--	--	--	--	--	50	50
IX	--	--	--	--	--	20	40	30	10	--	--	--

[a] Refer to table 2-5 for species listings and seasonal yield information.

Source: Adapted from The Penn State Agronomy Guide. 2002. *Eston Martz, Ed. College of Agricultural Sciences. The Pennsylvania State University, University Park, PA.* © *The Pennsylvania State University. Used with permission.*

Using methods previously described and reasonable estimates of forage nutritive value, one could calculate nutrient intake given the amount of DM made available.

SUMMARY

The basic nutrients of energy, protein, minerals, vitamins, and water are common to all livestock whether grazing pastures or offered diets in confinement. When grazing pastures, the ruminant has a wide variety of herbages from which to choose. Combinations of grasses, legumes, and nontraditional plants determine dietary quality, but selectivity by the grazing ruminant becomes an important factor in determining overall intake of nutrients. Environmental factors influence intake and physiological nutrients demanded by livestock, and also influence plant growth and nutritive value. Plant development and nutritive value are further influenced by pasture management. If forage growth is to be harvested for conserved feed, stage of plant maturity, harvesting methods, and postharvest storage interact to determine nutrient content. Concentration of nutrients, or more importantly the ratio of one nutrient to another nutrient in forages and feedstuffs, becomes an important issue in optimizing nutrient use and livestock performance, and also in determining supplementation needs. The energy to protein ratio becomes an important consideration to optimize protein use in ruminants, especially grazing livestock. Protein supplementation can be expensive, but knowing the basics of nutrition, nutrient interactions, and ration balancing techniques can help to maximize nutrient use by livestock and minimize inputs into livestock production systems for higher economic returns and profit margins.

CHAPTER 3
Beef Nutrition and Management

Edward C. Prigge, John B. Hall, and John W. Comerford

NUTRITION REQUIREMENTS OF THE HERD

Cows

The production of beef cattle worldwide has been successfully adapted to a wide variety of ecosystems. Within a production region, forage quality can vary according to season, soil types, local weather conditions, topography, and management. The seasonal variation in forage quantity and quality is a major consideration in the management of beef cattle. The nutrition requirements of the cow will vary throughout the year according to the demands of gestation and lactation. Consequently, in the northeastern United States one of the most obvious management considerations for beef production is to align the changing nutrition demands of the herd with the seasonal variations in forage growth and quality.

To accomplish this we must know the degree to which the physiological changes in the yearly production cycle of the cow affect her nutritional requirements. The nutrient requirements for a 1,200-pound beef cow (43) are outlined in table 3-1 (p. 44), in regard to months after parturition and level of lactation. The requirements for energy and protein are at their lowest when the cow is in midgestation and dry and are the greatest during the second month after calving, coinciding with peak milk production. Breeds of beef cattle known for high milk production, such as the Simmental, will require diets of greater nutrient densities when compared to breeds

with lower levels of milk production, such as the Angus or Hereford. Average milk production levels of various breeds of cattle common to the Northeast are reported in table 3-2 (p. 45). Considerable variation in milk production exists within a breed. Consequently, it is not unusual for herds selected intensely for high weaning weights and milk production to greatly exceed these averages.

The typical grazing season in most of the Northeast will range from late April or early May to November or December. Most cows will calve in February through April to use the high quality pasture during peak lactation and for the breeding season, when the extra energy could enhance success. The optimal calving time for a particular herd within the season (February–April) should vary based on level of milk produced by the cow and quality of winter feeds. Cows with high nutrient requirements and low quality winter feed would suggest that calving later in the season would be the most desirable for optimal reproductive efficiency and lactation. Conversely, cows with low nutrient requirements in conjunction with high quality winter feed would suggest that calving in the early part of the calving season would result in greater economic returns because of greater calf weight at marketing.

In general there is little risk that quality of forage will pose a problem in meeting nutrient requirements during the grazing season, although quantity of forage is sometimes a problem in the summer and fall. Conversely, winter feed quality often is insufficient to meet the nutrient requirements of the cow herd. Average energy (total digestible nutrients [TDN] and net energy for maintenance [NEm]) and protein values of mixed grass-legume and grass hays are reported in table 3-3. If we use the more optimistic TDN value rather than the NEm value for the hays, based on the standard deviations, about 17% of the hays analyzed will not meet the requirements for an average-producing beef cow in peak lactation.

First-Calf Heifers

The nutritional needs of first-calf heifers are even greater than those of mature cows because of the additional needs of growth. These requirements

| Table 3-1. Diet nutrient density requirements on a dry matter basis for beef cows. |

	Months since calving											
	1	**2**	**3**	**4**	**5**	**6**	**7**	**8**	**9**	**10**	**11**	**12**
1200 lb mature weight, 10 lb peak milk												
TDN[a] (%)	55.3	56.0	53.7	52.9	52.1	51.5	44.9	45.8	47.1	49.3	52.3	56.2
NEm	0.54	0.55	0.51	0.50	0.49	0.48	0.37	0.38	0.41	0.44	0.49	0.55
CP (%)	8.43	8.79	8.13	7.73	7.33	7.00	5.99	6.18	6.50	7.00	7.73	8.78
Ca (%)	0.24	0.25	0.23	0.21	0.20	0.19	0.15	0.15	0.15	0.26	0.25	0.25
P (%)	0.17	0.17	0.16	0.15	0.14	0.14	0.12	0.12	0.12	0.16	0.16	0.16
1200 lb mature weight, 20 lb peak milk												
TDN (%)	58.7	59.9	57.6	56.2	57.7	53.4	44.9	45.8	47.1	49.3	52.3	56.2
NEm	0.59	0.61	0.57	0.55	0.53	0.51	0.37	0.38	0.41	0.44	0.49	0.55
CP (%)	10.10	10.69	9.92	9.25	8.54	7.92	5.99	6.18	6.50	7.00	7.73	8.78
Ca (%)	0.29	0.31	0.29	0.26	0.24	0.22	0.15	0.15	0.15	0.26	0.25	0.25
P (%)	0.19	0.21	0.19	0.18	0.17	0.15	0.12	0.12	0.12	0.16	0.16	0.16
1200 lb mature weight, 30 lb peak milk												
TDN (%)	61.6	63.2	60.8	59.0	57.0	55.2	44.9	45.8	47.1	49.3	52.3	56.2
NEm	0.64	0.66	0.62	0.59	0.56	0.54	0.37	0.38	0.41	0.44	0.49	0.55
CP (%)	11.51	12.25	11.41	10.55	9.61	8.75	5.99	6.18	6.50	7.00	7.73	8.78
Ca (%)	0.34	0.36	0.34	0.31	0.27	0.25	0.15	0.15	0.15	0.26	0.25	0.25
P (%)	0.22	0.23	0.22	0.20	0.18	0.17	0.12	0.12	0.12	0.16	0.16	0.16

[a] TDN-total digestible nutrients. NEm-net energy for maintenance (Mcal/lb); 1 Mcal = 1 million calories = 1,000 kilocalories. CP-crude protein. Ca-calcium. P-phosphorus.

Source: Reprinted with permission from the National Academies Press, Copyright 1981, National Academy of Sciences.

are listed in table 3-4 (p. 46), and based on these requirements, more than half the hays submitted for analysis to the Northeast DHIA Lab (table 3-3) would not meet the requirements for lactating first-calf heifers. This suggests that these cattle should be managed differently from mature cows and that supplemental feed should be provided

in many cases. In addition crude protein in the hay is marginal for first-calf heifers, suggesting a need for protein supplementation.

Yearlings

In the production of yearling cattle (steers and heifers) the objective is often to produce the most economical gains possible. Unlike cows, in which additional energy intake beyond requirements is often not beneficial, yearling cattle will invariably respond to increased energy intake by increasing average daily gain, provided other nutrients are not limiting. For market cattle, because of the differential costs of feed inputs such as hay versus pasture, the cost of gain is often minimized when gain is less than maximum over the wintering period. With replacement heifers the objective is to achieve sufficient gain to be pubescent prior to breeding season.

Stockers

The stocker phase of production involves the growth of cattle from weaning until they enter the feedlot for finishing. Cattle that go into stocker programs must be of appropriate weight and frame size to be suitable for finishing in a feedlot, usually within 140 days after completing the stocker phase of production. The appropriate management of stocker cattle is influenced

Table 3-2. Peak milk production for various breeds.	
Breed	**Peak milk yield (lb/d)**
Angus	17.6
Charolais	19.8
Hereford	15.4
Limousin	19.8
Shorthorn	18.7
Simmental	26.5

Source: Adapted from National Research Council. 1996. Nutrient Requirements of Beef Cattle, 7th ed. National Academy Press, Washington, D.C. Appendix table 4, p. 214.

Table 3-3. Average nutrition value of hay in the Northeast.[a]			
Hay type	**Crude protein[b] (%)**	**TDN[b] (%)**	**NEm[b] (Mcal/lb)**
Grass – legume hay	12.1 ± 3.3	60.1 ± 2.9	0.52 ± 0.6
Grass hay	10.6 ± 3.2	61.2 ± 3.2	0.53 ± 0.7

[a] Mean of hay samples submitted to the Northeast Dairy Herd Improvement Association (DHIA) Laboratory in 1995.

[b] Dry matter basis ± standard deviation.

Table 3-4. Diet nutrient density requirements on a dry matter basis for 2-year-old heifers nursing calves, 10 pounds milk/day.				
Item	700 lb cow	800 lb cow	900 lb cow	1,000 lb cow
TDN (%)	65.1	63.8	62.7	61.9
NEm (Mcal/lb)	0.67	0.66	0.64	0.62
CP (%)	11.3	10.8	10.71	10.0
Ca (%)	0.36	0.34	0.32	0.31
P (%)	0.24	0.24	0.23	0.23

Source: Adapted from National Research Council. 1996. Nutrient Requirements of Beef Cattle, *7th ed. National Academy Press, Washington, D.C.*

by many factors, of which weaning weight is a major consideration.

Generally there are three production alternatives to consider when managing stocker cattle. Producers may retain the cattle for a wintering period only; this option is probably most suitable for heavier weight cattle (600 pounds plus at weaning). Producers using this type of program should feed an economical ration with enough energy to result in moderate gains. In most instances a properly balanced, corn silage-based diet will result in an average daily gain of about 2 pounds. A 600-pound steer entering this program fed for 120 days would be expected to weigh about 840 pounds at the end of the winter feeding period and would be a suitable weight to enter a feedlot for finishing rather than for grazing the following spring.

Many stocker producers elect to winter weanling calves at a low rate of gain (0.5–0.75 pound per day) and pasture them the following grazing season. A moderate quality hay will produce gains at this level. On well managed cool-season grass pastures a gain of about 1.8 pounds per day can be expected for steers averaged over the entire grazing season. For the winter feeding and spring/summer grazing, a total gain of more than 400 pounds can be expected. Lightweight steer calves less than 500 pounds at weaning would be most desirable for this type of program. Because of hay costs and the low rate of winter gain the costs per pound of gain for the winter feeding period are high; however, they are offset by low-cost pasture gains, which are enhanced by compensatory gain.

The third type of stocker program is to maintain cattle for the grazing season only. Because about 200–300 pounds of gain can be expected for steers over the grazing season, steers weighing about 600 pounds at the start of the grazing season are the most desirable. Heifers weigh about 20–40 pounds less at weaning than comparable steers and about 150 pounds less when finished to choice; consequently, adjustments in stocker cattle management must be made to accommodate heifers as finished cattle.

Replacement Heifers

The nutritional requirements of the replacement heifer from weaning to breeding depend on the weaning weight and the expected weight to achieve puberty for that particular breed and type of cattle. Puberty appears to be more a function of weight rather than age (55). Heifers with a 500-pound weaning weight would be expected to reach puberty at 650–700 pounds. These heifers need to gain 200-plus pounds each to ensure that the majority reach puberty by breeding season if they are to calve as 2-year-olds. If weaned October 1 and breeding is expected the following May 15 (226 days), the heifers must gain an average of 0.9 pound per day. If we assume the animals are grazing on pasture about 70 days and we allow for shrinkage during weaning, they should average at least 1.5 pounds per day on cool-season grass pastures with adequate herbage. This indicates that an average daily gain of 0.6 pound per day would have to be achieved during the winter feeding period to ensure puberty prior to the breeding season. Consequently, average quality hay with a TDN level in the high fifties and NEm (net energy for maintenance) and NEg (net energy for gain) levels of about 0.58 and 0.33 Mcal per pound, respectively, must be fed. If the weaning weight of the calf is lighter and/or the frame score of the herd is above average, a faster rate of gain is needed and a higher quality diet must be fed. Changes in hay management would be necessary to improve the quality of hay or supplement feed would have to be provided under these conditions.

BODY CONDITION SCORE

Scoring

Body condition scoring is an essential technique for nutritional management of the beef herd. By using body condition scores (BCSs), producers can evaluate if their nutrition program is meeting the needs of the herd. In addition, body condition scoring can identify individual animals that are not suited for the producer's environment or animals that are underproductive. These scores are highly correlated with animal productivity, especially reproductive performance.

Body condition scoring is a subjective assessment of energy reserves of the cow. It involves assigning numerical scores to cows based on their relative amount of body energy reserves, primarily fat. Although several different scoring systems have been developed around the country (4, 28, 50,), the 9-point body condition scoring system (50) is the most common. This system ranges from BCS 1 = emaciated to BCS 9 = obese. Ideal body condition for cows is BCS 5 or 6.

Producers can easily learn to assign BCSs. A basic knowledge of the key points of each BCS 1–9 is essential to understanding how to score cows. The primary areas of the cow to examine when condition scoring are the hooks, pin, tailhead, spine, spinous and lumbar processes, flank, ribs, and brisket. Working with a producer or extension agent who knows the BCS system is the easiest way to learn and refine your scoring abilities. Descriptions of BCS 1–9 are found in table 3-5 (p. 48).

Influence of Body Condition on Reproduction

Energy availability from the diet and energy reserves from body fat greatly affect reproductive efficiency in cattle. Research over the last 30 years has demonstrated that cattle that receive insufficient nutrients, especially energy, immediately before and after calving have poor pregnancy rates. Cattle that were thin at calving or lost weight between calving and breeding were later in returning to heat and had lower pregnancy rates during the breeding season.

Because BCSs are good indicators of body energy reserves, researchers have focused on

	Table 3-5. The nine-point cow body condition scoring system.
Score	**Description**
Thin	
1	Severely emaciated; starving and weak; no palpable fat detectable over back, hips, or ribs; tailhead and individual ribs prominently visible; all skeletal structures are visible and sharp to the touch; animals are usually disease stricken. Under normal production systems cattle in this condition score are rare. Cattle will contain about 3.8% body fat.
2	Emaciated; similar to BCS 1, but not weakened; little visible muscle tissue; tailhead and ribs less prominent. Cattle will contain about 7.5% body fat.
3	Very thin; no fat over ribs or in brisket; backbone easily visible, slight increase in muscling over BCS 2. Cattle will contain about 8–11% body fat.
Borderline	
4	Borderline; individual ribs noticeable but overall fat cover is lacking; increased musculature through shoulders and hindquarters; hips and backbone slightly rounded versus sharp appearance of BCS 3. Cattle will contain about 12–15% body fat.
Optimum	
5	Moderate; increased fat cover over ribs, generally only 12th and 13th ribs are individually distinguishable; tailhead full, but not rounded (about 17–19% body fat).
6	Good; back, ribs, and tailhead slightly rounded and spongy when palpated; slight fat deposition in brisket (about 20–22% body fat).
Fat	
7	Fat; cow appears fleshy and carries fat over the back, tailhead, and brisket; ribs are not visible; area of vulva and external rectum contains moderate fat deposits; may have slight fat in udder (about 24–26% body fat).
8	Very fat; squared appearance due to excess fat over back, tailhead, and hindquarters; extreme fat deposition in brisket and throughout ribs; excessive fat around vulva and rectum, and within udder; mobility may begin to be restricted (about 28–30% body fat).
9	Obese; similar to BCS 8, but to a greater degree; majority of fat deposited in udder limits effective lactation. Under normal production systems cattle in this condition score are rare (greater than 32% body fat).

Source: Encinias, A.M., and G. Lardy. 2000. Body condition scoring I: Managing your cow herd through body condition scoring. *Beef InfoBase, Version 1.2. Adds Center, Inc., Madison, WI.*

the relationship between body composition and reproduction. Several studies found that BCS at calving and BCS at the beginning of the breeding season were the most important indicators of reproductive performance (45, 56). BCS at calving has the greatest effect on pregnancy rate during a controlled breeding season (33).

Impact of BCS at Calving

Mature cows must calve at BSC 5 or above to maximize pregnancy rates in the following breeding. Cows calving in BCS ≤ 4 had a 9–29% lower pregnancy rate compared to cows calving at BCS ≥ 5 (40, 54). Pregnancy rates for cows of various BCSs are illustrated in figure 3-1. Changes in BCS between 4 and 6 have a greater impact on pregnancy rate than changes in BCS above 6 or below 4 (54). Little improvement in pregnancy rates is seen when cows calve in BCS above 6. Pregnancy rate does not get much worse below BCS 4.

In addition to the overall decrease in pregnancy rates, cows calving at BCS ≤ 4 that do conceive become pregnant later in the breeding season (table 3-6, p. 50). As a result, these cows calve later in the calving season. Late calving cows are more likely to fail to conceive during a controlled breeding season. Calves born late in the calving season will be lighter at weaning than calves born early in the calving season. At weaning, calves will be approximately 35 pounds lighter for every 21-day delay in calving (34).

First-calf heifers are even more sensitive to the effects of BCS at calving on pregnancy rates. Dramatic decreases of 40–50% (figure 3-2, p. 50) occur as heifers drop from BCS 6 to 4 (18, 56). In contrast to mature cows, heifers exhibit a significant decrease of approximately 16% in pregnancy rate between BCS 6 and 5. Therefore, the optimum BCS at calving is 6 or 7 in heifers.

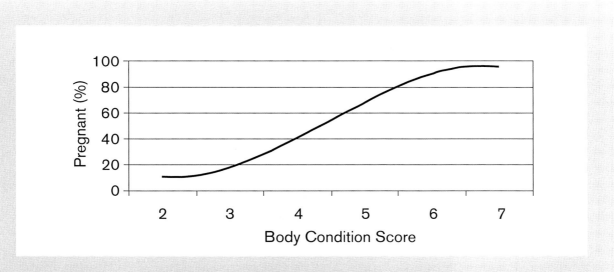

Figure 3-1. Relative influence of body condition score at calving on pregnancy rate.

Source: Adapted from Selk, G. E., R. P. Wettemann, K. S. Lusby, J. W. Oltjen, S. L. Mobley, R. J. Rasby and J. C. Garmendia. 1988. Relationships among weight change, body condition, and reproductive performance of range beef cows. J. Anim. Sci. 66: 3153–3159.

Animal class (reference)	BCS	Day of breeding season		
		20 d	40 d	60 d
Mature cows (50)		Cumulative % pregnant		
	4	41	67	84
	5	51	79	91
First-calf heifers (56)		Cumulative % pregnant		
	4	27	43	56
	5	35	65	80
	6	47	90	96

Table 3-6. Effect of BCS at calving on cumulative pregnancy rates.

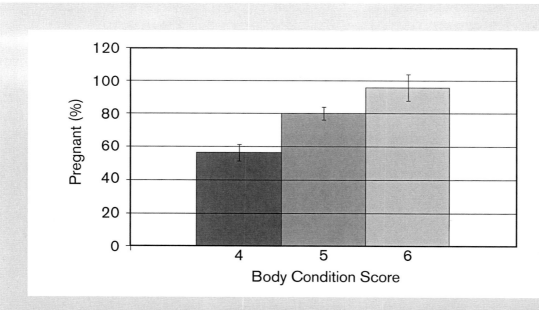

Figure 3-2. Effect of body condition score at calving on subsequent pregnancy rate in first-calf heifers.

Source: Adapted from Spitzer, J. C., D. G. Morrison, R. P. Wettemann, and L. C. Faulkner. 1995. Reproductive responses and calf birth and weaning weights as affected by body condition at parturition and postpartum weight gain in primiparous beef cows. J. Anim. Sci. 73: 1251–1257.

Limited data indicate that cows that calve at BCS ≥ 7 and heifers that calve at BCS ≥ 8 may have impaired reproduction during the breeding season (50, 28). Producers should be cautious in drawing any conclusions about "fat" cows, as the numbers of cows with BCS ≥ 7 in these studies were limited. In addition, it is not always clear if cows were in high BCS due to nutritional manipulation or physiological factors. For economic as well as reproductive reasons, producers should try to keep cows in the BCS 5–7 range.

BCS Changes from Calving to Breeding

Although BCS at calving has the greatest impact on cow reproduction, changes in body weight and BCS postpartum will also affect reproductive performance. Change in BCS postpartum dramatically affects cows that calve at BCS ≤ 4. Low BCS cows that continue to lose weight and BCS after calving are unlikely to become pregnant during the breeding season. Thin cows that continue to lose BCS have a longer interval from calving to first heat (postpartum interval). This means a low percentage of these cows (0–40%) are cycling by the start of the breeding season (28, 45). Often it may take more than 80–100 days for these cows to begin cycling. As a result of delayed cyclicity, thin cows losing BCS postpartum have low pregnancy rates, often 30–50% lower than their well-fed counterparts.

Cows that calve in BCS ≥ 5 are less sensitive to the effects of postpartum nutrition, but reproductive ability of cows losing weight after calving may be compromised. The interval from calving to heat lengthens and pregnancy rate decreases in fleshy cows that lose weight postpartum. Researchers in Oklahoma reported an increase of 22 days in postpartum interval and a reduction in pregnancy rate of 14% in cows that calved at an average BCS 5.4 but lost 1 BCS before the start of the breeding season (10).

Producers often hope that feeding thin cows to increase BCS and body weight after calving will solve their reproductive problems. Once a cow has calved, her metabolism shifts to support milk production. Therefore, only a portion of the additional energy fed to postpartum cows is available to combat the effects of low BCS. Cows that calve at BCS ≤ 4 and are fed high energy diets postpartum usually have a 10–20% reduction in cyclicity compared to moderate-flesh cows that maintain their weight (45). A reduction in the percentage of cows cycling diminishes the chances of high pregnancy rates. Occasionally, these refed cows have conception rates equal to cows maintained in better body condition (50, 28).

First-calf heifers (primiparous) are less responsive than cows that have previously calved to attempts to fatten them after calving. First, these primiparous cows have a longer postpartum interval and are more sensitive to the negative effects of poor body condition on reproduction. Because they are growing as well as lactating, enhancing dietary energy intake does not readily enhance reproductive performance. Most studies indicate that thin heifers that are adequately fed during early lactation have lower pregnancy rates at the end of the breeding season compared to heifers that calve at BCS ≥ 5 and maintain their body weight (33, 56). Distribution of conception is also affected, as thin, refed heifers tend to breed later in the breeding season.

Summary

BCS at calving is the most important factor influencing rebreeding success in beef cattle; changes in BCS postpartum may also affect reproduction. To maximize pregnancy rates, cows should be managed to calve in BCS 5–6, whereas heifers should calve in BCS 6–7. After calving, cows and heifers should maintain their ideal calving BCS. Cows or heifers calving in less than optimum BCS should be fed to gain weight during the postpartum period. However, good postpartum nutrition does not always repair the damage caused by calving in low BCS. Overconditioning cattle so BCS exceeds 7 at calving should be avoided.

Dietary Energy and Changes in BCS

The major factor influencing the BCS of a cow is the amount of energy she consumes. Cows consuming less than the required amount of energy will lose body condition; cows that consume more energy than required will gain in condition. Excess body condition of the cows can be used as a buffer for short-term dietary energy deficiencies. At times during the grazing season it may be difficult to increase the BCS of the cow herd. At the beginning of the grazing season when forage quality of cool-season grasses is at its greatest, the energy demands of the cow for lactation often prevent the accumulation of body fat. In midsummer when lactation demands taper off, pasture quantity and quality often limit the accumulation of body fat. In the fall, pasture growth and quality often increase because the cool temperature favors the growth of cool-season grasses. The energy requirement for the cow, after weaning of the calf, is also the lowest at this time. During this period it is not uncommon for cows to gain 3–5 pounds per day (2). At this rate of gain a cow can go from a BCS of 4 to 5 in about 20 days (43). Thus adequate pasture after weaning of the calf can be used to compensate for feeding hay of lower quality than requirements during the winter. In most areas of the Northeast, pasture can be grazed into November and in some cases well into December or January.

Another alternative to increase BCS if conserved forage for winter feeding is lower in quality than desirable is to wean early. This would lengthen the time period when it is relatively easy to increase BCS because of the difference between energy requirements of the dry cow and energy density of the feed supply (pasture).

According to the National Research Council (NRC) (43), the authoritative source for information on nutrient requirements for livestock in the United States, the nutrient requirements (table 3-1) for a dry cow increase to a large extent during the last third (months 10 to 12) of pregnancy. If we estimate the NEm intake from NRC (43) requirements for a 1,200-pound beef cow 2 months prior to calving, it would take approximately 160 days to increase BCS from 4 to 5 if the diet was 10% above the cow's energy requirement. Because most cow/calf producers rely on hay for winter and because of the narrow spread between the cow's energy requirements and the energy density of the hay, it may be impractical to try to increase BCS at this time. In many cases a relatively expensive high-energy diet involving supplemental concentrate would have to be fed. After calving, because of the demands of lactation, it may be even more difficult to increase BCS and therefore decrease postpartum interval (41). These results suggest that the opportunity to easily and economically increase BCS is greatest immediately after weaning.

If we fed a diet 10% lower in energy than required during the eleventh month postcalving, a cow with an initial score of 5 would lose a little more than 0.25 BCS. However, during the 10th month energy requirements of the cow are 10% lower and this diet would meet requirements at this time and exceed requirements during the ninth month. During the twelfth month (1 month prior to calving) this diet would be about 18% below energy requirements; however, most beef producers feed a higher quality hay (second cutting) at this time. In summary, if cows are in adequate body condition going into the season when conserved forage is fed, the cows can at times be fed diets that are lacking energy without major changes in BCS. This again emphasizes the importance of cows being in adequate condition prior to being fed their wintering diets.

CATTLE TYPE AND MANAGEMENT

Breed and Cattle Type to Optimize Grazing

The appropriate breeding program will be different for every farm that produces beef cattle. Beef cattle production is a highly segmented, diverse industry that exists under many environments and with many managers. Before planning any breeding program, the manager should answer four questions: (i) what are our goals, (ii) where are we now with regard to these goals, (iii) where would we like to be, and (iv) how can we get to where we want to be.

Setting goals for the production and breeding program is not concerned with the present; it is concerned with the future. Even the cattle owner who has beef cattle just to eat the grass so he does not have to mow it has a goal for production. Consequently, a breeding program to suit that low level of management is very important. Before any genetic improvement can be made, there must be some target for the process. This target may simultaneously be several things: improvement of production traits such as weaning weight, the merchandising of cattle from the herd, or the sale of forages and grain produced on the farm through the production of beef cattle.

To answer the goals questions, some basic tools are necessary. All cattle in the herd must be identified with ear tags, tattoos, or brands. Secondly, some recordkeeping system must be developed to measure changes in the performance of the herd. Finally, there must be equipment (e.g., scales) and facilities (e.g., working chutes) available to obtain the desired information. The equipment and facilities need not be elaborate or expensive, but they are essential to genetic improvement.

What are the resources available in the enterprise? These would include:

- land, buildings, and working facilities,
- capital,
- management skills and resources,
- the potential to endure risk, and
- commitment to a long-term program.

The physical assets are the place to start in the planning process. The determination of how many cows are appropriate for the land available is a fundamental task. This includes both the total acreage and the potential of the land to support cattle in pastures and crop production. Another fundamental task is to determine how much feed can be produced or made available for the cattle. This is important in a breeding program because all cows, both within and across breeds, do not have the same nutritional needs every day. The weight of the animal, the stage of production, and the environment will determine the nutritional requirements. Although the environment will not easily be controlled, the effects of the stage of production (e.g., the relative milk production potential) and the weight of the animal (e.g., yearling weight selection) can be highly related to the breeding program.

Management skills and resources are also part of the breeding system. The breeder must know how management of pastures, nutrition, reproduction, marketing, labor needs, and facilities will change as the program progresses. Additionally, the beef business requires large amounts of capital and has a relatively slow cash flow.

The breeder must determine how much risk is tolerable and find ways to manage that risk on a long-term basis. Finally, a breeding program with beef cattle is a slow process. The results of a single generation of selection will probably not be known for at least 3 years; a well planned rotational crossbreeding system will be disrupted for at least one generation of production by any deviation from the planned matings; and changing the goals of a selection program expands the time needed to reach any single production goal.

Market

Unfortunately, the question of market is often answered "after the fact." Determination of the market, how to meet the standards of the market, and the potential for marketing flexibility should be examined before designing a program. For example, in Pennsylvania and the eastern United States, there are many types of beef markets—purebred sales, high quality finished cattle, freezer beef, veal calves, cull cows, feeder calves, etc.—and each one has a unique set of standards that must be addressed by a well planned breeding and management program. The "cowboy logic" that a breeder should develop a cow herd to meet the restrictions of the environment and bull selection should meet the needs of the market has real value in planning a program. The selection of a market is closely related to the management and other resources available to the breeder. This is also a long-term activity, so markets should be stable, accessible, and competitive.

Genetics

The management program will dictate how outside genetics are used. In an artificial insemination program, for example, there are unlimited opportunities to incorporate new genetics into the herd. However, if artificial insemination is not part of the program, a close look at the genetics available locally will determine if there is sufficient quality (availability of performance information) and quantity (sufficient variation to select above-average animals) of potential breeding stock to meet the goals of the program. The latter will be true for breed selection as well as selection of individual animals.

The effect of cattle breed and type in a grazing system is a function of the maintenance and production requirements of the animal. The ability of grazed forages to serve the maintenance and production needs is determined by forage intake. Variations among, and probably within, breeds of cattle exist for both of these factors. Tables 3-7 and 3-8 (pp. 55 and 56) outline some of the production factors related to breed selection. For example, if a producer is interested in a breed with a large mature size, table 3-7 indicates that forage availability, as well as other factors listed as high, would have to be major considerations in selecting a breed such as Simmental, with a heavy mature weight.

Forage Intake

Forage intake on pasture is a function of forage palatability, grazing time, biting rate, sward density, and tensile strength of the forage (the dry matter [DM] available in each bite). Cattle will generally graze to satiation (fullness) when forage availability is ample; however, when forage availability is limited, grazing time becomes the limiting factor. Once forage height is less than 3 or 4 inches, grazing time becomes the major factor limiting forage intake. In theory, when forage availability is limiting, bite size and time available for grazing would be the same for both large and small breeds of cattle. If forage selectivity is limited, such as in a well managed rotationally grazed sward, the breed with the larger mature size would consume the same total amount of forage daily as a small breed. However, the large breed would consume

Market/resource	Mature size[a]	Milk production	Lean to fat ratio	Maternal traits	Hide color
Forage availability	H	H	L	MH	L
Labor availability	H	L	L	H	L
Purchased feed needs	H	H	LM	LM	L
Management ability	H	H	M	H	L
Grazing methods	M	M	L	H	L
Feeder calf production	M	H	H	H	MH
Finished beef	M	L	M	L	MH
Replacement heifers	M	H	M	H	H
Purebred production	H	H	H	H	H
Club calf production	M	L	H	M	H
Freezer beef production	M	L	H	M	L

Table 3-7. Relative emphasis for production factors and breed selection.

[a] Intensity of selection pressure: H-high, M-moderate, L-low.

a smaller proportion of its nutrient requirements than the small breed. On a sward of low height where forage selectivity of the cattle could be expressed, such as a patchy continuously grazed pasture, the large breed would compensate for a lower proportion of gut fill than the small breed by selecting for forage volume as opposed to quality to increase bite size. This would most likely result in the consumption of mature forage with a lower nutritive value. Consequently, under situations where forage availability on pasture is often limited, a smaller framed breed of cattle might be more desirable than large framed cattle.

Animal Maintenance

"Maintenance" is described as an animal at rest in a thermoneutral environment. This is usually determined as a function of animal weight. The usual calculation is animal weight raised to the 0.75 power. There is known variation among breeds of cattle from this calculation due to biological considerations of body composition, gut weight relative to total weight (guts have a higher maintenance requirement), and surface area. Environmental factors, particularly temperature, will also influence maintenance. As described by the NRC (43), heat production in cattle results from tissue metabolism and fermentation in the digestive tract. Cattle dissipate this heat through evaporation, radiation, convection,

Table 3-8. Breed crosses grouped by biological type for four production criteria.				
Breed group[a]	Growth/mature size	Lean to fat ratio	Age at puberty	Milk production
JerseyX	X[b]	X	X	XXXXX
Angus/Hereford	XX	XX	XXX	XX
Red PollX	XX	XX	XX	XXX
BrahmanX	XXXX	XXX	XXXXX	XXX
GelbviehX	XXXX	XXXX	XX	XXXX
SimmentalX	XXXXX	XXXX	XXX	XXXX
LimousinX	XXX	XXXXX	XXXX	X
CharolaisX	XXXXX	XXXXX	XXXX	X

[a] Sire bred with Hereford or Angus dam.
[b] Increasing number of Xs indicates relative difference between breeds.

Source: Adapted from Cundiff, L. V., K. E. Gregory, and R. M. Koch. 1984. Germ plasm evaluation program report No. 11. *Roman L. Hruska Meat Animal Evaluation Center, Clay Center, NE.*

and conduction. A constant body temperature is produced by regulation of heat production and dissipation. Lower critical temperature is reached when cattle cannot maintain body temperature by normal metabolic heat production, and maintenance feed needs increase. Conversely, when the upper critical temperature is reached, metabolism is slowed to reduce heat production by the reduction of feed intake. Cattle readily adapt to moderate to cool weather. Extremes in temperature, particularly when combined as heat plus humidity or cold plus precipitation, will increase maintenance feed needs. The most vulnerable to temperature extremes are newborns, market-weight feedlot cattle, and breeding females.

Various reports have documented variations in maintenance needs among breeds. Farrell and Jenkins (22) reported that Simmental cattle had a 19% higher maintenance requirement than Herefords. Comerford et al. (15) showed that Brahman cattle probably had a lower maintenance requirement than Simmental, Limousin, or Hereford cattle. House (29) and others have documented that Holstein steers have a 7–10% greater maintenance requirement than traditional beef breeds. In summary, it appears that Holsteins and some larger mature-size breeds may have greater maintenance needs (17). This implies that feed efficiency at some constant weight will be reduced in these cattle, and the stocking rate for grazing cattle will need to be adjusted (in addition to a weight adjustment) for some breeds. The most effective method to make these adjustments is by using animal BCSs.

Table 3-9. Estimated annual production cycle metabolizable energy needs for four sire breeds using the Angus/Hereford as a base for comparison.				
Sire breed	Maintenance (%)	Gestation (%)	Lactation (%)	Total
Angus/Hereford	73	8	19	100
JerseyX	76	7	23	106
CharolaisX	80	9	20	109
SimmentalX	96	8	24	128

Source: Adapted from Farrell, C. L., and T. G. Jenkins. 1982. Energy utilization by mature cows. Beef Research Program Report No. 1. Roman L. Hruska Meat Animal Evaluation Center. Clay Center, NE.

The relationships among breeds for various production traits shown in table 3-8 are combined with the information shown in table 3-7 (p. 55) to make breed selections. The combination of maintenance and production will determine the nutritional requirement. Production phases that increase nutritional requirements include lactation, growth, and reproduction. The variation in metabolizable energy needs for four sire breeds for Angus and Hereford cows is shown in table 3-9. These data indicate that lactation and maintenance are the principal forces of nutritional requirements of cows.

This information is useful in determining pasture needs relative to calf performance. Consider the following example:

- Angus/Hereford crossbred cow: daily metabolizable energy (ME) needs = 100%
- Simmental/Angus crossbred cow: daily ME needs = 120% (in relation to Angus/Hereford)
- Pasture cost = $40.00 per acre (assuming 1 acre annual pasture need for Hereford/Angus cow)

- Additional cost of pasture for 20% higher maintenance needs = $8.00/acre
- Sale value of Hereford/Angus calves @ 500 pounds = $400/calf ($80/hundredweight [cwt])
- Equivalent sale value for Simmental/Angus calves @ 500 pounds = $408/calf ($81.60/ cwt for a 500-pound calf or a sale weight of 510 pounds at $80/cwt.)

Both maintenance and production energy needs vary within breeds, due primarily to mature weight and milk production potential. Research has shown (44) that purchased feed is one of the most important factors reducing net returns to the cow-calf enterprise. It follows that cow size and milk production should closely follow forage quality and availability. This feature would tend to favor moderate-sized cows that are highly productive. Selection within and across breeds for milk production—using average breed effects and **expected progeny differences** (EPDs)—should tend to match forage quality and availability closely. EPDs provide an estimate of the genetic value of an animal as a parent. Differences in EPDs between two individuals of the same breed predict differences in performance

	Birth wt. (lbs.)	Weaning direct (lbs.)	Maternal milk (lbs.)	Combined weaning (lbs.)	Yearling wt. (lbs.)	Marbling score
Table 3-10. EPDs						
Bull A	6.4	29.2	−1.2	25.8	42.3	−0.03
Bull B	1.4	6.0	4.9	5.4	10.1	0.62
Bull C	3.6	16.2	8.4	23.1	36.3	0.47

between their future offspring when each is mated to animals of the same average genetic merit. A poor match of these resources will not necessarily reduce calf weight, but will reduce reproductive performance of the herd with a subsequent loss of net returns.

It should also be clear that there is as much variation in most production traits within breeds as there is across breeds. A single animal—or even a herd of them—will not necessarily represent the average value of some trait for that breed. Selection tools, such as EPDs and other performance data, will help to categorize many production traits within breeds. Using all of the information available within and across breeds will result in selections that more accurately match production targets.

Buying Yearling Bulls Using EPDs

No two beef breeders will have exactly the same needs for a breeding bull. However, all breeders must consider several management and marketing factors to make the most appropriate choice for a bull. Individual management skills, the quality and availability of feed, and marketing targets must be considered with genetic information to make the best choice. Let's look at an example.

Three different beef producers want to purchase a bull. Each is considering the same three bulls of the same breed (table 3-10).

Herd #1

Considerations:

- Cow age is mostly 5–9 years; no females less than 3 years.
- No heifers will be retained from the matings.
- Average weight of cows is 1,250 pounds.
- A full-time manager is employed for the herd year-round.
- The nutrition program is well managed and home-grown feeds are fed.
- All calves are sold at weaning, but the possibility exists to retain ownership through a backgrounding phase.

Bull of choice: Bull A

Reasons: The keys to this selection are (i) all heifers are sold, (ii) a full-time manager is employed, and (iii) the market is for feeder calves. The birth weight EPD of 6.4 pounds will probably result in "big" calves. However, because the cow herd is largely mature with no first-calf heifers, dystocia, or birthing difficulties should not be a problem. Because all females are sold, the maternal values of birth weight and milk are not weighed heavily. The direct weaning value and yearling weight are the important values to consider. Bull A has the most desirable data for these traits.

Herd # 2

Considerations:

- This is a small, part-time breeder.
- The breeder feeds out all calves for a local market that requires high quality beef and does not penalize feeders for heavy slaughter weights.
- The breeder may retain a small number of heifers as replacements.
- The cow herd is a "mixed bag" of ages and sizes.
- Some forage is home-grown (pasture), but additional hay and supplements are purchased.
- The operation experiences few breeding problems and is more concerned about eliminating calving problems.

Bull of choice: Bull B

Reasons: There are two keys for this breeder: his market and his feed supply. Bull B's marbling EPD of 0.62 is certainly in his favor, but, more importantly, the birth weight and yearling weight values are more desirable for this breeder. As a part-timer, special consideration should be given to birth weight in sire selection. Less time and expertise are available for calving management. Because the cow herd is composed of both young and old cows, the best course would be to select bulls suitable for the young cows. A birth weight EPD on a young sire should, in this case, be +2.0 or less to increase the probability that the bull will indeed sire calves with an average birth weight or less. Secondly, this breeder would probably not wish to increase the mature size of his cow herd. Because feed requirements are based largely on the weight of the animal, larger cows would imply more purchased feed and higher production costs. Selection of bulls with high yearling weight values would increase both cow size over time and the weight of the finished steer.

Both factors impose greater feed requirements and total feed costs, and slaughter weights of steers would increase to reach the same quality grade endpoint. Thus, an "average" bull on growth with light birth weight values suits the needs of this breeder.

Herd #3

Considerations:

- This is a small commercial herd.
- The replacement rate is 25% annually.
- The herd is well managed, but it is not a full-time enterprise.
- The operation uses home-grown forage, but the quality is variable.
- Feeder calves are usually sold at weaning, but retained ownership is possible when the market warrants.
- The breeder has occasional breeding problems with young cows.
- The average age of the cow herd is 4.5 years.
- The average cow weight is 1,350 pounds.

Bull of choice: None of the three.

Reasons: This herd is a well managed unit that has made some genetic progress and has some above average cows for growth and milk production. The restriction appears to be nutritional. The key comment that breeding problems happen occasionally with young cows implies that the part-time approach to harvesting forage and the bred-in growth and milking potential of the cows have clashed at some time. This breeder would need to select sires with strong, positive growth and milk values just to maintain his position. However, selection of sires that are well above average for yearling weight (+50 pounds or more) and milk production (+10 pounds or more) would intensify an existing problem. Secondly, although a sire like Bull C would be an acceptable choice for the high number of replacement heifers because of the good growth and milk values, the birth weight EPD of 3.6 in a young sire is risky due to low accuracy. The range for the bull's true breeding value for birth weight is too wide. The cows are mostly young, with at least 25% as first-calf heifers, and calving is not usually managed full-time. The risk involved with a young bull that is above the breed average for birth weight will be both for the current calving season and the perpetuation of high birth weights into future generations. This breeder should keep looking for a bull.

Nutritional Diseases of Grazing Cattle

Bloat

The incidence of bloat from grazed legumes is well documented. Bloat is caused by the rapid fermentation of legume plants that produce high levels of gas as a byproduct of the fermentation. This excessive gas cannot be eliminated fast enough, and the animal can die from the excess pressure on the internal organs. Its effect on reproductive efficiency is indirectly related to abortions or infertility in affected animals. The report by Majak et al. (39) provides a summary of the pasture management strategies to reduce bloat:

- Every cultivar of alfalfa tested caused bloat.
- Sainfoin, cicer milkvetch, and birdsfoot trefoil are legumes that did not cause bloat.
- Advanced stages of maturity of the alfalfa plant reduced the probability of bloat.
- Cattle susceptible to bloat have a slower passage rate in the rumen (allowing more time for gas production), and these cattle consume 18–25% less forage before bloating than nonbloaters.
- Mineral supplementation did not reduce the incidence of bloat.
- The only additive tested that consistently reduced pasture bloat was poloxalene.

- Seasonal weather conditions, including a killing frost, did not influence the incidence of bloat.
- Waiting until the dew was off alfalfa before grazing was substantiated as a method to reduce bloat.
- Cattle that had continuous access to alfalfa had less bloat than those that had access for shorter periods of time each day. Continuous access promotes continuous and rapid rumen clearance.

This research indicates that alfalfa and other legumes can be used safely in grazing systems, but management of supplementation and daily and seasonal timing of grazing is necessary. Pastures that contain trefoil in place of alfalfa or clover are less of a concern for bloat problems.

Grass Tetany

Grass tetany is a serious problem in many livestock herds. It is characterized by low blood serum levels of magnesium (Mg) from a dramatic deficiency of this mineral in forages and pastures. Symptoms of grass tetany (a.k.a. winter tetany, grass staggers, Mg tetany) usually first appear as extreme nervousness, an awkward gait, muscle spasms, and collapse. The symptoms may progress rapidly. Therefore, sometimes no clinical signs are observed and a cow may simply be found dead. Other symptoms may include grinding the teeth, violent convulsions, and coma. Cows suffering from grass tetany may often resemble those with milk fever and have low Ca as well as low serum Mg levels.

A positive diagnosis is difficult to obtain, but the status of the herd may be evaluated through blood samples. Serum Mg levels below 1.0 mg per 100 ml indicate Mg levels low enough to result in grass tetany.

Grass tetany can occur at almost any time of the year, but occurs most often in April and May in the Northeast. Other conditions favorable to the incidence of grass tetany include:

- warm temperatures in early spring followed by cool, cloudy weather.
- cows 6 years old or older nursing calves less than 2 months of age.
- grass pastures that contain few or no legumes.
- soil types that have a high level and availability of potassium (K).
- soils having low availability of P.
- pastures fertilized in the spring with nitrogen (N) and/or K.

Strategies for the prevention of grass tetany include:

- Make Mg additions to mineral supplements available during the latter part of the winter feeding period and the initial part of the grazing season.
- Wait until early spring grass growth reaches 8–10 inches before grazing.
- Graze grass-legume pastures first in the spring. Cases of grass tetany are seldom seen when legumes are included in pastures.
- Graze heifers, stockers, and dry cows on high-risk pastures.
- Identify cows that suffer from grass tetany; they tend to be more susceptible in following years.

Cows that suffer from grass tetany and go down for more than 12 hours seldom recover. Those in earlier stages should be handled gently and quietly. Stress and exertion will often cause affected animals to go down or die suddenly.

Early treatment involves preparing 200 ml of a saturated solution of Epsom salts (a soft drink bottle holds about 350 ml). The water and container should be very clean, and Epsom salts

should be added to the water until no more will dissolve. This solution should be offered as a drench using a stomach tube or given as an enema. It can also be injected under the skin of the animal in at least four sites (50 ml injected at each site) if an enema or drench cannot be administered. A veterinarian should be consulted to provide intravenous Mg supplements. All infected animals should be removed from the pasture and fed a legume or good grass/legume hay plus concentrate feeds.

Hardware Disease

Foreign objects that cattle may ingest (wire, nails, pins, screws, bolts, or glass) collect in the reticulum. The objects may puncture the wall of the reticulum, which can cause infection or damage to surrounding organs, especially the heart. Symptoms of hardware disease include loss of appetite, no cud chewing, swelling of the neck and brisket, and stiffness. The objects will normally have to be removed surgically. In some cases a magnet placed in the stomach may be used to remove metal objects (25).

White Muscle Disease

Deficiency of selenium (Se) in the soil can result in animal deficiency of this mineral. This problem may manifest itself as white muscle disease in young calves and reduced immune response in older cattle. Muscle damage results from lack of Se. Calves are born weak or dead. Selenium injections for newborn calves and mineral supplementation for cows will prevent white muscle disease. A mineral mixture with an average daily intake of 0.25 pound per day should contain 0.002% Se to provide the recommended intake of 0.2 ppm Se daily. Intake of Se at 5–10 times the recommended levels can result in toxicity.

Foot Rot

The bacterium *Fusobacterium necrophorum* has been reported to cause foot rot. However, researchers have not been able to reproduce typical foot rot lesions with this organism alone. Other organisms commonly isolated from animals with foot rot include streptococci, staphylococci, corynebacterium, bacteroides, and various fungi, all of which are common in the environment, especially where moisture is present. Cuts, bruises, puncture wounds, or severe abrasions permit these bacteria to enter the tissue of the foot to start an infection. The inability to cause foot rot in clinical trials has hampered the ability to recommend precise prevention and treatment procedures. Foot rot can become "seeded" in the soil, and it may persist for a long time. The incidence of foot rot may be variable in a given herd.

Symptoms include lameness followed by swelling of the foot, spreading of the toes, and reddening of the tissue above the hoof. In severe cases, the foot will abscess above the hoof with a discharge that has a characteristic foul odor. The animal usually has an elevated temperature with loss of appetite and body weight. If the infection is not stopped, it will invade the deeper tissues of the foot and may invade one or more joints, causing chronic arthritis.

Management practices that reduce hoof damage or avoid bruising will help decrease the incidence of foot rot. They include:

- Keep the hooves of heavy cows and bulls trimmed to help reduce stress on the soft tissue of the foot.
- Maintain drainage of lots and around water tanks to prevent mud accumulation, particularly when the mud freezes and causes the feet to bruise.
- Use walk-through foot baths in dairy operations. Place a copper sulfate (dissolve 2 pounds in 5 gallons of water) or formalin (1 gallon of 40% formalin in 9 gallons

of water) solution in the door or alleyway where the cattle come into the barn.
- Provide 50 mg per head per day of ethylene diamine dihydriodide (EDDI, tamed iodine) mixed in the feed or salt as a preventive measure. However, feeding EDDI has not been a very satisfactory control for foot rot. Overconsumption of the chemical can cause irritation of the respiratory tract. This may lead to pneumonia, hacking cough, depressed appetite, and watery eyes (30).
- Be sure that all cattle receive adequate Ca, P, and vitamin A for good bone and tissue health.

Early treatment is necessary to prevent animals from becoming chronically ill. Examine the feet of lame animals for foreign objects such as wires and nails and treat as soon as possible. Penicillin or the oxytetracyclines (terramycin, liquamycin, and oxy-tet) usually work well if given at the recommended dosage and treatment is started early. Sulfonamides (sulfapyridine, sulfamethazine, or triple sulfas) have been used successfully.

Feed additives containing chlortetracycline (aureomycin) or a combination of chlortetracycline and sulfamethazine can be used for treatment on a herd basis. To be effective, the minimum dose for calves should be at least 1 g of chlortetracycline per animal per day. Increase the amount of antibiotic for larger animals. Lower dosages may contribute to the production of drug-resistant organisms. When foot rot fails to respond to medication, thoroughly check the foot for foreign objects. A report by Hudson (30) provided an excellent summary of the causes and prevention of foot rot.

Johne's Disease

Johne's disease is a persistent, herdwide disease caused by *Mycobacterium paratuberculosis*. It is difficult to identify in a herd because all infected cows do not advance to clinical disease. The infection is long-lasting, and only 1–5% of infected cows may show clinical signs at any one time. Infected cows may shed the pathogen in manure for months to years before they develop clinical signs, and ingestion of infected manure and colostrum are the major methods of transmission. Clinical signs are diarrhea, weight loss, and "bottle jaw." Tests are available for cattle over 2 years of age to screen carrier individuals in a herd.

Prevention of Johne's disease in grazing animals includes the identification and elimination of carrier animals in the herd and closing the herd to any additions that are at risk for Johne's. Management strategies that may be used to control Johne's where potential infection exists include (26):

- Provide a clean, well drained area for calving.
- Clean calving pens between animals.
- Move cow-calf pairs from calving areas as soon as possible.
- Lower stocking rates.
- Raise heifers separately from mature cows.
- Cull progeny of infected animals.

Neospora spp.

Neosporosis is a cause of abortions in cattle. It is thought to arise from *Neospora caninum*, which is a protozoal parasite. Dogs have been identified as a definitive host of the pathogen, and cattle, deer, goats, and horses are intermediate hosts. Neospora has been identified in cattle in many areas of the world. Abortion is the usual symptom, and calves from infected cows that survive to birth have neurologic signs, are underweight, and may be unable to stand (19). Culling is the only method to prevent transmission from cows to calves. Prevention should focus on protection of cattle from feed and water sources that could be contaminated with dog feces.

Problems Related to the Ingestion of Legumes

Phytoestrogens

Legumes, particularly alfalfa and clover, can have a fairly high content of phytoestrogenic compounds. Because of a similar chemical structure at the binding site, these compounds mimic the effect of estrogen in the animal's body. In general, the plant must undergo some environmental dysfunction to be harmful. For example, alfalfa that has been attacked by aphids or fungal pathogens and suffers from foliar disease will have higher levels of phytoestrogens than normal, growing plants. Plants with genetic resistance to disease will have less estrogenic activity (35). Similar results have been reported for clovers that have suffered from foliar disease.

The extent of the problem of phytoestrogens in grazing cattle is not well known. Just by association, it appears that consuming legumes under some environmental stress may cause reproductive failure through abortions or poor estrous cycles due to estrogenic activity in the plants. It is not known if the effects of drought, animal intake levels, or other factors of grazing legumes may influence the estrogenic activity in cattle. Documentation of estrogenic content of various legumes under diverse growing conditions is needed to pinpoint grazing management decisions to avoid these circumstances. The value of legumes in pasture and hay crops for fixing N and enhancing animal performance outweighs the risk compared to not using legumes.

Blood and Milk Urea N

Cattle grazing pastures with relatively high amounts of legumes or highly fertilized with nitrogen can result in a reduced energy balance. Grazed high-quality forages contain dietary protein that is degraded primarily in the rumen, and there is a metabolic cost to this digestion compared to "bypass," or rumen-undegradable, proteins. When high levels of rumen-degradable protein are combined with relatively low levels of carbohydrate feeding, blood urea nitrogen (BUN) and milk urea nitrogen (MUN) values will increase. It has been shown in some studies that high (greater than 20 mg/dl) MUN can result in lower fertility of high-producing dairy cows (9). This may be the result of the negative energy balance in early lactation or the reduction of progesterone levels from the CL (corpus luteum) during later lactation (9). However, MUN values are not a good predictor of suboptimal fertility in dairy cattle (11).

It remains to be shown how levels of BUN or MUN in grazing beef cows may influence reproduction, particularly for early embryonic death. Forage variety, season, and plant maturity need to be evaluated for their effects on BUN and MUN, and these data can be related to fertility, particularly for measures of early embryonic mortality.

Slobbers Syndrome

A mycotoxicosis associated with *R. leguminicola* infestation of pastures results in slobbers syndrome. Two of the active alkaloids of the fungus are slaframine and swainsonine; the former is associated with a generally innocuous active salivation in infected animals, and the latter has been linked to more serious effects on the central nervous system in a condition called locoism (16).

Cool, wet weather that promotes fungal growth in legumes, particularly clovers, will often result in an incidence of slobbers syndrome. It has limited negative effects on grazing cattle. Swainsonine has been documented in red clovers, but is most often found in a plant known as locoweed in the western United States. The effects of

swainsonine ingestion are serious—staggering gait, depression, reduced sexual activity, abortions, and malformed fetuses (31). The regional existence of locoweed probably precludes attention to the effects of *R. leguminicola* in the Northeast.

Problems Related to the Ingestion of Grasses

Prussic Acid Poisoning

Most plants contain intact glucosides, but under certain conditions of climate, fertility, stage of growth, or retarded growth, a buildup of cyanide-containing compounds, called **prussic acid poisoning**, can result. This is particularly true for sorghum, sudangrass, and their hybrids, as well as Johnson grass. Some of the conditions that result in high levels of prussic acid in the plant include:

- a high N to phosphateratio in the soil,
- younger leaves, or regrowth,
- newly frosted leaves,
- extended drought preventing leaf maturity and growth, and
- regrowth of the plant following a frost.

Grazing management is the key to avoiding prussic acid poisoning in grazing cattle. Vough and Cassel (60) outlined some management steps to avoid prussic acid poisoning:

- Use certified seed.
- Select varieties low in prussic acid.
- Follow fertilizer application recommendations.
- Do not begin grazing until plants have reached a height of 18–20 inches.
- Allow frosted sudangrass to dry thoroughly before pasturing.
- Dilute intake of infected material with hay and other forages.

Nitrate Poisoning

Similar to prussic acid poisoning, nitrate toxicity can occur in grazing cattle and with the feeding of stored forages produced under specific environmental conditions, usually when high N fertilizers are applied followed by drought conditions. Quick tests for nitrate content should be used when nitrate poisoning is possible. Management of pastures, including drought-damaged corn, is necessary to avoid nitrate poisoning.

Mold (Aflatoxin)

Certain environmental conditions result in the formation of mold called aflatoxin. In pastures, this mold will most often be found when grazing infected corn aftermath. Aflatoxin contains an estrogenlike compound called zeralenone that can cause abortions in pregnant animals. Producers should know if mold is present in grain harvested from suspect fields. If so, pregnant cattle or cycling heifers should not graze these fields. Feeder cattle are generally unaffected. There is no known agent to eliminate the effect of aflatoxin in the field, but a field test for the presence of aflatoxin in stalk fields would be very useful.

Fescue Toxicosis

Since the 1970s we have known that fescue can be infected with a fungal endophyte called *Neotyphodium coenophialum* (formerly called *Acremonium coenophialum)*, and that intake of these infected plants by grazing cattle would result in a series of effects referred to as fescue toxicosis. These effects include:

- reduced feed intake,
- lower weight gains,
- lower fertility,
- "fescue foot" (a condition arising from reduced blood flow to the extremities that results in necrosis of the extremities [e.g., tail, feet, ears]), and
- elevated body temperature and others (3).

Reductions in cow weight gain and pregnancy rate from grazing highly endophyte-infected fescue are well documented (21, 24, 38, 57). In general, the pregnancy rate of cows grazing highly versus minimally endophyte-infected fescue can be reduced by about 40–50%. This results from both the reduction in body condition at calving and factors related directly to the fungal infection.

Endophyte-free and nontoxic endophyte-infected varieties of fescue are available and should be incorporated in a grazing system that includes fescue. The intake of infected fescue can be diluted by overseeding clover or other legumes in the fescue pasture or feeding grain or other feeds while cattle are grazing infected areas. There is tremendous potential for the use of stockpiled grasses, including fescue, to increase the grazing period and reduce the cost of production in the northeastern cow-calf enterprise. Careful variety selection and monitoring of endophyte infection is necessary to avoid the results of highly endophyte-infected fescue.

PASTURE MANAGEMENT STRATEGIES TO OPTIMIZE PRODUCTION FOR COW/CALF AND YEARLING CATTLE

Forage Quality and Availability

The quality of forage on pasture in the Northeast is a function mainly of its energy value (TDN, NEm, or ME) and secondarily of its crude protein content. Crude protein is often of secondary importance because it usually exceeds the requirements for cattle lactation or growth throughout the grazing season. In most instances energy is the limiting nutrient for beef cattle grazing cool-season grass pastures. Many factors, such as weather, forage species, aspect, soil type, and fertilization, influence the quality or energy content of pasture. However, grazing

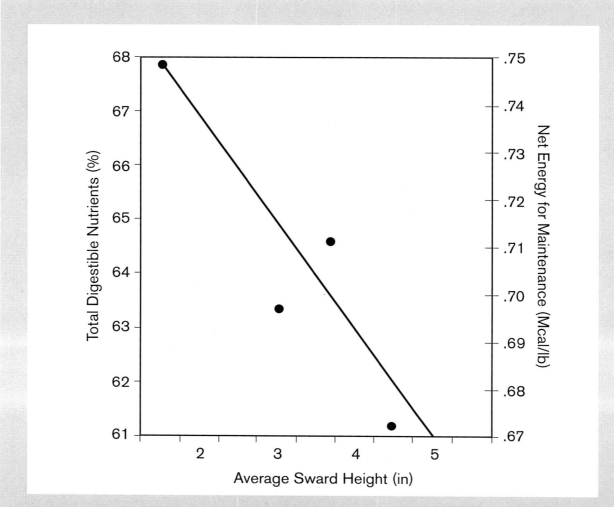

Figure 3-3. Sward fiber decreases and nutritive value increases with intensive or close grazing.

management is perhaps the most important controlled variable.

Intensive or close grazing regardless of the grazing management employed will result in a sward low in fiber and high in nutritive value compared to a sward that is lightly grazed. This relationship is illustrated in figure 3-3. As grazing intensity increases, the sward is kept in a more vegetative state and quality remains high; when grazing intensity is light, the sward matures, fiber content increases, and the energy value of the sward is low. With short-growing grasses such as Kentucky bluegrass and perennial ryegrass, yield or growth rate is also at its greatest when closely grazed. Results of studies in West Virginia (47) suggest that cattle cannot graze Kentucky bluegrass and possibly perennial ryegrass pastures close enough to influence yield. This is because cattle's mouth structure does not allow them to graze a pasture lower than 1.25 inches in height. Net herbage production of these forages (5, 47) may be maximized at heights lower than 1.25 inches. However, taller-growing grass species such as orchardgrass may not respond the same because grazing these pastures to 1.25 inches may depress growth rate. Results of studies in West Virginia (47) indicate that both quality and quantity of mixed bluegrass pasture are greatest when grazed to a 1.25-inch average height. However, grazing pastures very intensely to maximize quantity and quality of the harvest will not result in maximum gain per animal because quality and yield of pasture are not the only factors influencing performance of grazing cattle.

The availability of forage, as measured by height or pounds of forage per acre, is also a key element in determining the performance of grazing cattle because of its influence on forage consumption. When forage availability is high, the major factor limiting intake is forage quality. Forages high in fiber (58) and consequently low in energy will limit intake because of their influence on gut fill, and animal performance will be less than optimum. When forage availability is low, the sward is generally low in fiber content and high in energy; however, intake is limited because cattle are forced to take small bites of forage and the amount of time available for grazing limits forage intake (27). Studies with cool-season grass and legume pastures (47, 52, 63) indicate that the forage height that maximizes gain for yearling cattle is between 3 and 5 inches. However, if gain per acre is the major objective, it can be assumed that the optimal forage height would be lower than 3–5 inches because of the increased forage yield, at least for low-growing cool-season pastures. For cow/calf production, a height of 2.5–4 inches for Kentucky bluegrass might be ideal because calf response, due to contribution of milk from the cow, is less sensitive to the influence of sward height (47). In light of the seasonal variation in forage production of cool-season grasses in the Northeast, it is no easy task to maintain a sward at a height that would maximize gain per animal per acre regardless of the management. No experimental results are available for optimal heights to graze orchardgrass pastures. The information developed from grazing studies with bluegrass and perennial ryegrass swards is most likely not directly applicable to taller-growing grass species in the Northeast.

Grazing Systems

To develop appropriate grazing management systems the producer's goal must be considered; for example, is it to maximize production per animal, or is it to increase the carrying capacity of the farm. The producer also has to be aware of the factors limiting production with the present level of management. Most of the grazing management systems result in increased forage productivity. Thus if the advantages of these systems are to be realized, the number of cattle that can be maintained on a given area of land during the grazing season will have to increase. If the amount of winter feed is the major limiting factor, increased forage production from grazing management cannot be used unless excess forage is diverted to winter feeding. A producer grazing yearling cattle could take direct advantage of more forage by increasing cattle numbers during the grazing season.

A further complication to the development of the optimal grazing system is the variation of forage production between years as well as within the year (8). A producer must be flexible in managing pastures and plan to make adjustments as pasture or animal conditions warrant. The objective of a grazing management system is to maintain energy consumption of the cattle at adequate levels while keeping the forage in the sweet spot of its growth curve to ensure adequate yield. This sweet spot is the sward height which leaf area of the plant limits photosynthesis and below which maturity limits the growth of new tissue. As stated earlier, for yearling cattle grazing Kentucky bluegrass pastures, this optimum is when the average sward height is about 3–5 inches. Various grazing systems outlined below can help producers achieve this goal or enhance forage utilization on a farm.

Fixed or Rigid Paddock Rotational Grazing

Fixed or rigid paddock rotational grazing systems are popular in the British Isles and New Zealand for dairy production. Pastures are divided into numerous paddocks and the cattle are rotated to a new paddock every day. Thus, a paddock is grazed for 1 day and rested for approximately 3 weeks. These systems work very well where there is little variation in seasonal rainfall and temperature, and consequently a consistent pattern of forage growth. The system works on the assumption that an adequate amount of forage is available for the herd on a daily basis.

Flexible Paddock Rotational Grazing

This system is most likely more effective in the Northeast than fixed paddock rotational grazing when considering the variation in growth rate of forages over the grazing season in this area. This system is designed to provide at least a portion of the winter feed needs of a farm. It is similar to the previous system in that the grazing is divided into numerous paddocks. A limited number of paddocks are grazed in the spring when forage growth rates are high, and conserved forage is removed from the paddocks not initially grazed. As the growth rate of the forage decreases in the late spring, more paddocks and land area enter the grazing rotation, resting time between grazing of the paddocks increases, and the number of days grazing each paddock may change. A limitation of this system

is that on some sites' slope or soil type may limit the harvesting of conserved forage.

Buffer Grazing

Buffer grazing systems originated in Scotland. In this system a farm is generally divided into three portions. One portion of the farm is used entirely for grazing. A second portion is harvested as conserved forage and after suitable regrowth is opened to grazing along with the original grazing portion of the farm. After a second cutting of conserved forage in the third portion, it too is opened to grazing. The buffer system is designed so that any time there is a shortage of forage, portions of areas that have not been harvested can be opened to grazing to ensure ample forage availability for the cattle. Results from studies in West Virginia (7) seem to indicate that this system may be equal to rotational grazing systems on bluegrass-based pastures in terms of carrying capacity of the farm. It takes less effort to maintain a buffer grazing system than a rotational grazing system. Again, the ability to harvest conserved forage over two-thirds of the land area is necessary and site and soil limitations may limit the incorporation of this system on a particular farm.

Forward Grazing

Forward grazing is employed in a rotational grazing system when cattle with different nutrient requirements are maintained on the same farm. Consider, for example, yearling steers and lactating beef cows. Because the steers are more likely to respond in economic terms to a higher quality pasture it might be advisable to rotate them through the paddock prior to rotating the cows through the same paddocks. This system is quite common on beef farms in Great Britain and Ireland where young male calves and yearling males from dairy production systems are often on the same farms. The young calves proceed the yearlings through the pasture rotation.

Creep Grazing

Creep grazing is a method to provide the calf with ample or high quality pasture while relegating the cow to pasture of limited availability or quality. Studies by Vicini et al. (59) indicated that allowing calves to creep graze a Kentucky bluegrass/white clover pasture while cows grazed a tall fescue pasture resulted in a more than 10% increase in weaning weight of the calves. However, weaning weights with creep grazing were 10% less than when both cows and calves had grazed the same bluegrass/clover pastures. One would expect creep grazing to be more effective during the latter part of the grazing season when milk represents a smaller portion of the calves' diet.

Combination Systems

Combination systems often combine several aspects of the above systems. A study at West Virginia University incorporating components of both a rotational and buffer system seemed to indicate that the combination was more productive than either system alone. The combination system includes rotational grazing in the spring and continuous grazing in the summer on pasture in which conserved forage has been harvested. Approximately one-third of the land area is used strictly for grazing. In the early spring this area is divided into four paddocks. These paddocks are grazed for 1-week intervals in the spring after the initial grazing rotation, which lasts for 28 days. Two paddocks are grazed for 1 week following the initial grazing to compensate for reduced forage growth, and for the last week of grazing the entire area is open to grazing. During this time second-cutting hay was harvested from an additional area; this area as well as the initial grazing area was open to grazing. The cattle usually exclusively grazed the area from which the hay was removed for about 2 weeks and then expanded to the entire area, so rotational grazing was not needed. An additional area was harvested for hay twice and then opened to grazing in the same manner as the first area. There would be little advantage on bluegrass pastures to rotational grazing in the summer because the growth rate of the plant is less than consumption of the cattle and the yield of these pastures is generally greatest when they are intensively grazed. Study results showed that this combination system could obtain a 20% increase in carrying capacity over the buffer or rotational grazing system.

A producer's choice of grazing system must include consideration of land characteristics (e.g., portion of the farm from which winter feed can be harvested) as well as the producer's desires and economic situation. Consequently no one system is suitable for all farms.

Supplementation of Calves and Yearlings

Beef cattle are agriculture's great scavengers, and can meet most of their fundamental nutritional needs from crop residues, byproduct feeds, and pasture. Supplementation of grazing calves and yearlings may be needed depending on the following considerations:

- the production goals for the cattle,
- the metabolic and health issues related to pastures and cattle, and
- the method of delivering supplements to the cattle.

Production Goals

Production goals for calves and yearlings are determined by the level of animal growth that is desired, or, in the case of nursing calves, as both a growth and training device. Supplementation is designed to meet any nutritional deficiencies that exist in the pasture, and may also be necessary to reduce or eliminate the potential for nutritional disease or other health hazards.

Finally, supplementation may be needed as a method of estrous synchronization in heifers.

When setting goals for animal growth, it is necessary to determine:

- growth rate targets,
- the age and sex of the cattle,
- the price of delivering supplemental feed, and
- the availability and quality of forages.

The simple equation of supplementation for grazing cattle is animal requirements minus nutrients available in the pasture equals supplementation needed. Therefore, determining what supplement to use, or the need to use one at all, is predicated on knowing both the animal's needs and the availability of nutrients in the pasture. As shown in table 3-11, pastures generally can provide adequate nutrients to support a reasonable level of growth in young cattle.

However, care must be taken to recognize that forage quality is dynamic—the nutrients there this week will probably be different next week. Also, the factor limiting performance is forage intake.

Growth Rate Targets

Targeting average daily gain requires balancing biology with economics. It is important to know the efficiency of weight gain when giving a supplement to growing cattle. Faster growth is not always economical. A study by Comerford et al. (13) found that early-weaned calves must be supplemented with high energy feeds while on pasture to reach weaning weights equal to those of their unweaned, nursing contemporaries, but this weight gain was not cost-efficient. This result was also shown for grazing Holstein steers (14). Data from Roquette (53) in table 3-12 identify the optimum level of supplementation. The most effective supplementation rate for grazing cattle is at a relatively low

	Year	0–28 d	28–56 d	84–112 d	112–140 d
DM (%)	1994	21.3	31.3	25.4	17.7
	1995	18.9	24.9	31.8	25.9
CP (%)	1994	21.2	15.1	19.9	21.1
	1995	18.8	16.6	17.1	12.5
DM (lb/ac)	1994	1,355	1,488	1,209	1,522
	1995	2,427	2,189	1,790	861
NEg (Mcal/lb)	1994	0.53	0.41	0.45	0.48
	1995	0.49	0.40	0.39	0.56

Table 3-11. Quality and availability of pasture for 140 days.

Source: Adapted from Penn State University Haller Farm, unpublished data.

Table 3-12. Corn-based supplements for cattle grazing winter annual pasture.		
Supplement (lb/d/head)	Added gain (lb/d/head)	Supplement efficiency (supplement:weight gain)
0.74	0.38	1.9:1
1.43	0.77	1.9:1
2.44	0.45	5.4:1
4.06	0.45	9.1:1

Source: Roquette, F. M. 2000. Matching forage quality to beef cattle requirements. *Beef InfoBase, Adds Center, Inc., Madison, WI.*

level that does not force substitution of forage intake. Substitution rate depends on forage quality, level of nutrients in the supplement, energy sources, and feeding rate. This also implies that forage intake may be purposely reduced with supplementation, if necessary. Feeding 0.7–1.0% of body weight of corn supplements to grazing cattle results in a 1:1 substitution rate, and the stocking rate could be increased by 33% without changing animal performance. This result has practical significance during periods of drought or dormancy.

Supplements with high levels of readily degradable fiber, such as soy hulls, cottonseed hulls, wheat mids, and beet and citrus pulp, have shown the potential to improve the efficiency of supplement utilization on pasture over concentrates. Studies in progress at West Virginia University indicate that these feedstuffs do not have the same effects as high concentrate supplements, such as corn, in depressing fiber digestion. Although, these supplements have been quite effective in improving animal performance, they are often more expensive per unit of available energy than corn, and economic

evaluations of using supplements high in readily degradable fiber have not been completed.

Growth rate targets must reflect the market or "next user." Highly conditioned stocker cattle that will enter a feedlot after grazing (with all of the stress that goes with the relocation) can have in a negative growth rate due to weight loss and sickness in the early part of the feeding period. Conversely, healthy cattle that have significantly less condition will make economical compensatory gains in the early feeding period.

The considerations for growth rate targets are:

- Forage quality attributes. As shown in table 3-12, there are limited returns to increasing supplementation for cattle grazing reasonably good pasture. The higher the quality of the forage, the less economically efficient supplementation will be. Supplementing low quality forages (stalk fields and mature, dormant grasses) can be very economical and necessary to reach growth targets.
- Cattle market requirements. Condition may have a negative value in some feeder calf

markets, and the cost of production (gain) will be higher.

- Supplement costs and returns. Animal growth can be predicted fairly well if the forage quality, nutritional value of supplements, and feeding rate are known. The additional value of animal gain per unit of supplement fed must be compared to the cost to make sound production decisions.

Nutritional Requirements

The nutritional needs of young, growing cattle differ from those of mature animals (table 3-13). These needs are based on weight, growth targets, and sex of the animal. Protein needs are singularly different. Microbes in the rumen rapidly digest forage proteins, and undegraded proteins (passing through the rumen to the small intestine) may be needed to optimize animal growth. This fact may not be accounted for in the crude protein value of forage alone, and crude protein will need to be supplemented. For example, a 400-pound steer grazing orchardgrass pasture will have energy-allowable weight gain of 1.6 pounds per day, but only 1.3 pounds per day of metabolizable protein-allowable weight gain, in spite of a crude protein value of 16% for the orchardgrass pasture (43). A second method to

Table 3-13. Nutritional requirements of several classes of beef cattle.				
	ADG[a] (lb/d)	CP (%)	NEg (Mcal/lb)	TDN (%)
Steer calves 500 lb	1.5	10.5	0.38	63
	2.5	12.5	0.51	73.5
Heifer calves 500 lb	1.5	10.3	0.44	68.5
	2.0	11.4	0.55	77
Pregnant heifers	1.0	8.2	0.4	54.8
Cows in early gestation	0	7	0	48.8
Cows in late gestation	0	7.8	0	53.2
Young cows in lactation	0.5	10.2	0.37	62.3
Cows in heavy milking	0	11.5	0	63.7

[a] ADG-average daily gain.

Source: Adapted from National Research Council. 1996. Nutrient Requirements of Beef Cattle, 7th ed. National Academy Press, Washington, D.C.

determine protein supplementation is to evaluate the TDN:CP ratio. When the TDN:CP ratio is higher than 7, protein should be supplemented in the diet of growing cattle. Under conditions in the eastern United States, these energy:protein ratios will usually be found only in dormant grasses and when grazing crop aftermath. A third rule of thumb (32) is that if gain targets are more than 0.5 pound per day greater than can be achieved with grass alone, then supplemental protein is probably needed. As shown previously (table 3-12), the effective rate of supplementation may be limited. Forage quality will also determine the type of protein supplementation—"natural" proteins or nonprotein N (urea). A natural protein (32) should be used when supplementing medium- to low-quality forages (less than 50% TDN) because urea metabolism requires a higher level of energy intake than natural proteins.

Intake may be the first limiting factor for grazing cattle. Pasture availability, pasture quality, weather, and sward composition will all influence intake. Cattle will first eat the most desirable forage in the pasture, and they may eat certain forages only because nothing else is available. Pasture management and grazing systems will help dictate the quality and volume of grazed forage.

Mineral Supplementation

Mineral requirements of grazing cattle are highly variable due to variations in mineral content of soils, forages, and water. Only in rare cases in the eastern United States will mineral toxicity (usually from polluted water sources) be a problem, so the most effective mineral supplementation program is free-choice access to a good mineral mixture. Because of large variations in forage content and animal intake of mineral, the most effective program provides a daily mineral mixture that contains:

- 25% salt,
- 16% Ca,
- 8% P, and
- 0.002% Se.

This mixture can be made on site using products such as calcium carbonate, dicalcium phosphate, and sodium selenite. For a mixture designed for consumption at 2 ounces per head daily, the following mixture can be used:

- 48% salt,
- 34% dicalcium phosphate,
- 15% limestone, and
- 3% Se mix (0.06% Se).

Very small quantities of Se are fed daily, so Se supplements are usually prepared as 0.06% Se mixtures. It would be safer to use a commercial Se mixture because poor mixing of direct additions of sodium selenite could result in toxicity.

The source of salt for mineral supplements can be either plain salt or "trace-mineralized" salt. The latter salt is just what the name implies: there are only traces of other minerals in the mixture, and it is primarily just salt. Plain salt may be sufficient for these mixes in many cases; however, localized conditions may require the addition of trace minerals. For example, forages in West Virginia are often lacking in Cu and Zn, so these minerals should be included even beyond the levels found in some trace-mineralized salt mixes. In all cases proper mixing of ingredients is essential.

Another exception to this mixture may be the addition of Mg to prevent grass tetany. It is probably prudent to use a commercially available high-Mg mixture during spring and early summer. As with Se, small amounts of magnesium oxide are added as the Mg source in these mixes, and proper mixing is essential. On-site mixing of these small amounts of ingredients can be difficult.

Another addition to a mineral mixture could be iodine as a preventive against foot rot. Research on supplemental iodine to prevent foot rot has shown variable results. An iodine level of 0.016% in the mixture may be effective. Iodine and other minerals are available that are variously described as chelated, proteinated, or bypass. Several research trials have been conducted to evaluate the effectiveness of these mineral forms, with highly variable results. In most cases, when cattle have access to reasonably good forage and a free-choice "regular" mineral mixture, there is seldom any economic advantage to these mineral forms.

Vitamins A, D, and E may be deficient in cattle, but seldom when animals are grazing fresh, green forage. Grazing stalk fields and dormant grasses may require vitamin supplementation as part of the mineral mixture. Again, commercially available mixes may be the most effective and easiest way to deliver these vitamins to cattle.

Creep Feeding

Providing supplemental feed to nursing calves is termed **creep feeding**. This feed may be grain, or it may be access to grazing areas not available to mature cows. (This is accomplished with the use of a creep gate with openings 18 inches wide separating cows from the calf area, or with an electric wire placed 36–42 inches high to restrict cow access while allowing calves to enter under the wire.) As described by McCann (36), the following factors require evaluation prior to creep feeding calves:

- Use of creep feeding as a management tool rather than as an annual practice. The manager must first determine why creep feeding is necessary. Factors could include high

cattle prices relative to grain prices, lack of milk production in the cow herd, poor forage availability, calf marketing programs that require limited grain feeding prior to sale, transitioning calves in an early weaning program, and others.

- The cost of grain or additional pasture. As with any supplement, it makes no sense to supplement cattle for growth if the return will not pay for the supplementation. For creep grazing, the cost per acre of additional forage should be compared to additional weight gains, which have been shown by Wilson (1989) to be about 75% of those from grain creep. The efficiency of weight gain will vary from 3 to 12 pounds of feed per pound of gain due to the quality and quantity of other feeds available. Thus, the efficiency of creep feeding calves strictly to produce weight gain is questionable when cows have moderate milking ability and both cows and calves have access to reasonably good pasture. Conversely, drought conditions and poor pasture quality may enhance the effectiveness of creep feeding.

- The time of supplementation. Calves from spring-calving cows will usually be creep-fed during mid- to late summer. Local grain and additional pasture may not always be available.

- Inhibition of evaluation of herd performance. The optimum situation in the cow-calf enterprise is to have a cow use forages and pasture to produce milk and be reproductively efficient. Creep feeding a calf may distort the actual efficiency of a cow, particularly for milk production.

- Postweaning disposition of calves. Calves with excess condition that are to be marketed directly to feedlots or backgrounding programs will often have a discounted value because of the weight shrinkage they experience. Even when ownership is retained, the value of adding weight and condition from creep feeding may be lost in immediate postweaning shrinkage. Exceptions to these results include marketing programs such as preconditioned calf sales that require some grain feeding prior to sale, or transitioning calves to new feed sources in early weaning programs.

- Heifer reproduction. There is some evidence that young heifers that are overly fat from grain feeding will experience lower milk production in their lifetime.

Delivery Systems

Hand-Feeding

Supplemental feeds can be made available by hand on a daily to twice weekly basis, depending on the supplement and the class of cattle being fed. This will generally be the cheapest method of supplementation, and will generally provide the most consistent and uniform intake of feed. This is an especially important feature when supplements are used to deliver melangastrol acetate to heifers as part of an estrous synchronization program. This method also affords the opportunity to regulate and change intake as conditions warrant. The major disadvantage is the labor required to transport feed and to feed the cattle.

Salt-Containing Free-Choice Mixes

The process of providing salt-containing free-choice mixes was described for mineral mixes, and the same process may also be used to regulate intake of grain feeds for grazing cattle. Table 3-14 shows the ratios of grain and salt needed for various level of grain intake.

Regulation of the salt content of the mix is necessary to get correct grain consumption levels. Make small batches of the salt-grain mix at first

Cattle weight (lb)	Daily salt intake	Grain intake (lb/d)							
		1	2	3	4	5	6	7	8
		Salt in mix (%)							
400	0.4	29	17	12	9	7	6	5	–
500	0.5	33	20	14	11	9	8	7	6
600	0.6	38	23	17	13	11	9	8	7
700	0.7	41	26	19	15	12	10	9	8
900	0.9	47	31	23	18	15	13	11	10
1100	1.1	52	35	27	22	18	16	14	13
1400	1.4	58	41	32	26	22	19	17	15

Table 3-14. Salt needed to control intake of grains.

Source: ©MWPS.org (Midwest Plan Service), Iowa State University, Ames, IA. www.mwps.org. Used with permission: Beef Cattle Handbook, MWPS-CD-IP.

because the cattle will eat less of the mix when it is first offered and increase their consumption over a period of 3–4 weeks. Whenever possible wooden feeders or troughs under shelter should be used to protect equipment and prevent the salt from dissolving or caking in the feeder. It is usually a good idea to thoroughly clean grinder-mixers after mixing feeds with high salt content.

Plain, iodized salt should be used. In some trace-mineralized salt mixes it is possible to get toxic levels of certain minerals in a high salt feed. If it is necessary to provide trace minerals, keep the consumption level of the trace-mineralized salt at less than 0.1 pound per day for cows and 0.05 pound per day for yearlings. Provide plenty of fresh water at all times.

Molasses-Based Blocks

Free-choice access to molasses-based blocks can deliver various nutrients, particularly energy and protein. A report from Froetschel et al. (23) described blocks that contain cottonseed meal, molasses, and broiler litter. Others contain **anthelmintics** or bloat preventives. Most commercially available blocks have a high proportion of urea as the protein source. They are generally very palatable to cattle and often employ high salt content as an intake regulator. Intake by cattle may be excessive when first exposed to blocks, but will decrease with continued access. The major advantage of using blocks is convenience, and the disadvantages are inconsistent intake, excessive urea intake, and the cost per unit of nutrient delivered.

Lick Tanks

Molasses-based liquid supplements can be used to deliver energy, protein, and minerals with free-choice access to lick tanks. Various suspension agents are available that allow an array of products to be used, including blood and feather meal combinations (14). Similar to blocks, most commercially available liquid supplement formulations rely heavily on urea as the protein source, which again poses the problem of excessive intake. This is particularly true for initial access to the material. Lick tanks can be highly effective to deliver supplements in extensive management programs, when labor is costly, and when specific nutrients or minerals need to be delivered for cattle health or other reasons. Inconsistent intake, low efficiency of liquid feed use (14), and cost per unit of nutrient are the major disadvantages of this system.

WINTER FEEDING STRATEGIES FOR COWS AND YEARLING CATTLE

In many areas of the Northeast the production of winter feed is the major factor limiting herd size. Beef production enterprises are often on marginal lands where areas to harvest winter feed can be at a premium. The major result of intensive grazing management is to increase the carrying capacity of the grazing lands. If a cow-calf producer is unable to increase winter feed production to accommodate the increased stocking of grazing lands, the economical benefits of intensive grazing may be limited because of inability to increase herd size. Consequently, for producers who winter cattle, changes in grazing management should be accompanied by corresponding changes in the winter feeding program.

Hay

Hay is probably the most common source of feed for wintering beef herds in the Northeast. It has certain advantages over other alternatives for wintering cattle. The costs are often less for this commodity because it is produced on the farm. It can be produced on sites where slope limitations prevent other crops from being grown, and labor requirements for feeding are often less than for other feeds.

In the Northeast the predominant forage species used for hay production for beef cattle is orchardgrass with lesser amounts of timothy, bromegrass, and tall fescue. Legumes are generally found mixed with grass species in hay fields and include mainly alfalfa and red clover and other species to a lesser extent.

One of the greatest advantages of hay feeding is that large hay packages are generally placed in a wintering area to supply several days of feed, thereby reducing labor requirements. However, this can lead to a great deal of wastage during feeding. Cattle will remove hay from feeders and use it as a source of bedding. In addition, this waste hay can inhibit the subsequent growth of forage the following spring on large areas of land under this wastage. Beef producers should minimize this wastage by using specifically designed feeders and/or forcing the cattle to consume excess hay. Even under the best feeding conditions producers should plan for at least 15% more hay than expected consumption to allow for wastage. Based on a 1,200-pound cow's expected daily forage intake (43) of about 28 pounds of hay, one would need to feed more than 32 pounds of hay per cow daily to allow for minimal wastage. Producers also need to ensure that winter feeding of the herd is on land that is not highly desirable for production, because pouching by the cattle and excess hay will greatly inhibit subsequent forage growth, especially for the most desirable forage species.

Many beef producers feeding hay in the Northeast try to harvest two hay cuttings a year. The first hay cutting generally occurs in the late spring and the second cutting occurs in the mid- to late summer. Reported in table 3-15 are the nutritive values of hays summarized from the studies of Baker et al. (2) and Prigge et al. (46) and representing two grass species, orchardgrass and tall fescue, harvested as either pure grass stands or with legume incorporation. The results indicate the lower nutritive value and higher yields of first-cutting hay as opposed to second-cutting hay. Second-cutting predominantly grass hays generally have less stem and more leaf than first-cutting hays. These results are typical of other studies. Based on the increased nutrient requirements of the cow herd during the latter stages of pregnancy and early stages of lactation, it is recommended to feed the higher quality

Table 3-15. Protein and energy yield of first- and second-cutting cool-season grass and grass legume hays.

Harvest	TDN (% DM)	NEm (Mcal/kg DM)	CP (% DM)	Yield (lb/ac)
1st cut	54.9	0.59	8.2	3,361
2nd cut	56.0	0.61	12.0	2,360

Sources: Adapted from Baker, M. J., E. C. Prigge, and W. B. Bryan. 1988. Herbage production from hay fields grazed by cattle in fall and spring. J. Prod. Agric. 1: 275-279; and from Prigge, E. C., W. B. Bryan, and E. S. Goldman-Innis. 1999. Early and late-season grazing of orchardgrass hayfields overseeded with red clover. Agron. J. 91: 690-696.

second cutting during these periods of peak nutrient demand. Again it must be recognized that at times the hays may not be able to meet the nutritive requirements of the cows; when this occurs the body condition of the cows can be a buffer that protects the producer from the consequences of under-nutrition discussed previously. Alternative feed sources will also have to be considered at times as a supplement.

Hay can be used as the only energy source to winter yearling cattle provided gains can be maintained at least 0.75 lb per day and the cattle will be grazing the following spring. This would require hay with a TDN value of at least 57% and a crude protein content of about 8.5% (43). Most hays produced in the Northeast would meet these requirements. Cattle gaining at this restricted rate would be able to compensate during the grazing season and provide adequate returns for the producer when marketed.

For the development of replacement heifers in which the goal is to calve as 2-year-olds, gains in the wintering period would have to be greater than cattle destined for the market. The optimal gain for a replacement heifer depends on her weaning weight and estimated weight when puberty is reached. For a replacement weaning at 500 pounds and an expected weight at puberty of 700 pounds, a gain of 200 pounds would be needed by the time she is 14 months of age; this would be approximately 1 pound average daily gain from weaning. This would require hay with a TDN value of 59% and a crude protein of about 9.4%. Most hays sampled in the DHIA laboratory would meet these requirements, but more than 17% of the samples would not. Again if hay alone is used to develop replacement heifers, an analysis of the hay should be considered as well as the need to supplement with additional energy or protein. Another consideration if hay quality is a problem is wrapping the bales.

Ensiled bales should be of higher quality at harvesting and will not deteriorate during storage. The additional costs may be justified for at least a portion of the hay crop if quality is a problem.

Corn Silage

Corn silage as a winter feed has the advantage of greater DM yields per acre when compared to typical hay yields. Thus, acreage needed for winter feed production can be reduced significantly when corn silage is used. However, the feeding of corn silage has several negative aspects that limit its use as an alternative to hay for wintering beef cows. In addition to being more costly to produce, it does not lend itself to practical and low cost feeding systems and has to be fed daily to prevent spoilage.

The nutritive value of corn silage also presents problems for cow-calf producers. According to Northeast DHIA lab results in 1994, corn silage had an average TDN value of 70%; this exceeds the energy requirements of all beef cows even at times of peak energy demands (table 3-1, p. 44). Consequently, it has to be limit-fed if it is to be used efficiently. Also, the crude protein level of corn silage is lower than desirable; according to the Northeast DHIA summary for 1994, the average crude protein content was 8.3%. This is lower than requirements for the average lactating beef cow. The problem is exaggerated if one considers that corn silage is limit-fed to meet energy requirements. Additional protein sources must be fed with corn silage to meet the N requirements of the cattle. At the second month of lactation, the cows would require about 1.25 pounds of soybean oil meal daily assuming the corn silage has 8.3% crude protein. The economic returns from a beef production system may not justify the additional labor requirements for the use of corn silage as a winter feed for the cow herd if acreage for winter feed production is not limiting.

Corn silage can be used for backgrounding growing cattle. Growing cattle can take full advantage of the energy value of corn silage when supplemented with protein, and they are often fed in confinement situations, which lend themselves to the feeding of mixed corn silage-based rations. Protein sources such as soybean meal and/or urea can be used.

Extended Grazing

One option to lessen the need for conserved forage for winter feeding is to extend the grazing season. The production of conserved forage represents the greatest expense a cow-calf producer can incur in most cases. Because cool-season grasses do not grow in the winter, to extend the grazing season forage has to be accumulated or stockpiled in the summer and early fall for grazing during the winter months. The forage species that most readily lends itself to winter grazing is tall fescue. It produces a sod that withstands the punishment of winter grazing (2). Yields of tall

fescue in the autumn exceed those of other cool-season grasses (61), and its quality is maintained throughout the winter better than other grass species (2).

To effectively extend the grazing season using stockpiled forage one must know limitations and benefits of this practice. One benefit of extended grazing for cow-calf operations is that land with slope limitations can be diverted from grazing to winter feed production. Thus, extended grazing can enhance the carrying capacity and hopefully profitability of a farm in which acreage available for winter feed production is limiting.

Studies by Rayburn et al. (49) (table 3-16) showed the influence of date when stockpiling of forage for winter feeding was initiated on DM yield per acre. These results suggest that the earlier stockpiling is started, the greater the herbage available for grazing. However, late applications of N seem to enhance production

Table 3-16. Yield of tall fescue accumulated by December from several spring–summer dates averaged over 2 years.

Date of analysis 100 lb N/ac	Stockpiling periods			
	June to Dec.	July to Dec.	Aug. to Dec.	Sept. to Dec.
	lb/ac[a]			
No N	2,606	1,316	803	431
June	4,078	2,431	899	521
July	4,114	2,990	1,303	574
August	3,319	2,863	1,574	808
September	3,347	2,571	1,692	1,381

[a] DM basis.

Source: Adapted from Rayburn, E. B., R. E. Blaser, and D. D. Wolfe. 1979. Winter tall fescue yield and quality with different accumulation periods and N rates. Agron. J. 71: 959–963.

N application Rate (lb/ac)	Stockpiling period [b]					
	July to Dec.		July to Jan.		July to Feb.	
	TDN (%)	CP (%)	TDN (%)	CP (%)	TDN(%)	CP (%)
No N	47.1	7.8	43.8	7.2	42.9	7.2
107	47.4	8.6	43.2	8.5	41.5	8.8

Table 3-17. TDN and CP content of stockpiled fescue sward given different harvest dates.[a]

[a] DM basis.

[b] TDN and CP values at end of stockpiling period. TDN values were calculated from in vitro digestible DM determinations.

Source: Adapted from Collins, M., and V. A. Balasko. 1981. Effects of N fertilization and cutting schedules on stockpiled tall fescue. I. Forage yield. Agron. J. *73: 803–807.*

at the later dates for initiating stockpiling. In this study the crude protein content of the forage also increased at the later stockpiling times independent of N fertilization time, suggesting an improvement in energy content of the forage as well.

It appears from the research of Collins and Balasko (12) (table 3-17) that the TDN content of tall fescue deteriorates significantly with increasing stockpiling time. The TDN values as calculated from the in vitro digestibilities reported by Collins and Balasko are below the requirements for all cows at the mid-January and mid-February harvest dates. The mid-December TDN values are suitable for dry cows that are not in the last trimester of pregnancy. These results suggest that extended grazing is a viable alternative to other winter feeding programs only through December. Grazing can

probably be extended beyond December if the cows are still in good condition or if a late calving season is used. Other possibilities to enhance the use of extended grazing would be providing supplemental feeds such as soy hulls and other alternatives, especially during the later part of the wintering period. From personal experience, cows can graze tall fescue under at least 12 inches of snow. We generally strip graze tall fescue and provide a fresh paddock to graze after a snowfall.

Extended grazing can be a viable alternative or addition to the feeding of conserved forage for the cow herd; however, producers must be aware of the limitations of this practice. The major limitation is the low quality of the available forage, which negates extended grazing as an option for some classes of cattle, including yearling cattle production.

CHAPTER 4
Dairy Nutrition and Management

Lawrence D. Muller, Carl E. Polan, Steven P. Washburn, and Larry E. Chase

In chapter 2, the basics of nutrition, including nutrients and nutrient requirements, were discussed. This chapter expands on the principles discussed in previous chapters as they relate to the nutrition of dairy cattle. The focus will be on the grazing dairy animals.

NUTRITION BASICS FOR DAIRY CATTLE

The basic classes of nutrients that apply to all species are carbohydrates, proteins, lipids, vitamins, minerals (ash), and water. Figure 4-1 illus-trates the basic nutritional components of every feedstuff and the nutritional components and terminology that are used in formulating diets for dairy cattle.

Carbohydrates

The carbohydrates in forages and feedstuffs can be categorized into nonstructural (NSC) or nonfiber carbohydrates (NFC) and structural carbohydrates (figure 4-2, p. 82). Structural carbohydrates consist of the cell wall components, including cellulose, hemicellulose, and lignin. The unique digestive system of ruminants allows

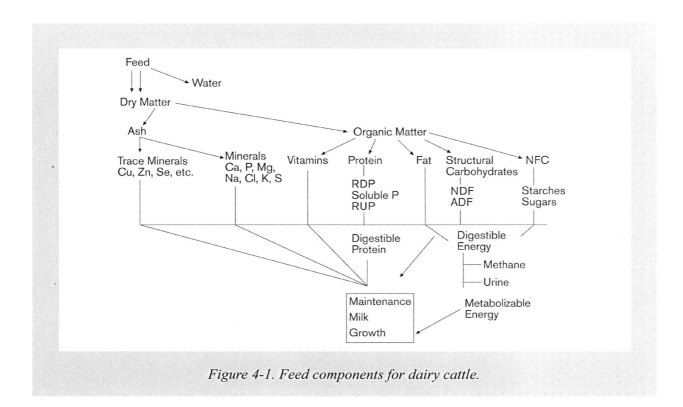

Figure 4-1. Feed components for dairy cattle.

digestion of fibrous carbohydrates, which provide energy for rumen microorganisms and for the cow. The two fractions of structural carbohydrates (figure 4-2) used in evaluating feedstuffs and in dietary formulations are:

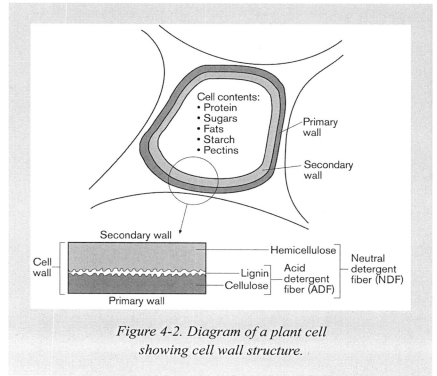

Figure 4-2. Diagram of a plant cell showing cell wall structure.

- acid detergent fiber (ADF)–The residue remaining after boiling a forage sample in acid detergent solution. ADF contains cellulose, lignin, and ash, but not hemicellulose.
- neutral detergent fiber (NDF)–The residue remaining after boiling a forage sample in neutral detergent solution. NDF represents the indigestible and slowly digestible components in plant cell walls (cellulose, hemicellulose, lignin, and ash).

Structural carbohydrates are a major source of energy because most forages are high in fiber; stimulate production of saliva, which is high in sodium bicarbonate, a rumen buffer; are needed for rumination and normal rumen function; and slow down rate of passage of feed in the digestive tract.

NFC, sometimes referred to as NSC, is the soluble carbohydrates found in the plant cell contents. The NFC consists of sugars, starch, pectin, and the fermentation acids in ensiled products. Pectin is part of the cell wall but is soluble and readily available for digestion in the rumen. NFC is highly and rapidly digestible in the rumen and is the major source of energy

in the animal's diet. In addition to the sugars, starch, and pectin, cell contents also include protein and fat.

When formulating diets for dairy cattle, a balance of NSC and NFC must be maintained for optimal rumen function, animal health, and productivity. This balance is illustrated in figure 4-3.

The targets in the *total ration* (forage and concentrates) for the high-producing dairy cow are:

- NFC - 35–40% of total ration dry matter (DM),
- NDF - 28–34% of total ration DM,
- forage NDF intake - > 0.85% of body weight, and
- ADF - > 19% of total ration DM.

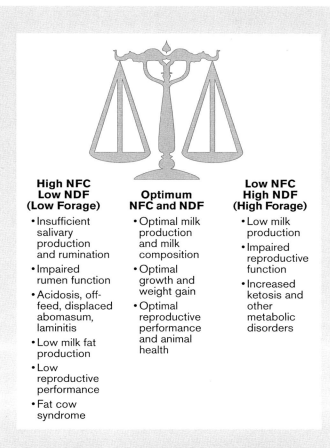

**High NFC
Low NDF
(Low Forage)**

- Insufficient salivary production and rumination
- Impaired rumen function
- Acidosis, off-feed, displaced abomasum, laminitis
- Low milk fat production
- Low reproductive performance
- Fat cow syndrome

**Optimum
NFC and NDF**

- Optimal milk production and milk composition
- Optimal growth and weight gain
- Optimal reproductive performance and animal health

**Low NFC
High NDF
(High Forage)**

- Low milk production
- Impaired reproductive function
- Increased ketosis and other metabolic disorders

Figure 4-3. Feed components for dairy cattle.

Energy

All cows need energy to survive and function. The cow requires energy for:

- maintenance,
- activity,
- milk production,
- growth,
- body condition, and
- reproduction/pregnancy.

Daily requirements for energy are second only to water in terms of actual quantity needed. But unlike other nutrients, energy is not a specific chemical substance or compound. Energy is derived as a product of the digestion, absorp-tion, and metabolism of various feedstuffs ingested by the animal (figure 4-1, p. 81). The carbohydrate components of feeds (structural and nonstructural) are the primary energy sources in a ruminant ration. An energy deficiency in the dairy cow will result in lower milk production, body condition loss as the cow draws on body reserves of fat and protein, lower weight gain or loss of body weight, lower reproductive perfor-mance, and more metabolic disorders. However, feeding excess energy and allowing the cow to become too fat and overconditioned can also be detrimental to animal performance. Cows on pasture consistently have lower body condition than cows in confinement. Pasture feed-ing is likely related to low energy intake in relation to energy requirements. It is nearly impossible to avoid some thin cows on pasture-based systems.

Grazing cows have a higher energy requirement for activity compared with nongrazing cows. The 2001 National Research Council (NRC) report (15) stated that for every mile walked to and from pasture and for eating activ-ity, a cow needs 12% or 1.2 megacalories of net energy added to the maintenance requirements. This extra energy need can be met by feeding about 2 pounds of concentrates. Cows grazing hilly topography and walking about 2 miles per day may require up to a 25–35% increase in maintenance energy compared to cows graz-ing relatively flat pastures. This extra energy may come in the form of nearly 4–5 pounds of concentrates per cow daily. The negative impact on milk production if extra dietary energy is not provided with relatively flat pasture land is shown in figure 4-4 (p. 84) (15). Hilly pastures and increased walking distances would result in even larger decreases in milk production.

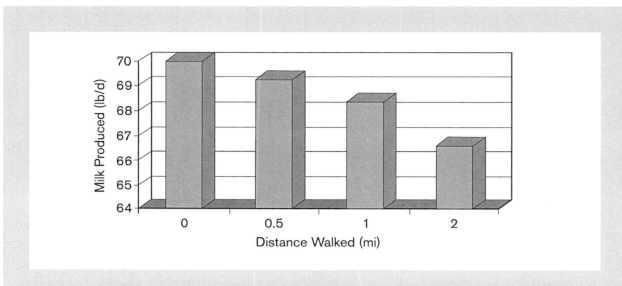

Figure 4-4. Energy cost of walking and impact on milk production.
Source: Adapted from National Research Council. 2001. Nutrient Requirements of Dairy Cattle,
7th ed. National Academy Press, Washington, D.C.

Nutritionists use the net energy system of the NRC (15) to describe energy requirements for dairy cattle and the energy content of feedstuffs. There are three net energy values for each feed because animals use feedstuffs with different efficiencies, depending on how the energy is being used. Net energy for gain (NEg) is the least efficient and will have the lowest value. Net energy used for maintenance (NEm) and lactation (NEl) are similar and more efficient than the energy used for gain. Net energy values for forages are best for figuring ration formulation because they most accurately reflect the energy available for milk production.

Laboratory digestibility and net energy values are not produced from digestion trials or metabolism studies. The feeding value of forages is negatively associated with the cell wall contents (as the ADF and NDF values increase, energy values decrease). Because of this, microbial

energy values, estimates of digestibility, and relative feed values reported on laboratory analysis are often estimated using the fiber content in the forage. NDF content is often used to estimate the amount of forage an animal will be expected to consume. The use of NDF values to generate many of the relative feeding values further emphasizes the importance of cell wall content for animal performance. Laboratories are now estimating digestibility and energy values by in vitro dry matter digestibility techniques. The digestibility of NDF is now used as an indicator of the feeding value of forages.

Protein

Protein makes up most of the cow's body (muscles, skin, organs, blood). It is also part of milk, ranging from 3.0 to 4.0% of milk, depending on breed and diet. Protein is needed for maintenance, growth, reproduction, and milk production. Protein and urea or other nonprotein nitrogen sources are needed to meet the rumen microbial nitrogen requirement for growth and protein synthesis. The rumen microbes cannot synthesize enough protein to meet the cow's protein needs for high milk production, so dietary protein, which escapes degradation in the rumen, may be needed.

Protein nutrition in ruminants is quite complex and requires examining protein fractions in addition to crude protein (CP). Certain levels of the various protein fractions must be present in a dairy cow's diet to meet the needs of the rumen microbes as well as to provide essential amino acids to the small intestine. Therefore, developing rations to meet the cow's requirement for protein requires balancing rations for protein fractions in addition to CP.

Soluble intake protein (SIP) is the protein in feedstuff that is readily soluble in the rumen. This protein is rapidly degraded in the rumen to ammonia and other simple nitrogen compounds. About 30–35% of the total protein in the ration for lactating cows should be SIP. Rumen-degradable protein (RDP) is the fraction of protein that is degraded to ammonia by rumen microbes and is available in the rumen. The optimum amount of RDP for the total ration is about 60–67% of the total protein. Rumen-undegradable protein (RUP) is the fraction of protein that escapes ruminal degradation and is available for digestion in the small intestine. The optimum range of RUP in the total ration is about 33–40% of CP. The amino acid content of RUP is important when meeting the nutrient and amino acid needs of high-producing cows. Unavailable or bound protein, often referred to as ADF-N or acid-insoluble nitrogen or rumen-undegradable protein, is the protein fraction that escapes ruminal degradation and is not digestible in the small intestine.

The protein in every forage or feedstuff is degraded and used differently in the digestive tract of the ruminant and has different values for these protein fractions. A partial listing of the protein and carbohydrate content of different pastures is shown in table 4-1 (p. 86). A schematic diagram of the protein fractions typically found in feedstuffs is shown in figure 4-5 (p. 86). A more in-depth discussion is offered below.

Dry Matter Intake

Animals must eat the proper amounts of all required nutrients to produce milk and/or gain weight and remain healthy. The NRC (15) has established guidelines for dry matter intake (DMI) requirements for dairy cattle. There are many factors affecting dry matter requirements and DMI. We need to understand these factors because the goal when feeding lactating dairy cattle, regardless of the system, is to maximize DMI. A more specific goal is to maximize DMI of pasture, the lowest cost source of nutrients. As a rule of thumb, 1 pound more DMI will provide the energy for the early lactation dairy cow to produce about 2 pounds more milk.

The major factors affecting DMI are:

- Animal factors:
  — Body weight
  — Milk yield and composition
  — Body condition gain or loss
  — Stage of pregnancy
  — Stage of lactation

- Pasture and feed factors
  — Pasture quality
  — Pasture availability
  — Sward density
  — Supplement type and amount
  — Time allowed to graze

- Environmental climate, temperature, humidity, and management

Table 4-1. Average nutrient composition for cool-season grass pasture and legumes over a grazing season.

Nutrient	Predominantly grass (cool-season)		Grass with legumes	
	Spring	Summer	Spring	Summer
CP (% DM)	21–25	18–22	22–26	20–24
RUP (% of CP)	20–25	25–30	20–25	25–30
Soluble P (% of CP)	35–40	25–30	30–35	25–30
ADF (% DM)	24–28	28–34	21–35	25–30
NDF (% DM)	40–45	48–55	30–36	35–45
Hemicellulose (% DM)	17–21	21–25	12–16	15–19
Cellulose (% DM)	16–20	21–26	16–20	18–23
NE, Mcal/lb (% DM)	0.72–0.78	0.66–0.72	0.74–0.80	0.70–0.74
NFC (% DM)	15–20	12–15	18–24	15–20
Fat (% DM)	3–4	3–4	3–4	3–4
Ash (% DM)	7–9	7–9	8–9	7–9
Ca (% DM)	0.40–0.60	0.40–0.60	0.60–0.80	1.1–1.3
P (% DM)	0.25–0.30	0.25–0.30	0.30–0.35	0.30–0.35
Mg (% DM)	0.15–0.20	0.15–0.20	0.18–0.24	0.18–0.24
K (% DM)	2.0–3.5	2.0–3.5	2.0–3.5	2.5–3.5
S (% DM)	0.16–0.22	0.16–0.22	0.18–0.24	0.18–0.24

Sources: Summarized from several sources, including Muller, L. D., and S. L. Fales. 1998. Supplementation of cool season grass pastures for dairy cattle. p. 335, in Grass for Dairy Cattle. *J. H. Cherney and D. J. R. Cherney, eds. CAB International, Oxon, UK.*

Figure 4-5. Schematic diagram of protein fractions in feeds.

Source: Adapted from National Research Council. 2001. Nutrient Requirements of Dairy Cattle, *7th ed. National Academy Press, Washington, D.C.*

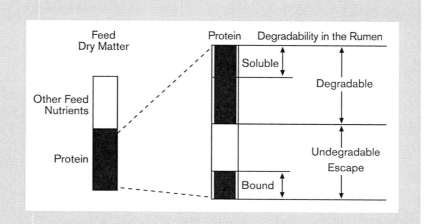

Lactation Cycle of a Dairy Cow

The typical lactation cycle of the dairy cow is as follows (15):

- Cows peak in milk production at about 5–6 weeks after calving. Milk yield typically declines at 6–8% per month after the peak milk, but is influenced by nutrition and management.
- Peak feed intake occurs about 8–10 weeks after calving and typically declines as milk yield declines.
- Cows typically lose body weight and body condition during the first 6–8 weeks after calving. The tissue that is mobilized is used to produce milk during the first 8–10 weeks of lactation.
- Cows typically regain body weight and body condition during the last half to two-thirds of the lactation.
- The goal is to begin breeding a cow at 60–90 days after calving with pregnancy occurring between days 100 and 120. For seasonal calving, cows should be pregnant by 85 days after calving to ensure a 12-month calving interval.
- Dairy cows need a 50–60-day dry period.

The energy status during a typical lactation cycle of a dairy cow is an important consideration and influences the lactation curve. A high-producing dairy cow in early lactation has a higher output of energy in milk than intake of energy through feed. Thus, she is in negative energy balance and mobilizes body tissue to use as energy to produce milk. During the latter half to two-thirds of lactation, milk production begins to decline in relation to energy intake. A cow is in positive energy balance and the goal is to restore body weight or condition that was lost during the first 8–10 weeks of lactating. Managing energy intake and body reserves is one of the major challenges in feeding dairy cattle, and the balance of carbohydrates in the diet is a key factor

in this management. This balance appears to be even more critical with cows managed in a pasture-based system. Several studies and farm experiences have shown that high-producing dairy cows on pasture-based feeding programs tend to decline faster in milk yield and lose more body condition than cows managed in confinement. Imbalances and deficiencies in dietary protein, carbohydrates, and energy likely contribute to these responses in milk yield and body condition loss.

QUALITY OF INTENSIVELY MANAGED PASTURES

Pasture *quantity* and *quality* are the two most important factors in maximizing intake and the amount of nutrients obtained from low cost pastures. Many factors influence the quality or nutrient composition of pasture, but when compared to stored forages, well managed pasture is generally higher in total protein and RDP, lower in fiber, and higher in estimated energy. A typical nutrient analysis of pastures over the grazing season in the northeast United States is presented in table 4-1. Well managed cool-season spring pastures may have up to 25–30% CP with NDF concentrations of less than 40%. Nitrogen fertilization usually increases total protein and soluble protein content in grass pastures. Ryegrass usually has lower NDF than orchardgrass, and inclusion of alfalfa and legumes with grasses will usually lower the NDF content of the pasture compared to grass alone.

Given a grazing height of 6–9 inches, pasture quantity and quality can be high during the entire grazing season. However, protein concentration and energy usually decrease in the summer and fiber content tends to increase. In general, pastures containing some legumes will be higher in nutrient value (lower fiber and higher digestibility) than grass pastures. These

From Confinement to Grazing to the Future

Adapted from an article by D. Forgey, "From Confinement to Grazing to the Future," in
Hoard's Dairyman, *Fort Atkinson, WI. Used with permission.*

Dave Forgey has become well known among dairymen across the country because of his articles about his grazing dairy in *Hoard's Dairyman*. He lives at Logansport, Cass County, Indiana. Dave is a third-generation dairyman on the farm he took over in the mid-1970s by purchasing land, livestock, and machinery. He milks about 150 cows with the same number of replacement heifers. To survive, it was necessary to have an efficient operation. The devaluation of the 1980s put a constant strain on the operation and net profit per cow was declining each year. Increases in milk production per cow came with nearly equal increases in cost of production until net profit was less than 5% of gross sales. A drought in August 1988 put the operation near collapse.

Then Dave learned of the Mahoning Project, a grazing study at Ohio State University. In 1991, he started rotationally grazing his heifers. He continued to read about rapid rotational grazing. He visited Wisconsin and met Alan Henning, a Wisconsin-based grazing consultant. He invited Alan to lay out his farm for a complete grazing system.

Grazing began in April 1992 and no additional forage was supplemented until October, when Dave began supplementing an 18% CP pellet in the parlor. But cows were losing condition and not settling well. CP in the pastures was in the high 20s, so Dave changed to a 12% CP. By July, cows were gaining well and pregnant. Milk peaked at 73 pounds in August.

In 1993, Dave fed a maximum of 16 pounds per day of 12% protein mix of rolled corn, oats, and soybean meal with added mineral. The rolling herd average slipped about 1,000 pounds. In 1994, Dave returned to a pelleted mix to get more intake in the parlor, up to 24 pounds per day.

Body condition seemed to improve, and the 1,000 pounds in production was regained.

In 1995, grain prices were high. All supplemental protein was cut and grain was fed at about 1% body weight. This practice was followed through 1999. Production was excellent and fertility higher than in the past.

The Forgeys started by grazing 120 acres, but now have 260 available. They practice a **leader-follower** system in their rotational grazing. When the forage is 8–10 inches in height, the lactating herd is allowed to graze the top 3–4 inches in a 12-hour period. Bred heifers follow for the next 12 hours, leaving about 3 inches of residue. Calves are kept on a separate grazing system to ensure excellent quality for younger animals. With seasonal grazing, calves are born in early spring, so three groups of similar animals are managed: calves, bred heifers, and the cow herd.

A number of other factors have changed in the operation. Fuel and repair costs decreased, because the cows were doing the harvesting. Overall herd health improved. Rumen disorders did not occur. Milk fever and retained placentas declined greatly. Hoofs did not need trimming. Breeding has continued to be a challenge.

The most important change: operating expense was 78% of gross sales in 1990, the last year of total confinement. Since, it has been in the 65% range, and only 57% in 1997.

The Forgeys are doing well now and enjoying life. They believe the only way young people can get started in dairying is to develop a low cost, low input system that allows a good return on investment. Grazing could be that type of system in the high rainfall areas of the United States.

high quality pastures will have an estimated net energy of lactation value of 0.70–0.78 mega-calories per pound of DM, depending on season, climate, and grazing management. Because of the high quality of pastures, passage through the digestive tract may be faster than with stored forages such as hay or silage. This rapid passage rate will reduce residence time in the rumen and may lead to reduced digestibility. In addition, key nutrients such as protein and carbohydrates (structural and nonstructural) in pasture forage may be degraded in the rumen at a more rapid rate than with stored forages.

Forage Testing

Forage testing is advisable to monitor the chemical composition and changes in composition within and between years. At minimum, analyses should include CP, soluble protein, ADF, NDF, an estimate of net energy, and macrominerals (calcium [Ca], phosphorus [P], potassium [K], magnesium [Mg]). In conventional dairy ration programming, lab analyses are used to determine the nutrients in forage and estimate the nutrient intake by the animals. This allows diets to be as consistent as the person who does the feeding. Rations cannot be formulated with the same accuracy in a grazing system as with nongrazing. Visual assessment of pasture does not discern changes in nutrient and mineral components that are nutritionally important. In addition to helping with ration formulation, grazed forage analysis provides information for the manager and nutritionist about how changes in grazing management affect the quality of the forage.

To be useful, pasture samples must represent what the cows are eating. The first step in sampling a pasture is to observe how the cows graze the paddocks. Samples should be plucked by hand from 20–30 sites to represent the grazing patterns and grazing height of the animals.

Samples should be kept cold before sending them to the laboratory. As much air as possible should be squeezed from the sample bag.

Wet chemistry analyses will likely be the most accurate (especially for minerals). Many labs have NIR calibrations to analyze samples of fresh grasses and legumes. Besides the usual protein and fiber analyses, soluble protein should be analyzed because it can be quite variable. Mineral analysis is important because values may vary by 20–30% during a grazing season and will be influenced greatly by the amount of legumes in the pasture. In addition, minerals have a major effect on several metabolic diseases and on the environment.

A sound grazing management program will provide the necessary forages and environment for good milk production. Proper forage sampling and analyses are management tools that allow the producer to formulate rations and maximize the return from each dollar spent on feed.

ENERGY/PROTEIN RELATIONSHIPS

Nutrient Imbalances/Deficiencies in Pasture

High-producing dairy cows require a well balanced diet to optimize production and profitability. The basics of dairy nutrition published in the NRC guide (15) are useful for developing balanced rations to complement pasture, including CP and the protein fractions, NDF, NEl, minerals, and other nutrient components (figure 4-1, p. 81). Although some believe that feeding cows on pasture is an "art," science is needed to develop the most profitable total feeding program. Application of knowledge about nutrient utilization in the rumen and by the animal will help ensure optimal animal performance.

High total protein and energy and low fiber content, particularly in spring pastures, indicate a quality of forage that is higher than most stored forages. However, there are nutritional limitations and imbalances with pastures when attempting to develop feeding programs to meet total nutrient needs and achieve profitable milk production.

- Total nutrient intake may be inadequate to support high levels of milk production.
- Total protein in pasture is high. It is highly degradable in the rumen but not efficiently used there. Adding NFC from grains will help "capture" this degraded protein in the rumen. RUP intake may be inadequate for high-producing cows.
- The fermentable carbohydrate content, which is the major source of energy for rumen microbes and the cow, is usually low in pasture compared to the needs of the cow.
- High quality pasture may be too low in "effective fiber" to stimulate adequate cud chewing and rumination, especially when NFC is included in the diet via grains. This may result in reduced milk fat content. The rate and extent of fiber digestion may be altered with grazing due to the high quality and moisture of pasture.
- The amount of minerals, including Ca, P, Mg, sulfur, copper, zinc, selenium, and salt, is often inadequate in pasture to meet the cow's needs, whereas K may be too high, depending on the legume content of the pasture. Ideally, these minerals should be analyzed in pasture, added to a concentrate mixture, and force fed to more precisely meet the mineral needs.

Nutrient utilization by the rumen microorganisms, and thus by the cow, is not optimal with only pasture. The above limitations, if not properly addressed, may lead to high nutrient passage through the digestive tract (and loose manure) and less than optimal milk production. The inadequate fiber can contribute to low milk fat. Inefficient utilization of the high protein content of pasture and the energy cost to the cow to excrete this excess protein can lead to losses in milk production and is an environmental concern. The energy needed to excrete the excess urea may be equal to 2–4 pounds of milk per day. These limitations suggest the need to strategically supplement pasture to maximize the utilization of the high quality, low cost forage that the cows are harvesting.

A goal of most dairy producers should be to maximize the intake and utilization of pasture, which is the lowest cost forage on the farm. Maintaining intake of high quality pasture through proper grazing management offers the best opportunity to maximize pasture intake and to reduce total feed cost. A system with several paddocks and a well-planned rotation provides the opportunity to have high quality forage available to cows at the correct time.

Pasture as the Only Forage and DMI

One of the basics in feed programming and feeding dairy cattle is estimating both total and forage DMI. Intake is influenced by many environmental, diet, and animal factors, many of which are well understood in nongrazing systems when total mixed rations (TMRs) are fed. A basic difference in feeding management between grazing and confinement is that adequate amounts of pasture must be provided in the "field feedbunk," similar to what is provided in the confinement feedbunk with nongrazing farms. In the field, estimating total DMI using NRC (15) values and subtracting the amount of grain and supplemental forage fed from total expected DMI can provide a rough estimate of expected DMI from pasture. This method may be satisfactory if pasture availability is adequate.

The maximum expected DMI of pasture alone is about 36–40 pounds per cow per day for Holsteins when *high quality* pasture is the *only* feedstuff. The NDF content is related to DMI and can be used as a guideline for expected pasture intake based on forage NDF (F-NDF) intake expressed as a percentage of body weight. Average forage NDF intake of cows on pasture ranges from about 1.1 to 1.3% of body weight. For a 1,300-pound cow, this is about 14–17 pounds of F-NDF intake per day. This amount of pasture DMI may support about 45–55 pounds of milk per day in early lactation. However, body condition loss can be high at higher milk production levels. When supplemental concentrates are fed, pasture DMI decreases, so total DMI increases. A summary of several studies (1, 14) indicates that *when adequate high quality pasture is available* and when concentrates are fed in amounts "typical" for these milk production levels, total DMI is comparable to the DMI expected with nongrazing cows.

If one of the goals is to maximize the intake of low cost, high quality pasture, then understanding the factors that influence pasture DMI, and in turn managing the pastures and cows to maximize this intake, become key factors. Pasture intake by the grazing dairy cow is largely determined by how effectively the cow harvests the available pasture in the field. Pasture intake is primarily a function of:

biting rate (bites/min) × grazing time (min/day) × intake (g DM)/bite = pasture intake

Most detailed grazing behavior studies have shown that both biting rate and grazing time per day are influenced by hunger drive or genetic merit. High-producing dairy cows have higher biting rates (up to a maximum of 60 bites per minute) than low-producing cows and graze for longer periods each day (up to a maximum of about 600–650 minutes per day) because they require more nutrients and have more bites per day. This is illustrated in figure 4-6, which shows that some high-producing cows had more than 40,000 bites per day compared to 25,000 bites per day for low-producing cows. If intake per bite declines, as it inevitably does with short and less dense swards, the behavioral constraints on biting rate and grazing time often mean a reduction in daily forage intake and the need for more available pasture and/or supplementation. Cows may try to compensate by increasing grazing time; however, this may represent an energetic loss.

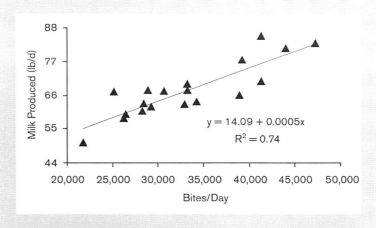

Figure 4-6. Relationship between milk production (lb/d) and number of bites per day with grazing cows fed pasture and grain supplements.

One of the keys to achieving high intakes of grazed pasture is to manipulate and manage the sward to *maximize intake per bite*. Recent studies illustrate the overriding importance of intake per bite in achieving high daily pasture intake rates, with intake being maximized on moderately tall, dense, leafy swards. The sward bulk density has a major effect on intake rate. Similar intakes have been observed at sward surface heights of 6–8 inches with high, medium, and low density swards. As surface height decreases, DMI decreases most for low density swards. When properly managed, rotational grazing systems provide the opportunity to maintain the high quantity and quality of the sward that can maximize intake per bite and total DMI.

In a recent Penn State study (2) with Holstein cows, the allocation of 90 pounds of pasture DM per cow per day clearly resulted in more pasture DMI (41 pounds) than the allocation of 55 pounds of pasture DM per cow per day (35 pounds of DMI) when pasture was the sole feedstuff (table 4-2). This higher pasture allowance resulted in more intake per bite, more bites per day, 6 pounds more DMI, and 7 pounds more milk per cow per day.

Correcting Nutritional Limitations/ Imbalances of Pasture

Energy

Energy is the most limiting nutrient for profitable milk production and normal reproductive performance when pasture is the major source of nutrients. The NFC content of grass pastures tends to be low, with a range of 15–25% of DM. This compares to the total ration needs for high-producing cows of about 33–38% of DM. The NFC content (starches and sugars) in grains must be the major source of energy for the rumen microbes. The amount of concentrates or NFC fed to increase the total energy intake on a pasture-based system can have long-term effects on energy balance, milk production, body weight and condition changes, reproductive performance, and profitability. Research and producer experiences indicate that supplemental energy from concentrates is beneficial and profitable with high-producing cows in grazing systems in the United States.

	Pasture allowance (lb DM/cow/d)	
Item	**55**	**90**
Pasture DMI (lb DM/d)	35	41
Milk yield (lb/d)	42	49
Grazing behavior		
Grazing time (min/day)	609	626
Rate of intake (bites/min)	56	58
Intake/bite, (g DM/bite)	0.55	0.60
Total bites/day	34,400	35,200

Table 4-2. Performance of lactating Holstein cows grazing two pasture allowances (no supplement).

Source: Adapted from Bargo, F., L. D. Muller, J. E. Delahoy, and T. W. Cassidy. 2002. Milk response to concentrate supplementation of high-producing dairy cows grazing at two pasture allowances. J. Dairy Sci. 85:1777–1792.

Without adequate NFC in the diet from concentrates, protein is used for energy and excess nitrogen is excreted in urine, which is an energy cost to the cow. The result is lower milk production and a higher nutrient loading into soil and water. Providing concentrate supplements rich in NFC and lower in protein can help to "capture" more of the protein in pasture and convert it to milk protein. Research and producer experiences show milk production responses of 0.6–1.2 pounds for each pound of concentrates fed with high-producing cows. One pound of milk produced per 1 pound of concentrates fed is the average response. The milk response to concentrate supplementation is usually highest with the first amounts fed, depending on the initial milk yield, and diminishes as additional concentrates are fed to high-producing cows. The response in milk yield to supplemental concentrates is generally higher when pasture quality is low and pasture availability is limited.

The end result is that feeding concentrates to high-producing cows on pasture causes higher DMI, which translates into higher, more profitable milk production and improved body condition. Australian researchers suggested that the greatest benefit to grain (energy) supplementation may be the improvement in body condition, and in turn, reproductive performance. Based on research and producer experiences, the concentrate feeding guidelines below (table 4-3) can be helpful for dairy farmers. Responses will vary, depending on the composition, quality, and quantity of pasture.

4% fat-corrected milk	Spring		Summer		Fall	
Production (lb/day)	lb	G:M[c]	lb	G:M[c]	lb	G:M[c]
> 80	20	1:4–1:5	21–24	1:3.5	20	1:4–1:5
70	16–18	1:4–1:5	18–20	1:3.5–1:4	16–18	1:4–1:5
60	12–14	1:5	15–18	1:3.5–1:4	12–14	1:5
50	8–10	1:5–1:6	10–12	1:4–1:5	8–10	1:4–1:5
> 40	6–8	1:6–1:7	8–10	1:5–1:6	6–8	1:6– 1:7

Table 4-3. Concentrate (DM) feeding guidelines for a grass-based pasture system.[a, b]

[a] Assume 1300-pound body weight (average for Holsteins).

[b] These guidelines are based on high quality grass pasture available in adequate quantities assuming the approximate DMI. Lower quality forages may require more grain. Maximum grain DM fed should be equivalent to about 20 pounds per day. Some adjustment of grain should be made based on body condition scores and stage of lactation. Lower amounts can likely be fed when the pasture contains legumes.

[c] Concentrate or grain (G) fed (DM basis) to milk (M) yield on a pound for pound basis.

Economics of Supplemental Grain Feeding

A commonly asked question relates to the optimum amount of concentrates to feed to achieve the most profitable milk production and maintain good health and reproductive performance. The total profitability of concentrate feeding depends on: (i) the price of milk and the price of grain (milk:grain price ratio; M:G), (ii) the substitution rate of concentrates for pasture, (iii) the expected milk yield response to supplemental concentrates, and (iv) production of milk fat and protein. The economics of supplemental feeding are usually positive with *high genetic merit* cows in early lactation when 0.7–1.3 pounds of milk are produced for each 1.0 pound of concentrate fed and when M:G is 1.5:1 or greater. Targeting concentrates to the high-producing cows will likely reduce body condition loss and improve reproductive performance.

A summary of the research with lactating cows in pasture-based systems such as New Zealand is that the increase in milk yield per pound of grain supplement ranges from about 0.4 to 0.8 pound. This research may not be directly relevant to the United States because of differences in cow size, genetics, and type of grasses. This marginal return in milk yield to supplementation is usually not profitable given the low M:G in New Zealand and several other countries with pasture-based systems.

The marginal response of milk per unit of concentrates fed follows the law of diminishing returns. The first units are most profitable, but each extra unit gives a lower return. Table 4-4 summarizes the expected milk response of high-producing cows in early lactation with increasing increments (4-pound increments) of grain feeding. This information is based on research in other countries and at Penn State with *high-producing* cows (> 60–70 pounds of milk per day). As concentrate feeding increased from 0 to 20 pounds of grain, the milk yield per unit of concentrates fed tends to decrease from about 1.3 to 0.6 pounds of milk per 1.0 pound of concentrate. Overall the average milk response to feeding 20 pounds of concentrates is about 20 pounds of milk or 1 pound of milk per 1 pound of concentrates fed. The substitution rate is defined as the decrease in pasture DMI for each increment

Table 4-4. Expected milk yield response of high-producing cows to increasing increments of concentrate feeding.

Supplemental concentrates fed (lb/cow/d)	Expected lb milk//lb concentrates fed[a]
0–4	1.2–1.4
4–8	1.0–1.2
8–12	0.8–1.0
12–16	0.65–0.8
16–20	0.5–0.65

[a] Average about 1 pound milk yield/1 pound concentrates fed when 20 pounds are fed.

of concentrates fed. The substitution rate influences intake and profitability. With a substitution rate of 0.5 pound (0.5-pound decrease in pasture DMI per 1.0 pound DMI from concentrates) and an average milk response of 1 pound milk to 1 pound grain, we would expect cows to have adequate energy intake to produce about 60–70 pounds of milk with minimal change in body condition. This grain:milk ratio is about 1:4 (18 pounds concentrates fed:70 pounds milk produced).

There is another advantage to feeding more concentrates that is often not considered. With less pasture consumed per cow as more concentrates are fed, it is possible to graze more cows on the same land area. Pasture can also be extended during periods of reduced growth such as drought by supplementing with additional concentrates.

To calculate the income over feed cost, the M:G price ratio now must be considered. Table 4-5 is a summary table incorporating the numbers discussed in this section. As concentrates are fed, pasture DMI decreases and total DMI and feed costs increase. However, milk yield increases at about 1 pound per pound of concentrates fed, thus total milk income increases. For example, when 12 pounds of concentrates are fed compared to no concentrates, total feed cost is increased by $0.78 from $1.14 to $1.92 per day. However, the expected milk response is 13 pounds with a total value of $7.54, or a $1.69 ($7.54–5.85) greater income minus feed costs when no concentrates are fed. The income minus feed cost is $0.91 greater ($5.62–4.71) than when feeding no concentrates. Even at the highest level of concentrate feeding (20 pounds), the marginal response to the last 4 pounds of concentrates fed (16–20 pounds) is $0.13 ($5.88–5.75).

Table 4-5. Expected marginal economic response (early lactation) to feeding supplemental grain.

Pasture[a]		Grain[b]		Total		Milk[d]			
DMI (lb)	Cost ($)	DMI (lb)	Cost ($)	DMI[c] (lb)	Feed cost ($)	(lb)	($)	Inc.[e,f] ($)	Marginal income
38	1.14	0	0	38	1.14	45	5.85	4.71	
36	1.08	4	0.32	40	1.40	50	6.50	5.10	0.39
34	1.02	8	0.64	42	1.66	54.5	7.08	5.42	0.32
32	0.96	12	0.96	44	1.92	58	7.54	5.62	0.20
30	0.90	16	1.28	46	2.18	61	7.93	5.75	0.13
28	0.84	20	1.60	48	2.44	64	8.32	5.88	0.13

[a] Pasture cost = 3 cents per pound DM.
[b] Grain cost = 8 cents per pound DM.
[c] Assume substitution rate of 0.5 (1 pound grain with 0.5-pound decrease in pasture DMI).
[d] Milk price = 13 cents per pound of 3.5% fat milk.
[e] Income = total milk price – total feed cost.
[f] Does not consider long-term benefits on body condition and reproduction, or the response on milk fat and milk protein.

With the typical M:G price ratio in the United States between 1.5:1 and 2.0:1 (1.6 was used in this example), it makes economic sense to feed up to 16–20 pounds of supplemental concentrates for high-producing cows. When the M:G price ratio approaches 1.0 or less, as it often does in New Zealand and some other countries, then concentrate feeding is not profitable, except perhaps when targeted for early lactation, high genetic merit cows. In our example, if the M:G price ratio decreases to below 1.5:1, as it does occasionally with high grain prices and low milk prices, then we may want to reduce concentrate feeding to 10–14 pounds or less of grain per cow per day and target feeding to selected early lactation cows. Table 4-6 provides a summary of expected profit responses with changing milk: feed price ratio.

The bottom line is that feeding concentrates to high-producing cows on pasture results in higher DMI, which translates into higher, more profitable milk production and improved body condition.

Energy Sources

Energy is the most limiting nutrient for grazing dairy cows. In addition to the amount of energy, the type of supplemental carbohydrate (grain) fed and the method of providing energy must be considered. Energetic uncoupling may occur with pasture diets from an undersupply of ruminal available energy (or carbohydrate) relative to the degradation of the pasture protein or nitrogen in the rumen. Synchronizing the supply of N and energy-yielding substrates to ruminal microbes has been suggested as a means to improve the capture of RUP and improve animal performance with nongrazing systems. Feeding a TMR to cows under confinement housing is an excellent method to synchronize the supply of N and NFC to the rumen. With grazing conditions, the largest time spent grazing during a 24-hour period is after milking. In many situations, including grazing, cows are fed grain rations twice daily at milking time. The synchrony of providing protein-N and NFC to the rumen does not exist with grazing to the same extent as with a TMR under nongrazing.

Frequent feeding of concentrates to reduce **slug feeding**, or excess concentrate feeding, has long been practiced and has been shown to be beneficial with confinement feeding. Feeding concentrates more than twice daily may reduce the risk of slug feeding. If concentrates are fed twice daily, feeding before the cows eat pasture may provide rumen-fermentable carbohydrates before cows consume a high protein pasture.

Table 4-6. Milk and profit response to concentrate supplementation.		
	Early lactation	**Mid-lactation**
Expected milk response/lb of concentrate supplement	1.0	0.6–0.8
Milk:grain price ratio		
1.50:1	Profit	Profit
1.25:1	Profit	Break even
1.00:1	Break even	Loss

In addition to feeding strategy and frequency of feeding, studies have compared different types and sources of carbohydrates for the availability of the carbohydrates in the rumen. The effects of grain processing, different grain sources, fineness of grind, pelleting, and heat processing have not been studied extensively with grazing cows. Based on knowledge of confinement feeding, feeding grains that are high in starch and readily fermented in the rumen would be desirable. Corn would be a logical choice because of availability and price, and because it is high in readily fermentable starch. Research at Virginia Tech did not find differences in milk response when feeding coarse or finely ground corn, although milk fat test was lower with the finely ground corn. High moisture corn and dry corn did not give a different milk response. Research at Penn State found no change in milk response between feeding finely ground dry corn and steam-flaked corn. Steam-flaked corn increases the ruminal degradation rate of starch. Feeding the steam flaking of corn showed the expected trend of increased milk protein and reduced milk urea-N, indicating an improvement in dietary protein utilization.

The research cited above and other research from around the world does not show a strong advantage for a specific grain source or method of processing. However, our knowledge of nutrition suggests that a diet high in rumen-degradable starch would be desirable. Providing a variety of grain sources that differ in rumen availability of carbohydrates may help in utilizing pasture nutrients. Availability of economical grains that are high in starch should serve as the basis of which grains to feed.

Fiber

High quality pasture is often marginally low in rumen "effective fiber" for optimal rumen health. Lush spring pasture is 80–85% moisture,

which can lead to fast passage of feedstuffs through the digestive tract. This can contribute to reduced milk fat content. Feeding a few pounds of forage such as long hay will add some effective fiber and slow the rate of feed passage, and may help maintain DMI and milk fat content. In the United States, an abundance of byproducts often referred to as NFF (nonforage fiber) sources are economical in some areas. Many of these sources are high in NDF, and the fiber in these ingredients is often highly fermentable in the rumen compared to stored forages. In Virginia Tech research with Holsteins, feeding high amounts (20 pounds) of fiber-based supplement resulted in increased milk yield compared to feeding 12 pounds. Increased yield was not seen with Jerseys. In other Virginia Tech research, a fibrous supplement (18.3% ADF) containing pressed brewer grains, cracked corn, soy hulls, cottonseed meal, and brewers liquid yeast was compared to one containing 85% ground corn and 11% fishmeal. Both were supplemented at 14 pounds daily air-dry basis. Cows fed the high fiber supplement decreased more sharply in milk production, then paralleled production of the corn-fishmeal cows, but produced about 5 pounds less milk daily. Beginning at week 5, 6 pounds of ground corn daily were top-dressed onto the fibrous supplement. Milk yield advanced to equal and finally exceed by 2 pounds yields with the corn-fishmeal. This demonstrates the importance of a fermentable starch source.

In Penn State research, a comparison of feeding a "fiber" supplement containing soyhulls and beet pulp and a corn-based supplement resulted in similar milk production but a higher milk fat percent in the fiber supplement.

Research in Northern Ireland with feeding fiber-based supplements to high-producing cows (> 75 pounds) grazing high quality ryegrass pasture resulted in higher milk yields and milk fat content than feeding corn-based supplements. Producer experiences and research suggest that the addition of NFF sources, such as soy hulls, cottonseed hulls, beet pulp, brewers grains, distillers grains, citrus pulp, and other byproducts, to the grain ration may be beneficial in providing fermentable fiber in the rumen and should be considered if economical and readily available. Feeding NFF sources along with starch-based grains such as corn provides a blend of rapidly and slowly fermentable carbohydrates, and shows promise to improve milk yield and milk fat content. Feeding a grain mixture that contains about half the DM as starch, primarily corn, and the other half from NFF sources such as soy hulls, citrus pulp, and wheat mids will likely optimize the rumen environment and performance in the grazing dairy cow, particularly when high quality pastures are fed.

Corn silage can be an excellent supplemental forage and fiber source to complement pasture because it provides needed NFC as a source of energy for the rumen microbes and also "dilutes" out the high protein in spring pasture. Corn silage is a highly palatable feed, is an excellent carrier for the grain, and allows lower amounts of grain to be fed. One management problem is that adequate amounts must be removed daily from a silo to maintain good quality silage. Stored forages are needed to maintain DMI when pasture availability is low, particularly during the summer.

Fat Supplementation

The common practice of fat supplementation of diets provides a more energy-dense diet for high-producing cows and may lead to less body condition loss and improved reproductive performance. A review of 18 experiments with fat supplementation to cows fed pasture indicated that milk yield averaged 2.5 pounds higher with fat supplementation (19). Milk yield increased in

80% of the studies, and was associated primarily with the feeding of saturated fats. Supplementation with unsaturated fat sources did not increase milk yield, and may be related to the fact that grasses and legumes have 70–80% of the total fat as unsaturated fats. Feeding too much unsaturated fat can affect the rumen environment and decrease fiber digestion. With added fat, milk protein percent decreased slightly, but milk protein yield did not. Feeding 0.4–0.8 pound fat (saturated preferred) per cow per day should be considered, particularly for cows in the first half of lactation. Feeding low amounts of unsaturated fat such as 1–2 pounds of roast soybeans or whole cottonseed will provide 0.3–0.5 pound of added fat. Changes in body condition and reproductive performance were not reported in many of these research studies but are likely to improve with the feeding of supplemental fat.

Protein

Although the total protein in well-managed pastures is high—often more than 25% in the spring—the true protein is often highly degradable in the rumen (table 4-1, p. 86). Often, 70–80% of the protein in pasture will be degraded in the rumen (table 4-5, p. 95). Providing ruminally available NFC, and rumen-fermentable carbohydrates, primarily from concentrates, will improve the utilization of the high levels of RDP in pastures, increase rumen microbial protein production, and increase milk yield. If NFC or energy are lacking in the diet and rumen, the high RDP in pasture will result in high levels of rumen ammonia, which is then converted to urea. This urea then appears in blood and milk, and much of it is eventually excreted in the urine. Thus, the high protein in pastures is often wasted by the cow if fermentable carbohydrates are not fed. High levels of urea in blood and milk have also been linked to lower reproductive efficiencies. Because formation and excretion of urea by the cow requires energy, this process wastes protein and uses energy that could be used for milk production. The energy required to excrete urea, often called the urea cost, may be equivalent to 2–4 pounds of lost milk production per day. This can also have a negative impact on the environment.

The amount of CP needed in the grain ration to supplement high quality pasture is typically 12–14% of the DM. With high ruminally degraded protein in the pasture, the supply of protein to the small intestine depends largely on rumen microbial protein. This suggests that cows may be deficient in dietary RUP and in specific amino acids. Research studies and farmer experiences do not report a consistent response to increasing the RUP in supplements. Research at Penn State (6) reported that multiparous cows producing 80 pounds of milk produced more milk protein when fed high RUP diets. Virginia Tech research on feeding high RUP sources also reported some benefit with high-producing cows. A 4-pound milk response was found when dried brewers grain and corn gluten meal replaced soybean meal. Feeding a treated soybean meal, which increased rumen bypass protein, resulted in 4 pounds more milk than soybean meal. Protein sources such as brewers or distillers grain, corn gluten meal, and roasted or cooked soybeans are good sources of RUP. Producers should consider feeding 0.5–1.5 pounds of RUP to grazing cows producing more than 70 pounds of milk. Costs must be considered, but energy is the first limiting nutrient and must be supplemented to the rumen to optimize the capture of nitrogen from pasture and to optimize rumen microbial production.

Milk urea nitrogen (MUN) is used as a diagnostic tool to monitor the dietary protein and carbohydrate balance. The nutritional basis for using MUN testing is either that excess dietary protein in relation to dietary NFC and ruminal available carbohydrates will result in elevated MUN, or that low dietary protein in relation to high dietary NFC will result in low MUN. Normal levels of MUN are 10–16 mg per 100 ml.

A field study in New York did not find MUN values excessive with pasture-based systems. Values averaged 11–15 mg per 100 ml. A study in Pennsylvania with three grazing herds reported 14–15 mg MUN per 100 ml. A field study in Pennsylvania during which MUN was monitored monthly on seven grazing and seven nongrazing herds for 18 months found no major differences in MUN during the grazing season compared to the nongrazing season. Levels of MUN in milk averaged 13–15 mg per 100 ml for both groups. However, in a Virginia study, MUN averaged 22.5 mg per 100 ml when 12 pounds of supplement (12% ADF) were fed and was similar (21.7) when 20 pounds were fed. When 14 pounds of a ground corn-fishmeal was supplemented, MUN averaged 18.2, compared to 15.7 for comparable cows fed a well balanced TMR.

MUN in grazing cows is likely most influenced by two factors, the lushness and quality of the forage and the amount of fermentable starch grains supplemented. Lush, actively growing forage contains greater quantities of rapidly degradable protein. Fermentable starch aids in incorporating the nitrogen compounds into microbial protein. As more grain is fed, less forage is grazed, resulting in less degradable protein intake and less MUN. Analysis of MUN can be a useful tool to monitor the dietary protein and energy with a pasture-based system.

Supplemental Forages and TMRs

Supplementing lactating cows on pasture with additional forages is a common practice of dairy producers. Stored forages are needed to maintain DMI during times when pasture availability is low, particularly when high stocking rates are used. When feeding other forages, the hay or silage will replace the DMI from pasture on a one-to-one basis. There are pluses and minuses to feeding different forages. As previously discussed, corn silage can be an excellent supplemental forage to complement pasture. Grazing herds in New York that supplemented with corn silage (4–7 pounds of DMI per cow per day) had more net farm income per cow than nonsupplemented herds. Forage supplementation with adequate pasture has advantages and disadvantages, and the manager's goals should guide decisions about supplementation.

Advantages of Supplemental Forage

- Supplemental forage provides a more uniform ration throughout the year, with less chance for disruption of rumen function, particularly with fermented forages.
- Adequate DMI is more easily provided compared to pasture alone.
- Supplemental forage may complement pasture quality for better nutrient use. For example, corn silage is a high energy, low protein forage that complements pasture.
- Feeding a small amount (2–4 pounds per head per day) of long, dry hay can provide effective fiber to grazing cows and decrease nutrient passage rate with high quality pastures.
- Milk production and milk fat concentration per cow may be higher.
- More cows can graze per acre.

Disadvantages of Supplemental Forage

- Extra time and labor is involved in feeding stored forages.
- When feeding fermented forages, adequate amounts of silage must be removed daily from the silo.
- Supplemental forage reduces the intake of low cost, higher quality pasture.
- Feed costs are higher than when pasture is the only source of forage.

- With less pasture intake, stocking rates and manure loading of pastures increase environmental risks.

Supplementation with a pTMR

Many dairy producers using pasture and feeding supplemental forage are supplementing grazing cows with a "partial" TMR (pTMR) (partial because pasture is not physically part of the mixed ration) using a mixer wagon. Many of these producers have the feeding equipment because a TMR is fed during the winter months. Feeding a pTMR to grazing cows offers more control over the entire feeding program compared to offering forage or grain separately. An estimate of the DMI from pasture should be part of the ration formulation process to develop a balanced ration. As with all pasture supplementation strategies, feeding a pTMR has advantages and disadvantages that must be considered, along with the goals of the dairy producer.

Advantages of a pTMR

- Forage is fed with the concentrates rather than separately, so there is less chance for rumen digestive problems due to slug feeding of grain.
- A pTMR provides a more uniform ration with less chance for disruption of rumen function.
- Feeding a pTMR makes monitoring DMI of pasture easier.
- Feeding a pTMR likely contributes to higher milk production per cow. A study at Penn State (3) found that grazing cows supplemented with a pTMR produced 8 pounds more milk per day than grazing cows supplemented with just concentrates (70 vs. 62 pounds of milk), and had higher milk fat and milk protein content and improved body condition (table 4-7).

Disadvantages of a pTMR

- Extra labor is involved with feeding a pTMR.
- Feed costs are higher than with pasture as the only source of forage.
- Feeding a pTMR requires management of changing formulations or amounts to feed.
- A feeding area is needed for a pTMR.

Using a pTMR

Because much of a pTMR is stored forage, some similarities exist between supplementing grazing cows with pTMR and with stored forages. However, the nutrient content and feed ingredients in the supplemental grain ration differ. The quality of the forage in the pTMR will influence DMI and milk response, but substitution of pasture for pTMR may differ from substitution with stored forage alone. Because the grain is included in the pTMR, cows will replace less pasture with pTMR supplementation than with supplementation of forage alone. The quality and amounts of stored forages in the pTMR will affect substitution rate, DMI, and milk production.

How much pTMR to feed depends on the cows' requirements as well as the quantity and quality of available pasture. Time of feeding also affects intake of both pTMR and pasture. In general, feeding pTMR before grazing encourages more

Table 4-7. Results of feeding a pTMR to supplement pasture for high-producing Holsteins (22-week study).

Item	Pasture + 19.1 lb concentrate	Pasture + pTMR[a]	TMR (nongrazing)
Intake:			
Pasture intake (lb DM)	28.4	16.5	--
TMR intake (lb DM)	--	38.9	58.7
Concentrate intake (lb DM)	19.1	--	--
Total intake (lb)	47.5	55.4	58.7
Total intake (% of body weight)	3.58	3.99	4.15
Milk yield (lb/d)	62.7	70.4	83.3
Milk fat (%)	3.13	3.35	3.30
True milk protein (%)	2.82	2.95	2.99
MUN (mg/d)	14.9	12.0	10.6
Expenses/cow/d ($/d)	2.69	3.30	4.20
Milk income/cow/d minus feed costs ($)	5.07	5.75	6.35

[a] Cow grazed half days.

Sources: Adapted from Bargo, F., L. D. Muller, J. E. Delahoy, and T. W. Cassidy. 2002. Performance of high-producing dairy cows with three different feeding systems combining pasture or total mixed rations. J. Dairy Sci. 85:2959–2974.

consumption of pTMR, and offering pTMR after an initial period of grazing encourages greater consumption of pasture. Generally, cows adjust intake of pTMR based on how much pasture is available, but quality and palatability of forage species in the pasture also affect how much pTMR is left in the bunk. Many dairy managers and nutritionists adjust feeding practices and ration formulation based on the amount of pTMR left in the feed bunk and the amount of milk in the bulk tank. These must be monitored closely and the amount of pTMR fed must be adjusted frequently.

Formulating a pTMR

Balancing a pTMR for cows on pasture is similar to balancing a pTMR for nongrazing cows except for two factors. Because feed ingredients and sometimes feed refusals are weighed in a pTMR system, it is much easier to estimate pasture DMI when supplementing with a pTMR compared to feeding forage and grain separately. Using estimated DMI from NRC recommendations and forage analysis from spring, summer, and fall pasture, it is possible to formulate a reasonably balanced ration for grazing cows. Changing quantity of pasture greatly affects the ration, even more than changes in pasture quality. Having flexibility in the formulations is the key to maintaining optimal feed availability for the cows.

Although pasture quality can fluctuate rapidly, it is not necessary to attempt to reformulate rations to adjust to these fluctuations. Formulating rations every few weeks during the grazing season may be adequate. Planning ahead for changing pasture DMI based on pasture inventory and budgeting can help to minimize difficulties that changes may cause. It is important to the success of a pTMR with pasture-based systems to record daily the levels of the bulk milk tank and pTMR refusals. The basic principles of nutrition

and nutritional management still apply: monitor DMI, monitor forage availability, monitor forage to concentrates ratios, monitor protein and carbohydrate ratios, and monitor daily milk yield.

The economics of feeding a pTMR to grazing cows suggest an improvement over grazing cows fed primarily concentrates. Feeding a pTMR resulted in a $0.68 higher income over feed cost when compared to pasture and concentrates. Soder and Rotz (21) evaluated the long-term economic impacts of feeding either a pTMR or concentrates with grazing dairies. The net return per cow was $223 higher per year for the pTMR system.

EFFECTS OF ENERGY/PROTEIN RELATIONSHIPS ON MILK PRODUCTION AND COMPOSITION

Milk Composition

Dairy producers receive additional payments for increased milk components, primarily milk fat and milk protein. The increased demand for cheese has increased the value of the protein and fat in milk. Therefore, dairy producers are searching for feeding strategies to enhance the production of milk fat and protein. Milk composition is very important to dairy producers with pasture-based systems because milk volume is often lower than with confinement systems. Focus must be on protein and fat yields as well as the percentages. Many factors affect milk fat and milk protein, including pasture type and quality, diet, and climate. Milk fat is altered more by nutrition than is milk protein. Altering milk protein by 0.2% is about the maximum change expected through nutrition.

With pasture as the only feedstuff, research studies report that the milk fat concentration for Holsteins ranges from 3.1 to 3.7% fat. These values are on the low end of the Holstein breed average of 3.7%. As higher amounts of concentrates and high starch diets are fed, milk fat percent tends to decrease. However, with the increased milk yields, the yield of milk fat is often higher as more concentrates are fed. Nutritional management may modify these trends. Inclusion of highly fermentable starch sources in the grain mixture has been reported to decrease milk fat percent in some studies; however, milkfat yield is often increased. Processing of grains such as steam flaking or adding heat tends to decrease milk fat percent, but may increase yield. Including fiber such as long hay or nonforage fiber sources will help maintain milk fat percent with high quality pasture in the spring. Feeding a TMR will often increase milk fat and protein concentration and yield (see table 4-7, p. 101). Milk protein percent from Holsteins fed only pasture tends to be low compared to the breed average, and can range from 2.6 to 3.0%. Feeding concentrates high in starch increases milk protein about 0.1–0.2 percentage units over pasture-only diets. Including bypass protein sources for high-producing cows will likely maintain or increase milk protein percent and yield.

Milk Fatty Acids—Conjugated Linoleic Acid

Reduction of dietary fat intake has been a focus of human health concerns over the past several years. One method to reduce fat intake is to minimize the consumption of fat in dairy products. Consumers have the choice of many low-fat dairy products, including milk, cheeses, and ice cream, and can also switch from butter to margarine.

Research indicates that **conjugated linoleic acid (CLA)**, a fatty acid found in dairy products, can be beneficial to human health. Current research has revealed a positive role for CLA in the fight against cancer; CLA is the only fatty acid shown to inhibit carcinogenesis in animals. Inclusion of

CLA in the human diet may elicit anticarcinogenic effects by inhibiting some types of cancer, including skin, prostate, and mammary cancers. Recent research reports that CLA can prevent the onset of diabetes in laboratory animals.

The dairy cow's unique digestive system includes billions of microorganisms in the rumen to break down feed for use by the cow. Through this action of microorganisms a common dietary fatty acid, linoleic acid, is changed to CLA. This fatty acid is incorporated into the milk fat of cows and ultimately dairy products for human consumption. Milk fat is the richest natural dietary source of CLA, which suggests an enormous marketing opportunity, particularly if levels of CLA can be increased in dairy products.

Researchers are evaluating several nutritional factors that can increase CLA content in milk. Feeding fresh pasture increases the level of CLA in milk. Research at Cornell University (8), in cooperation with Penn State, found a twofold increase in CLA in milk when cows were fed pasture as the sole forage compared to a TMR. Wisconsin and Virginia researchers reported a similar finding. A summary of research with grazing cows indicated that pasture in the diet increased CLA in milk by two- to fourfold. Supplementation of the cows' diet with various dietary fat sources increased CLA in milk. Virginia Tech research (11) indicated that CLA content of milk from pasture-fed cows was further increased by supplementation of mechanically extracted soybean meal compared to solvent-extracted soybean meal.

There are still unanswered questions as to how best to deliver CLA in the human diet. Pasture-based forage systems and direct supplementation of various fat sources show promise for increasing the content in dairy products. CLA provides a promise for prevention of several types of cancers. In addition, CLA may improve the public's perception of dairy products.

Gradually Introduce Pasture in Spring

In the spring, dairy cows should be adjusted gradually to a change from stored feeds to pasture. This is primarily an adjustment of the rumen microbes to a change in feedstuffs. As with any feeding changes, more gradual adaptation will help the rumen microbes to adjust. There is not a lot of "science" on the adjustment procedures. Fresh forage is highly palatable and animals will readily consume spring pasture. One suggestion is to gradually adapt over a 10–14-day period using the following guidelines.

- Graze for a few hours initially when pasture is 3–4 inches tall. This will help stage different paddocks to different growth rates, maintain the nongrazing feeding program, and keep forage growth under control.
- Gradually increase grazing time to one-half day as pasture grows. This will likely increase pasture intake.
- Continue to decrease other forages fed over a 10–14-day period.
- Watch bulk tank and manure, and make a decision when to switch to predominantly pasture or all pasture as the total forage.
- Provide some forage as long hay to help reduce the risk of bloat.

Using Models to Predict Limiting Nutrients and Animal Performance with Pasture-Based Diets

As we reflect on managing the supplemental feeding program for grazing dairy cattle, we need to move away from the "art of feeding" and toward the "science of feeding," similar to our nongrazing nutrition programs. One attempt to accomplish this is to adopt simulation models such as the Cornell Net Carbohydrate Protein System (CNCPS), CPM, and the NRC model to identify limiting nutrients and to predict animal performances with pasture-based systems. Simulation models are increasingly being used to understand the many interactions involved in animal production responses. The CNCPS has primarily been used to predict performance and evaluate feeding programs with diets based on hay and silages when fed as a TMR.

Research at Penn State (9) tested the predictive ability and potential to use the model for diets based on pasture. Data were obtained from eight pasture studies in the United States and New Zealand in which DMI and animal performance provided reasonably good estimates of changes in body condition score (BCS), estimated energy balance, blood urea nitrogen, and milk production under grazing conditions. Milk production was first limited by the supply of metabolizable energy (ME) when only pasture was fed, but specific amino acids may be limiting milk production when more than 20% of the diet consists of a grain supplement. However, in a study at Penn State, milk synthesis did not appear to be limited by the supply or profile of amino acids, and the supply of protein did not appear to be limiting until a daily milk production of 80–85 pounds was reached. Preliminary evaluations with fiber addition from forages or byproducts to pasture-based diets indicate an increase in ruminal pH and microbial protein synthesis, and higher milk yield potential. This model was developed primarily under nongrazing conditions, and in contrast to TMR feeding in which cows eat several 5 to 6 times per day, cows on pasture usually have 3 to 4 major grazing meals per day. Clearly, more use of models, which integrate the results of many research studies, must occur to improve the on-farm supplementation strategies with pasture-based systems.

USING PASTURE IN VARIOUS MANAGEMENT SYSTEMS

The practice of seasonal dairying using intensive rotational grazing as a forage source is gradually increasing in this country. This has largely been modeled from the New Zealand grazing system. This practice maximizes on the conversion of high quality pasture by cows into milk. To be very efficient at this requires the application of science and art to the management of both plants and animals. Seasonal dairying has the potential for the lowest investment of any form of dairying. Other than a milking parlor, not much other high-cost equipment and facilities are involved.

The stored feed may be corn silage, hay crop silage, or often hay or round bale silage. Hay crop silage or hay often comes from harvesting the excess forage from pastures not grazed in the spring. Ideally, this is custom-made preserved forage, but that is not practical for many farm locations. Some seasonal producers have corn silage custom grown and stored.

Good grazing practices must begin early in the growing season to develop proper conditions for high quality grazing material. Sound rotational grazing practices must be followed. With so much dependence on pasture quality, new forage to graze should be available once or twice daily. Typically, supplemental feeds are purchased concentrates or byproducts.

Conditioning the animal prior to or during the dry period for the next lactation may be important for a healthy transition to lactation, may help maintain body condition, and may improve reproductive function. A dry cow feeding program is important. A period of intense animal and pasture management begins with calving. Within a period of 90 days, the cow calves, reaches a point of peak milk production that coincides with considerable loss of body weight and body condition, and is expected to cycle and potentially conceive.

Timely reproductive function or conception is a major problem for producers who change to seasonal grazing from conventional herds or those who begin seasonal dairies with cows from conventional feeding programs. These cows generally milk well the first season, but 30–40% are culled because they do not conceive to fit the calving window for the next season. Two major physiological concerns adversely affect reproduction: large losses of body weight and high levels of blood and milk urea. Inadequate supplemental carbohydrates are the result of excessive CP in high quality forage. Both of these are addressed by energy supplementation, covered in the preceding section, "Effects of Energy/Protein Relationships."

Some very successful graziers are semiseasonal, but still emphasize producing milk from grass. Often, these producers breed as many cows as possible to calve at the proper time for spring grazing. The cows that do not breed back in a timely manner continue to milk on a winter diet and provide income throughout the year. This style of dairying offers the opportunity to keep expenses less than a conventional operation by planting fewer crops, requiring less storage facility and less equipment, and purchasing less feed. Contract harvesting of corn silage or other crops is an option where available.

Some producers with confinement operations could probably add to their bottom line by using grazing. Many have land nearby in pasture species or land that could be easily converted to good pasture. This provides the opportunity to use pasture partially or as the total forage source during the flush growing season.

Research has shown that cows can adjust to grazing with proper supplementation without large losses in milk yield. Getting the cows off concrete can be a healthful benefit. The advantages to the producers are that daily mixing of rations is not required and handling manure and cleanup is minimized. In addition, if 15–20% of stored feed is replaced by pasture, that reduces storage space needs and depreciation on equipment by 15–20%.

Although often not considered high quality, pasture is a good source of protein. The excess of nitrogen and the cost of eliminating ammonia from the body as urea is a negative aspect of grazing when it is the only forage source. However, that could make a positive contribution to the nutrition of the animal, especially if a carefully balanced grain-based silage diet is fed that depends on pasture for the diet protein contribution. If properly managed, this could reduce the costs of protein purchases considerably.

REPRODUCTION IN PASTURE-BASED DAIRY SYSTEMS

Dairy Reproductive Performance Has Declined

We have been concerned that reproductive performance has declined markedly over the past 25–30 years in the United States. Dairy herd summary data from several states in the southeastern United States were examined at North Carolina State University (2000). All breed groups increased in herd size, milk, fat, days open, and services per conception over time. The regional data show similar trends in declining reproduction across breed groups, although in a controlled study, Jerseys had higher conception than Holsteins (24). Increases of services per conception of 1.81–1.85 to 3.08–3.12 for both Jerseys and Holsteins and 1.97–3.08 for other breeds are a major concern. Conception rates decreased from about 50–55% in the mid-1970s to 32–33% in the late 1990s. Also of concern is

the trend for increasing days open. Decline in fertility occurred at a greater rate beginning in the late 1980s. Fertility of heifers has been more stable.

In countries where seasonal breeding and pasture systems are routine, fertility of dairy cattle has remained quite high. Daughter fertility and survivability are factors used in evaluating sires. Although milk production in these countries is not as high as in the United States, their production has increased modestly without severely affecting dairy cow fertility. New Zealand milk production has slowly increased over the years, but conception rates typically remain above 60%. In the United States, slower breeding cows milk more than their contemporaries, so our system may be biased against fertility of a sire's daughters.

In Ireland, a study comparing high genetic merit cows to medium genetic merit cows by use of U.S./Dutch Holstein semen showed that high genetic merit cows had lower conception rates (48 vs. 56%) and much higher culling rates for infertility (23 vs. 6%) after a 13-week breeding season (5). However, with selection of the most fertile cows each year, it may be possible to improve reproductive efficiency in seasonally calving dairy herds in the United States. The Mahoning County dairy grazing project in Ohio reported that involuntary culling for infertility tended to decrease over time. Similarly, observations from the Dave Forgey herd (see sidebar, p. 107) in Indiana indicate a trend toward improved fertility over time.

Seasonal Calving Versus Year-Round Calving

Approaches on reproductive management of pasture-based dairy herds include single calving seasons, split calving seasons, or year-round calving. There are potential advantages and disadvantages to each approach. A single seasonal breeding and calving strategy in the spring allows dairy systems to effectively match the lactation period to the availability of rapidly growing, high quality pastures. Seasonal management results in very predictable annual work cycles and cattle that can be managed basically in three groups: (i) the cow herd, either lactating or dry as the year progresses; (ii) heifers greater than 1 year of age; and (iii) heifer calves less than 1 year of age. In seasonal systems, labor requirements are significantly higher from the beginning of calving until calves are weaned and the breeding season ends. After that, workloads are less demanding. Later, when cows are dry,

From Confinement to Grazing to the Future

Adapted from an article by D. Forgey, "From Confinement to Grazing to the Future," in Hoard's Dairyman, *Fort Atkinson, WI. Used with permission.*

Forgey said: "Our selection process for the past five years has been to keep only the cows which breed back in a six-week window. Cow numbers declined slightly the first couple of years, but we have maintained numbers with heifers from our herd. The 1998 breeding season was a challenge with very high temperatures and humidity during the breeding season. Early embryonic death occurred in about 10% of our early-bred animals. Many heifers that freshened in 1999 were the first from New Zealand genetics from bulls selected for survivability of daughters. Initial milk production figures from those outcrosses show slightly lower milk production at peak but perhaps more persistency."

Forgey's herd reproduction is managed intensively with use of estrous synchronization and an electronic heat mount detection system. Although the herd's reproduction has improved generally, a seasonal breeding program is very susceptible to adverse breeding conditions such as the 1998 summer heat stress. Interestingly, Forgey's herd reproduction continued to improve in 1996 and 1997 when the daily grain ration was decreased from 24 to 12 pounds. The ration amount was further reduced to 6 pounds per head in 1999 while maintaining reproduction. Forgey continues to breed cows after the desired breeding season, but such cows are marketed to nonseasonal dairies after their current lactation.

Table 4-8. Reproductive performance of Dave Forgey's seasonal Holstein dairy herd.

Year	1995	1996	1997	1998	1999
Days to first service	68	75	77	76	--
Services per pregnancy	2.4	1.8	1.6	2.0	1.8
Breeding season (weeks)	12	10	8	8	8
Cows pregnant (%)	75	70	68	72	80
Calved by April 15 (%)	68	76	78	87	84

renovations in facilities and/or family vacations can easily be scheduled. Potential disadvantages of a calving season include the increased stress level on dairy workers at calving time, uneven distribution of income, potential loss of seasonality with disease or fertility problems, and the optimal milk market not matching seasonal production. An aggressive breeding program must be maintained to be successful in seasonal management. However, because most dairy herds in the United States are not seasonal, a market exists for productive cows bred outside the desired season. In the northern United States, seasonal milk production in pasture systems involves late winter or early spring calving with rebreeding during late spring or early summer when pasture supply and quality are high. In warmer areas of the United States, inability to breed in hot weather would necessitate fall or winter calving.

Split seasonal calving management keeps some efficiencies of a single season in that several tasks on the dairy farm can be concentrated in two times of the year. Two calving seasons would produce six versus three groups of animals to manage, and part of the herd would be milking every day. A split season does allow flexibility for keeping productive cows that fail to rebreed within the desired breeding period. Such cows could be shifted to the other seasonal group and rebred with them. However, that would reduce selection pressure on fertility, and reproductive efficiency would likely not improve. The cow group earliest in lactation at any time could have priority for the highest quality pastures. Income and workloads would be more consistent throughout the year.

Year-round calving is most common in the United States. This allows reasonably uniform monthly milk production and labor requirements. Management of pasture-based systems is complicated because of cows in all stages of lac-

tation and heifers of all ages at any given time. Numerous small groups make pasture rotation schemes difficult to manage, and large groups may result in over- or underfeeding of some animals. With strategic use of supplements, many dairy graziers have managed year-round calving systems. Although the producer's "grazing lifestyle" would exist, individuals would have to plan more to schedule vacations, etc. Lower reproductive efficiency would be expected because short-term risk of fertility failure is less. Twelve-month calving intervals would not be as critical with year-round calving.

BCS and Reproduction

Differences in BCS have been reported between pasture and confinement-fed dairy cattle. In a 4-year study across total lactations, White (24) at North Carolina State reported that cows fed TMRs in confinement had higher BCS compared to cows on pasture fed supplemental grain and hay or haylage. Lower BCS was also observed for cows fed pasture compared with confinement-fed cows at Penn State (10) and Virginia Tech (Polan et al., 1997) (18). During a 24-week trial at Penn State of grazed cows fed different supplements researchers found that BCS did not exceed a score of 3.0. Lower BCS for grazing cows is likely due to lower energy consumption and greater energy expenditure due to walking. There is also energy cost in processing excess nitrogen from high protein pastures into urea.

Lower BCS in pasture-fed cows may not be a problem if effects on production, reproduction, or health are not severe. Pastured cows milked 7.9% (Holsteins) or 12.9% (Jerseys) less than respective confinement cows (24). Among Holsteins, fertility tended to favor pastured cows over confinement cows in spite of lower BCS. Jerseys had similar fertility in both systems, higher than Holsteins. Grazed cows had less mastitis and did regain body condition later in lactation. As mentioned earlier, supplementation with readily fermentable carbohydrates may help use excess soluble nitrogen in pasture, thereby enhancing production, body condition, and perhaps reproduction.

Summer Heat

Heat stress can occur for relatively short durations in northern areas or for extended periods in southern areas. When the environmental temperature nears the cow's body temperature and relative humidity is high, the cow's cooling mechanism is impaired, body temperature rises, panting increases to more than 80 breaths per minute, and the cow eats less, is less active, and milk yield declines.

Heat stress is believed to have both acute (immediate) and chronic (delayed) effects on reproductive performance of dairy cattle. Cows are less active and estrous is difficult to detect when it is hot. Elevated uterine temperature at or near insemination has acute negative effects on fertility. Chronic effects of heat stress include an adverse environment for initial growth of follicles expected to ovulate much later. Conception rates continue to increase for 2–3 months after temperatures have declined in the fall. In northern areas, depressions of fertility are shorter, but summer breeding during periods of heat stress could still be difficult. Lower conception and increased loss of embryos would be more critical for seasonal herds. Although virgin heifers may have slightly lower conception rates if bred in summer, conception rates of lactating cows in hot, humid weather may drop below 10%.

Management practices may reduce heat loading on cows. Cows can be kept inside during the hottest part of the day under permanent shade structures with eave heights of 12–16 feet, open ridge vents, and roof slopes at least 4–12%. In chronic heat-stress environments,

fans and sprinklers might be beneficial. In pasture systems, trees used for shade can be both advantageous and harmful. In a pasture with a few shade trees, many cows will cluster, lounge, and deposit manure nutrients there rather than throughout the pasture. Eventually the trees will die. However, six or eight designated shade paddocks with numerous trees will allow rotation and reduce the clustering effect. Shade paddocks should be used only when cows are stressed and then for only the hotter parts of the day. Portable shades (12–16 feet high) with at least 50 square feet per cow can be used but may be difficult to manage for large herds. Some producers have used irrigation equipment or even simple sprinkling systems to cool cows while on pastures, although evaporative cooling is most effective with lower humidity. In the South, some producers have successfully used cooling ponds for lactating dairy cattle. However, proper construction and maintenance is critical to avoid mastitis and other health problems.

Other approaches to heat-stress issues include adjustments to milking times, calving seasons, and cow selection. Adjustment in milking times to allow cows to graze at cooler times of the day and have shade when it is hot can help maintain intake levels. Dry cows or those in late lactation can tolerate moderate heat stresses slightly better than cows in early lactation. Therefore, producers with seasonal herds in areas with long periods of hot weather could use fall or early winter calving to reduce impacts of heat stress. Small- to moderate-sized cows work best in pasture systems in general and particularly in warmer environments. All dairy breeds can be managed successfully in hot environments, but many graziers use colored breeds or crosses of them with Holsteins (see next section). Continual access to plenty of fresh, cool water near paddocks is a must for managing pastured cows in the summer.

Genetics of the Grazing Cow for Reproduction and Performance

Dairy remains the only animal industry in the United States that depends mostly on purebreds, of which about 93% are Holsteins. Even though heritability of reproductive traits is low, Weigel and Rekaya (23) concluded that genetic improvement of reproductive performance is possible. McDaniel et al. (1995) recommended use of cow (daughter) reproduction in sire selection decisions. Both direct and indirect measures of fertility have been used in selection indices in countries where fertility is critical in seasonal breeding. Such practices appear useful in maintaining cow herd fertility, but milk yields have improved more slowly than in the United States. Also, most dairy breeds in the United States have experienced increases in the average inbreeding coefficient—up to about 4–6% in 1999—which may negatively affect reproduction.

White (24) reported higher average conception rate in Jerseys (60%) than in Holsteins (50%). During the 75-day breeding season, almost all (97%) Jerseys were detected in estrous and inseminated, but only 86% of Holsteins were inseminated. Because of these differences, the average pregnancy rate within the 75-day period was only 58% (63% on pasture) for Holsteins compared to 78% (80% on pasture) for Jerseys. Percentages for Holsteins were only slightly better in Dave Forgey's herd for breeding periods of 8–12 weeks annually (see sidebar, p. 107). A herd with pure U.S. Holsteins may struggle to maintain seasonality without high culling rates, although some producers have been successful in this. Jersey genetics could increase efficiency of seasonal breeding. Given our historical decline in dairy cow fertility, perhaps genetic selection of dairy cattle should include fertility indicators along with milk production. Michael Murphy of Ireland has indicated that some dairy cattle were totally unsuited for seasonal dairy-

ing due to their inability to maintain a 365-day breeding cycle.

There is interest among dairy graziers in use of crossbred dairy cows to obtain smaller cow size, increased heat tolerance, improved fertility, and increased herd life. There is not much recent research on dairy crossbreeding. One study comparing Ayrshires and Holsteins with crosses of those breeds found that crossbred females had a 21-week longer median estimated herd life than the pure lines. McDowell and McDaniel (13) reported economic aspects of purebred and crossbred cows using Ayrshire, Brown Swiss, and Holstein genetics, including veterinary treatment, death losses, and dry cow maintenance. Crossbred combinations with 50% Holstein genetics had greater economic worth than purebreds or crosses with only 25% Holstein breeding. Dickinson and Touchberry (4) reported that 86% of crossbreds survived from birth through two lactations compared to only 69% of purebred Holsteins and Guernseys. McDowell (12) concluded that crossbreds may not exceed the best purebreds for any single trait but that net economic merit of crossbreds may be superior when all important economic traits are considered, particularly in "poor or hot environments."

A planned criss-cross mating system using Jersey and Holstein crosses could lead to very productive and reproductively efficient cows for pasture-based dairy systems. Several dairy farmers have used crosses or have imported semen within breeds for outcrosses. More data should become available. Use of crossbreeding without performance information on breeds or animals being considered is not recommended.

Artificial Insemination or Natural Service

Bulls for natural service breeding are used in many dairy herds. Bulls are excellent at catching cows in heat and may reduce labor requirements during the breeding season, but use of bulls raises several issues of concern. Bulls are a safety problem for people, and facilities must be adequate to allow safe handling and/or escape. *Every year workers are killed or injured by dairy bulls*. Safety concerns are greatest with mature dairy bulls. Therefore, younger bulls (less than age three) are typically used; any aggressive behavior should result in immediate culling regardless of age. A breeding soundness examination should be done on all bulls used, but some bulls may still fail to breed cows if they get injured or have low libido. Fertile bulls may not solve a herd infertility problem. Bulls can also introduce biosecurity concerns and should be purchased from reliable sources and isolated initially to minimize disease.

Genetic selection, ease of calving, and crossbreeding can be more difficult to manage if bulls are used for natural service. It would be difficult for dairy producers to purchase young, unproven bulls for natural service with the genetic reliability of proven artificial insemination (AI) sires for traits of economic importance. Calving difficulty could be an issue for some breeds of bulls used on heifers.

Dairy graziers have successfully used natural service in breeding programs, but long-term genetic effects are not known. For seasonally bred herds, plenty of bull power would be needed. With use of yearling and 2-year-old bulls, a ratio of no more than 20 cows per bull should be used, and if estrous synchronization is used, there should 10 cows or fewer per bull. With a split calving season, the same bulls could be used for both seasons. With year-round calving, fewer bulls are needed, but they should be rotated in and out of the herd frequently.

In contrast to natural service, AI gives dairy producers access to sires from all over the world. Such sires are often of proven merit for various traits, including estimated relative conception rates, through the Dairy Records Management Services. AI sires also provide flexibility for potential use of crossbreeding. There are labor costs associated with an AI program, and the program must be managed well to avoid a reproductive disaster. Virgin heifers should always be bred via AI because heifers have consistently high conception rates and are easily synchronized for breeding. Because fertility is usually good in seasonal calving herds in Ireland and New Zealand, AI is used for short breeding periods (3–6 weeks) and clean-up bulls are used for the balance of 45–80-day breeding seasons.

Reproduction is critical to seasonal dairying success. An aggressive program for ensuring reproductive success is essential. The tighter a seasonal breeding and calving program is maintained, the more successful it is likely to be. If cows calve in periods of 90 days or greater, then cows are still calving at the time to start rebreeding. In contrast, a calving period of 50 days means that all cows will have calved about 5 weeks before starting the breeding season. With shorter calving seasons, all cows have time for uterine involution and will be ready to rebreed at the start of the season.

Success of AI programs requires attention to detail and high rates of submission of animals for insemination and subsequent conception. The estrous detection rate multiplied by the conception rate gives the pregnancy rate or proportion of the breeding herd that becomes pregnant in a 21-day period. Typically, many dairy herds have both estrous detection and conception rates below 50%, resulting in 21-day pregnancy rates below 25%. In herds that breed and calve year round, delayed breeding has less obvious initial consequences but long-term efficiency is reduced. In contrast, seasonal calving herds must have 21-day pregnancy rates above 40% to ensure that a high proportion of the herd conceives on time. This requires high estrous detection efficiency (> 80%) and conception rates (> 50%) during short breeding seasons.

Improving reproductive efficiency requires improved detection of estrous, including more frequent observations, tail-head paint or other mount detectors, or use of electronic detectors. Estrous synchronization products are also available that can help concentrate the breeding in fewer days. Specific strategies will vary for heifers and cows. Conception rates are often lower after synchronization but pregnancy rates may improve if higher percentages of cows are inseminated.

DRY COW CONSIDERATIONS

Evaluating pasture for the dry cow has not been a high research priority, so the value of it is not well validated. However, pasture has traditionally been a major source of forage in many and perhaps most of the dairy operations in the East. In other cases, grassed land has served as a means to get cows off concrete during the dry period even though stored forage is the primary source of feed. Usually, either of these options work well for dry cows; all nutrient requirements are easily supplied by pasture forage of medium quality. Nutritional demands are greater in the close-up dry cow and may require supplementation to maintain body weight in preparation for lactation.

With current concepts of cation-anion balance, risk for development of milk fever upon freshening increases considerably when the diet is high in K during the prepartum period. Pasture forages are typically quite high in K and may predispose susceptible animals to milk fever. However, this is more a cautionary statement than one with substance because of the lack of information on grazing cows, especially with known differences in dietary ionic balances.

Decreasing the cation-anion balance of the diet of cows grazing prepartum did not affect DMI, herd health, or postpartum milk yield (20). The benefits of exercise and improved rumen health of grazing animals may offset any problems encountered with cation-anion imbalance.

PASTURE FOR REPLACEMENT HEIFERS

The current intensive grazing areas of the world—parts of Australia, New Zealand, and Ireland—begin grazing heifer calves at a very early age, in some cases while they are still suckling from the nipple. From that point forth, rotational grazing begins and high quality forage is made available. Some limited grain is fed early on to develop the rumen. But these producers have shown that heifers can successfully be grown with very little concentrates while grazing high quality forage.

In this country, our grains and commercial feed sources cost much less, so our producers have a choice to either practice early conventional calf rearing or choose the intensive grazing route. The conventional method may be less demanding for the first 3–4 months, i.e., calves don't eat so much and are relatively easy to care for. After that, it is usually economical for them to graze as much as possible.

The 300–400-pound heifer has protein and energy demands per 100 pounds of body weight approximately equal to a cow producing 50 pounds of milk. With high quality pasture, a cow can produce 50–55 pounds of milk per day, as mentioned earlier. Likewise, a calf can grow at adequate rates. However, in the transition to pasture, the calf probably needs 2–4 pounds of a 16% CP supplement until adequate adaptation occurs and forage intake capacity develops.

Work with the CNCPS at Penn State predicted that high quality pastures should support daily gains of 1.7–2.0 pounds per day, which is within the range of desirable growth rates for prepubertal heifers. The CNCPS also predicted, however, that because more than 75% of the protein is ruminally degraded, pasture may not provide enough undegraded protein to meet the needs of lightweight heifers.

Numerous studies have shown that 400–500-pound heifers will gain in excess of 2 pounds per day grazing high quality pasture alone (7). As with lactating cows, energy intake is limiting production (growth) for heifers; protein is more than adequate.

Researchers at Virginia Tech (16) compared 550-pound Holstein heifers raised in confinement to pasture-reared heifers without supplement, with 200 mg lasalocid (Bovatec), or with 4.4 pounds of a 19% protein grain mixture. Respective daily gains were 1.65, 1.84, 1.92, and 2.05 pounds over a grazing season. Tall fescue, orchardgrass, and white clover were the predominant forage species. Fat content was slightly higher in confinement-reared heifers based on ultrasound and urea space techniques. Acceptable gains were obtained with minimal concentrate supplementation, and the inclusion of ionophores resulted in gain approaching that provided by 4 pounds of concentrate supplementation.

Grazing trials were conducted with Holstein heifers in two successive years at Cornell University to determine the performance possible with heifers of different weights (light or heavy) when intensive grazing was used (Fox, 1991). Initial weight (352 or 478 pounds) and the response to supplemental protein (none, or 0.4 pound protein from soybean or fish meal) were evaluated. Daily gains for unsupplemented, soybean-, and fish meal-supplemented groups were 2.02, 2.19, and 2.23 pounds, respectively.

Total protein of the diet averaged 21.5% across years, but varied considerably between months. Total protein in May averaged 20.4%, declined to 13.5% in June, then increased continuously to a peak of 27.6% in October.

Pasture-available carbohydrates followed the same trend, with the highest cell wall content in June, likely due to seed heads. Forage NDF averaged 44% in May, increased to 54% in June, then decreased to a seasonal low of 39% in October. Protein solubility and degradability also varied as total protein and fiber changed. Protein degradability averaged 80% over the 2 years, but was highest in the spring and fall (80–85%) and was lowest in summer (76%).

All of the response to supplemental protein occurred in May and June; these were the months when forage protein had the highest degradability and when the heifers were at their lightest weight. Daily gain was lowest for July; forage availability was also lowest for that month. The results indicate that dairy heifers can meet both energy and protein requirements for desirable growth rates when intensive rotational grazing is practiced along with animal management practices that include deworming and fly control.

In another study at Virginia Tech, Novaes et al. (1991)(17) compared growth of heifers receiving pasture plus 3 pounds per head per day of ground corn, corn and soybean meal (26% protein), or dried brewer's grains as a source of RUP. Daily gains averaged 1.89 pounds, 1.95 pounds, and 1.96 pounds for the respective diets, indicating little benefit to supplementation above that provided by ground corn.

As with beef cattle, ionophores, such as Bovatec and Rumensin, have been widely used in dairy heifer feeding programs since their approval by the U.S. Food and Drug Administration.

In a 1995 study, a better apparent response to ionophore occurred when it was force-fed in corn-soybean meal than when it was a part of a salt-based mineral supplement free-choice. The authors assumed that daily intake of free-choice mineral varied greatly.

Generally speaking, dairy heifers' growth response on pasture to forage quality and ionophores is comparable to that of beef cattle. Tall fescue generally has not been a forage of choice for lactating dairy cows, but tall fescue and stockpiled tall fescue are useful for growing dairy animals.

In summary, when grazing heifers, provide plenty of quality pasture. In the transition to pasture, provide some energy supplementation. Consider providing an ionophore, and measure growth progress of the animal.

MANAGING THE FEEDING PROGRAM

To obtain optimal milk production and profitability, even in times of high grain costs, high-producing dairy cows need to be fed supplemental nutrients and feeds to complement the nutritional content and limitations of pasture. The basic supplemental nutrients needed to complement pasture are *energy* from the *NFC* in grains, *RUP* and *NFC* to maximize the rumen-undegradable protein and the rumen microbial protein synthesis, *fiber* from forages and/or high fiber feed ingredients to increase the effective fiber in the total ration, and deficient *minerals*. There is no one magic grain mixture for pasture. The variation in pasture availability and quality during the year suggests that we may need to have several ration formulas available during a grazing season. Feeding strategy or allocation is important, and a pTMR is likely to improve nutrient utilization and rumen health. It is challenging to successfully manage the feeding program with grazing herds. This requires even greater management skills in evaluation and monitoring than in total confinement systems, including:

- Daily evaluation of pasture availability
- Seasonal analysis of pasture quality
- Adjusting stored forage and amount of supplemental grain according to pasture availability and quality
- Daily monitoring of milk production per cow based on the bulk tank
- Monitoring milk fat and protein percent and MUN
- Monitoring body condition scores of the herd on a regular basis
- Observing cows and manure consistency and making necessary ration adjustments
- Monitoring feed costs.

CHAPTER 5
Sheep Nutrition and Management

Scott P. Greiner, Mark L. Wahlberg, and Bill R. McKinnon

Sheep are very efficient converters of forage to meat and fiber, and can produce an acceptable carcass from forage alone. The topography, climate, and forage resources of the eastern region make it very well suited for sheep production. The close proximity to large lamb consumption areas of the United States results in a variety of viable marketing opportunities for lambs from diverse production systems. Historically, sheep production has been profitable from year to year largely due to extensive use of cheap forage and competitive market prices.

Sheep production may serve as a primary enterprise on both large and small, part-time farms. Sheep work well as a secondary enterprise in combination with existing enterprises such as beef cattle due to differences in grazing behavior. Sheep will consume forages that cattle will not, resulting in increased returns per acre and improvement in pasture quality over that grazed by one species alone.

REPRODUCTIVE BIOLOGY AND ITS INFLUENCE ON PRODUCTION

Reproductive efficiency, measured as percent lamb crop raised and marketed, is the major factor determining profitability of the sheep enterprise. Given that approximately 70% of the total costs of the sheep enterprise can be attributed to the ewe flock, maximizing receipts for total weight of lambs marketed per ewe exposed is of primary importance. Pounds of lamb sold is influenced by several factors, but is most

directly related to percent lamb crop weaned. Therefore, strategies that optimize reproductive efficiency are critically important to profitability.

The seasonality of reproduction is inherent to the biology of sheep and has a major influence on the production, management, and marketing decisions and practices for the operation. Sheep are referred to as short-day breeders, meaning that they exhibit estrous and ovulate as days become shorter. Typically, fertility in ewes is highest and most efficient (in terms of embryo survival) from September through November. Conversely, many breeds are anestrous from April through July when days are long. Therefore, pregnancy rates and lamb crop percentages typically favor fall breeding versus spring breeding. Much like the ewe, fertility in the ram is also affected by season of the year. The length of the estrous cycle in ewes averages 17 days, with a typical range of 14–19 days. The duration of estrous (heat) in ewes ranges from 24 to 36 hours.

Several factors affect reproduction in sheep. Genetics is a primary influence, as there are large differences both within and between breeds for reproductive traits. Of most concern are differences for litter size (number of lambs born) and seasonal fertility. Breed differences and their implications are discussed in the Genetics and Selection section of this chapter (pp. 135–140). Age is a second factor affecting reproduction. Reproductive rates in ewes four to eight years of age are generally greater than those of young ewes. Lifetime production of the ewe is affected by age at first lambing. Ewes lambing for the first time at 12 months of age have been shown to have greater lifetime production than ewes lambing for the first time at 24 months of age. For most breeds, ewe lambs reach puberty between 4 and 8 months of age, with a body weight between 100 and 130 pounds. Proper nutrition and management is critical in these young ewes to optimize both growth and reproduction.

Strategies to synchronize estrous in the ewe flock can be advantageous in shortening the breeding season, resulting in a large percentage of the lamb crop being born in a short period. Short lambing seasons concentrate labor resources, which may result in reduced lamb mortality. Additional benefits include the potential for a more uniform lamb crop for targeted marketing as a result of less weight variation due to age differences.

Currently, there is an absence of approved products available for synchronization of ewes in the United States. The "ram effect" is commonly used to induce ovulation in anestrous ewes that have been previously isolated from rams. The ram effect is an effective, inexpensive, practical means to increase the percentage of ewes lambing out of season or early in a lambing season. Use of the ram effect requires isolating the ewes from rams for a minimum of one month, and preferably longer. Isolation from rams must be complete. Avoid fenceline contact and any association with rams (sight, smell, touch). Upon joining rams with ewes that have been previously isolated, ewes will ovulate within 7 days after introduction of the rams. However, less than 20% of the ewes will be in heat during the first 7 days. Active estrous and ovulation will occur 16–20 days after introduction of rams, resulting in pregnancy. Vasectomized teaser rams are frequently used during the first two weeks because there is a delay in estrous with

the ram effect. Aggressive teaser rams (from breeds that are active out of season) with high libido are most effective in eliciting response in the ewe.

Breed of ewe is an important factor in response to the ram effect; more response is typically seen in breeds that have out-of-season capability. Additionally, ewes with long anestrous periods will be more responsive to the ram effect as they reach the end of anestrous.

EWE NUTRITION

Ewe nutrition is a very important aspect of total flock management. Proper nutrition of the ewe is necessary to optimize productivity and profitability, as ewe feed costs are the largest single cost of maintaining the flock.

Five factors affect the nutritional needs of the ewe, specifically:

- age. Because young ewes are still growing, their requirement for nutrients is higher.
- size, or more importantly, body weight.
- body condition (amount of body fat).
- stage of production (maintenance, early gestation, late gestation, or lactation).
- level of production (how much milk, how many fetuses carried, litter size).

Additionally, health status (including parasite load), activity level, weather, and other environmental factors may influence nutritional requirements and management. However, consideration of the following questions should allow the shepherd to make decisions relative to nutritional management.

- Is the ewe pregnant?
- If so, in which stage of pregnancy is she?
- If lactating, how many lambs is she nursing?
- When will the lambs be weaned?

Tables of Requirements

To determine when and how much to feed the flock, the animals' requirements must be known. These requirements (tables 5-1 and 5-2, pp. 118 and 119) are affected by the five factors listed above. There are four key nutrients of concern in feeding the ewe flock. Those are energy (expressed as total digestible nutrients [TDN]), protein, calcium (Ca), and phosphorus (P). Vitamins A and E are important, but as long as the ewe is eating green forage (hay or pasture), these vitamins are usually consumed in adequate amounts.

Tables 5-1 and 5-2 both provide information about requirements for nutrition, but the information is expressed in different ways. Table 5-1 lists quantities of specific nutrients needed by the animal each day. Heavier animals require more nutrients, thus requirements vary with weight. Table 5-2 depicts the nutrients as concentration in the diet dry matter. This table is most useful for animals that are able to consume all the feed they want, such as those grazing pasture or consuming unlimited hay. Table 5-2 allows comparison of the quality of the diet (hay composition) to the animal's requirements without having to estimate how much feed the animal is consuming.

The remainder of this chapter examines nutrient needs at the various stages of production.

Maintenance

The animal's requirements for maintenance are the amounts of dietary nutrients it must consume daily to neither gain nor lose weight. Maintenance is generally associated with the dry period, or the period between weaning and the breeding season. Maintenance requirements for four weights of ewes are found in table 5-1. These weights reflect prebreeding weights for ewes in average body condition. A 175-pound ewe has a maintenance energy requirement of 1.6 pounds TDN per day and a maintenance protein requirement of 0.27 pound per day. From table 5-2 we see that for this animal the diet must contain at least 55% TDN and 9.3% protein. Often ewes are grazing pastures during this stage of production and would have no trouble meeting these requirements. In fact, during spring and early summer, grazing lush pastures would allow the ewe to far exceed her maintenance requirement, resulting in some weight gain. This weight gain is desired and necessary, because most ewes will lose body condition during lactation.

Flushing

Flushing is the practice of increasing ewes' energy intake, and therefore body condition, during the 10–14 days prior to breeding. This practice is effective in increasing ovulation rates and thereby increasing lambing percentage by 10–20%. The response to flushing is affected by several factors, including body condition of the ewe. Ewes in poor body condition will respond most favorably to the increase in energy, whereas fat ewes will show little if any response. With ewes on pasture, flushing is most easily accomplished by providing 0.75–1.25 pounds corn or barley per head per day from 2 weeks prebreeding through 4 weeks into the breeding season. Because corn grain is approximately 80% TDN, providing 1 pound per day would provide 0.8 pound of additional TDN to the ewe (1 pound corn x 80% TDN = 0.8 pound TDN). This additional energy would approach the additional energy requirement shown in table 5-1. Alternatively, flushing may also be accomplished by movement of the flock to a high quality forage paddock (pasture or small grain). Flushing should not continue for an excessively long period, because overfeeding is costly. Additionally, ewes that become very fat and are subsequently placed on a lower plane of nutrition following flushing may be subject to increased prenatal mortality and lower lambing rates.

Table 5-1. Daily nutrient requirements of mature ewes.

Stage of production	Body wt. (lb)	Wt. gain or loss (lb)	DM[a] intake/day (lb)	Energy TDN (lb)	Protein (lb)	Ca (g)	P (g)	Vit. A (IU)[b]	Vit. D (IU)	Vit. E (IU)
Maintenance	130	0.02	2.4	1.3	0.23	2.3	2.1	2,820	333	16
	150	0.02	2.6	1.5	0.25	2.5	2.4	3,290	378	18
	175	0.02	2.9	1.6	0.27	2.7	2.8	3,760	441	20
	200	0.02	3.1	1.7	0.29	2.9	3.1	4,230	505	22
Flushing (2 wk. prebreeding & 1st 4 wk. breeding)	130	0.22	3.7	2.2	0.34	5.5	2.9	2,820	333	26
	150	0.22	4.0	2.3	0.36	5.7	3.2	3,290	378	27
	175	0.22	4.2	2.5	0.38	5.9	3.6	3,760	441	28
	200	0.22	4.4	2.6	0.39	6.1	3.9	4,230	505	29
1st 15 wk. gestation	130	0.07	2.9	1.6	0.27	3.2	2.5	2,820	333	20
	150	0.07	3.1	1.7	0.29	3.5	2.9	3,290	378	21
	175	0.07	3.3	1.8	0.31	3.8	3.3	3,760	441	22
	200	0.07	3.5	1.9	0.33	4.1	3.6	4,230	505	24
Last 4 wk. gestation (130–150% lamb crop)	130	0.40	3.7	2.2	0.40	6.0	5.2	5,100	333	26
	150	0.40	4.0	2.3	0.42	6.2	5.6	5,950	378	27
	175	0.40	4.2	2.4	0.44	6.3	6.1	6,800	441	28
	200	0.40	4.4	2.5	0.77	6.4	6.5	7,650	505	30
(180–225% lamb crop)	130	0.50	4.0	2.6	0.45	6.9	4.0	5,100	333	27
	150	0.50	4.2	2.8	0.47	7.6	4.5	5,950	378	28
	175	0.50	4.4	2.9	0.49	8.3	5.1	6,800	441	30
	200	0.50	4.6	3.0	0.51	8.9	5.7	7,650	505	32
Lactation (1st 8 wk.) Nursing single	130	-0.06	5.1	3.3	0.70	9.1	6.6	5,100	333	34
	150	-0.06	5.5	3.6	0.73	9.3	7.0	5,950	378	38
	175	-0.06	5.7	3.7	0.76	9.5	7.4	6,800	441	39
	200	-0.06	5.9	3.8	0.78	9.6	7.8	7,650	505	40
Nursing twins	130	-0.13	5.7	3.7	0.89	10.7	7.7	6,000	333	39
	150	-0.13	6.2	4.4	0.94	11.2	8.4	7,000	378	42
	175	-0.13	6.6	4.7	0.98	11.4	8.8	8,000	441	45
	200	-0.13	7.0	5.0	1.01	11.6	9.2	9,000	505	48
Nursing triplets	150	-0.20	6.5	4.9	1.04	12.2	9.0	8,000	378	47
	175	-0.20	7.2	5.2	1.08	12.4	9.4	9,000	441	50
	200	-0.20	8.0	5.5	1.11	12.6	9.6	10,000	505	53

[a] DM-dry matter. To convert dry matter to an as-fed basis, divide by percent dry matter.
[b] International units.

Source: Adapted from National Research Council. 1985. Nutrient Requirements for Sheep, 6th revised ed. Washington, D.C., table 1, pp. 45–47.

Table 5-2. Daily nutrient concentration in the dry matter for mature ewes (175 lb. body weight).[a]

Stage of production	DM intake/day[b] (lb)	Energy TDN (%)	Protein (%)	Ca (%)	P (%)
Maintenance	2.9	55	9.3	0.19	0.21
Flushing	4.2	60	9.0	0.31	0.19
1st 15 wk. gestation	3.3	55	9.4	0.25	0.21
Last 4 wk. gestation					
(130–150% lamb crop)	4.2	57	10.5	0.33	0.32
(180–225% lamb crop)	4.4	66	11.1	0.41	0.25
Lactation (1st 8 wk.)					
Nursing single	5.7	65	13.3	0.37	0.28
Nursing twins	6.6	71	14.8	0.38	0.29
Nursing triplets	7.2	72	15.0	0.38	0.29

[a] Values converted from table 5-1 by dividing requirement by DM intake.
[b] To convert dry matter to an as-fed basis, divide by percent dry matter.

Source: Adapted from National Research Council. 1985. Nutrient Requirements for Sheep, *6th revised ed. Washington, D.C., table 1, pp. 45–47.*

Early Gestation

Table 5-1 shows that there is a relatively small increase in ewe nutrient requirements for the first 15 weeks of gestation compared to maintenance. During this time winter- and spring-lambing ewes will make the transition from pasture to a diet of harvested feedstuffs. Ewes on fall pastures should consume enough forage to meet their nutritional requirements during this early gestation stage. When feeding hay becomes necessary, it is important that the quality and quantity of hay be carefully considered. Assuming that the available hay is 50% TDN and 12% crude protein on an as-fed basis, a 175-pound ewe eating 3.3 pounds per day of this hay would consume 1.7 pounds TDN and 0.40 pound crude protein. The requirements for this ewe in table 5-1 are 1.8 pounds TDN and 0.31 pound protein daily (55% TDN and 9.4% protein). Note that her protein intake exceeds the requirement. Additionally, a ewe given the opportunity to consume as much of this hay as she desired would consume considerably more than 3.3 pounds per day (ewes can consume 3.5% of their body weight [6.1 lb, in this case]), and easily meet her requirements for both energy and protein. This emphasizes the importance of using poorer to average quality hays during the early gestation period, when ewe nutrient requirements are low compared to late gestation and lactation. If high quality hays, such as alfalfa, are fed during this period, it is important to limit intake. Overfeeding during this period is costly, and may result in overconditioned ewes leading to complications later in the production cycle.

Late Gestation

Approximately two-thirds of the birth weight of a developing fetus is gained during the last six weeks of pregnancy, resulting in 10–20 pounds of weight gain during this time period. As a result, the nutritional requirement of the ewe for both energy and protein increases. Table 5-2 shows that TDN requirements increase to 57–66%, compared to 55% for maintenance and early gestation. Similarly, the protein requirement increases to around 11% compared to 9% for maintenance. The most critical difference is the increase in energy requirement. Inadequate nutrition during this period may result in pregnancy ketosis, low birth weights, weak lambs, and lower milk production. Supplementation of 1–2 pounds corn per ewe per day, in combination with average to good quality hay (> 11% CP) should provide adequate nutrition. An important consideration during this period is the number of fetuses the ewe is carrying (see table 5-1). As the ewes approach lambing, the size of the uterus increases and limits intake. Therefore, feeding nutrient-dense rations is important to ensure adequate nutrition. Although corn silage is an excellent feed for sheep, its high moisture content and bulkiness prevents its use as the sole energy source during late gestation. Additionally, corn silage is low in protein and Ca, so these nutrients must be added to the diet for balanced nutrition.

Lactation

The growth rate of lambs from birth to weaning is largely determined by milk production of the ewe, which emphasizes the importance of good nutritional management during this period. Lactation provides an opportunity to control feed costs by feeding ewes according to the number of lambs nursing. During lactation, the ewe's nutritional requirements for both energy and protein are at the highest level of the production cycle. As mentioned previously, the highest quality hays or pastures should be used during this time. Alfalfa hay is an excellent feedstuff during lactation due to its high energy and protein density relative to other forages. In most cases, a grain-protein supplement (such as corn-soybean meal) will need to be fed in addition to the highest quality hay available. The needed protein content of this grain mix will vary depending on the quality of the hay used. Generally, total rations should be formulated to contain 70% TDN and 14% protein for lactation. Table 5-1 demonstrates the significant differences in nutrient requirements of ewes nursing singles versus twins versus triplets. Splitting ewes by number of lambs nursing is an excellent management technique to minimize feed costs. Ewes rearing single lambs will require less grain supplementation than twin-rearing ewes. Similarly, triplet-rearing ewes could be provided the extra nutrition needed if they are separated from other ewes. When all ewes are fed together, single-rearing ewes are likely overfed, which can be costly. Facilities and labor will dictate the feasibility of this management practice. As mentioned previously, milk production of the ewe is influenced by nutrition. Feed intake is a critical nutritional factor affecting milk production. Therefore, diets that are nutrient dense and highly palatable will enhance milk production. High quality grass-legume pasture can satisfy the requirements for both energy and protein of ewes in early lactation. Management to ensure adequate forage availability is crucial, along with free-choice availability of a properly formulated mineral supplement.

Ewe Lambs

Females lambing for the first time at around their first birthday require special nutritional consideration during all stages of production. In addition to the requirements for pregnancy and lactation, ewe lambs also require additional nutrition for growth, because they have not yet reached mature body size. Also, ewe lambs consume less feed per day than mature ewes of the same body weight. Daily nutrient requirements of ewe lambs are presented in table 5-3 (p. 122). Because ewe lambs are frequently managed as a separate group from mature ewes, providing extra nutrition during gestation is easy. Maintaining ewe lambs as a separate management group during lactation is also critical, particularly for ewe lambs nursing multiples, so they can receive proper nutrition to maintain adequate body condition for future growth and productivity. Ewe lambs must replenish body condition quickly following lactation to be prepared to breed and lamb as 2-year-olds.

Monitoring Body Condition

Body condition of the ewe is an important consideration in nutritional management. If ewes are getting fat, they are consuming more energy than they need, and are likely being overfed. On the other hand, if they are thin, they are not receiving adequate energy (or they have a health-related problem). Table 5-1 (p. 118) lists requirements for ewes in average body condition; these requirements may be above or below the requirements for your flock. Proper body condition is essential for optimum productivity, and is most critical during the breeding season and late gestation. Ewes that need to improve body condition should be separated from the rest of the flock and supplemented.

Forage Quality

An important aspect of nutritional management is knowing the quality of forages that will be used, most importantly hay. To properly balance rations and formulate diets, an accurate forage analysis should be conducted on all harvested feeds (hays and silage). There can be significant variation in hays harvested from the same field from one year to the next, and from one cutting to another. Having accurate feed analysis may save feed costs and will improve the ability to adequately manage the nutrition of the flock.

In summary, ewe flock nutrition is an important aspect of the profitability of the sheep enterprise. Proper nutrition is key to getting optimum production from a sheep operation. Forages, both harvested and especially grazed, should be fully used to provide adequate, cost-effective nutrition for the ewe flock.

NUTRITION OF GROWING LAMBS

There are many different strategies for feeding growing lambs. There is no one right way. As with any producing animal, the amount of energy (TDN) consumed is directly related to how fast the lamb grows. However, maximum rate of gain may not always be the objective. The following factors influence the choice of feeding program for the growing lamb:

- Intended use—A market lamb is fed differently than a lamb to be used as a breeding animal.

- Growth potential and mature size—Smaller framed and earlier maturing sheep fatten at light weights, especially when fed a high energy diet. Large framed and late maturing sheep may need a high energy diet to finish at desirable market weights.

- Market specifications—Such factors as desired market weight, amount of finish (fat), and seasonality of demand are economically important.

Table 5-3. Daily nutrient requirements of ewe lambs.

Stage of production	Body wt. (lb)	Wt. gain or loss (lb)	DM Intake/day[a] (lb)	Energy TDN (lb)	Protein (lb)	Ca (g)	P (g)	Vit. A (IU)	Vit. D (IU)	Vit. E (IU)
1st 15 wk. gestation	110	0.30	3.3	1.9	0.35	5.2	3.1	2,350	277	22
	130	0.30	3.5	2.0	0.35	5.5	3.4	2,820	333	24
	155	0.28	3.7	2.2	0.36	5.5	3.7	3,290	389	26
Last 4 wk. gestation	110	0.35	3.5	2.2	0.42	6.3	3.4	4,250	277	24
(100–120% lamb crop)	130	0.35	3.7	2.4	0.42	6.6	3.8	5,100	333	26
	155	0.33	4.0	2.5	0.43	6.8	4.2	5,950	389	27
(135–175% lamb crop)	110	0.50	3.5	2.4	0.45	7.8	3.9	4,250	277	24
	130	0.50	3.7	2.6	0.46	8.1	4.3	5,200	333	26
	155	0.47	4.0	2.7	0.46	8.2	4.7	5,950	389	27
Lactation (1st 8 wk.)	110	–0.10	4.6	3.3	0.62	6.5	4.7	4,250	277	32
Nursing single	130	–0.10	5.1	3.6	0.65	6.8	5.1	5,200	333	34
	155	–0.10	5.5	3.8	0.68	7.1	5.6	5,950	389	38
Nursing twins	110	–0.22	5.1	3.7	0.71	8.7	6.0	5,000	277	34
	130	–0.22	5.5	4.0	0.74	9.0	6.4	6,000	333	38
	155	–0.22	6.0	4.3	0.77	9.3	6.9	7,000	389	40

[a] To convert dry matter to an as-fed basis, divide by percent dry matter.

Source: Adapted from National Research Council. 1985. Nutrient Requirements for Sheep, 6th revised ed. Washington, D.C., table 1, pp. 45–47.

- Facilities—Availability of feeding equipment and feed storage, animal housing, manure handling, and other factors must be considered.

- Feedstuffs available—The type and price of grains that are continuously available affect feed choices, as does pasture quality and quantity.

Growth and Development

The change in body weight of a lamb over time can be described in three phases:

During phase 1 the young lamb does not rapidly gain weight. Weight gain is bone and muscle. The lamb has not yet developed a functioning rumen, so its diet consists primarily of milk.

Phase 2 sees a much faster rate of gain. There is a rapid increase in the amount of muscle the animal possesses, but little fat is deposited during this phase. Sometime during this phase the lamb develops a functioning rumen, is weaned, and no longer consumes milk.

Phase 3 begins when the weight gain curve plateaus. This relates to physiological maturity, when skeletal and muscle growth ceases and change in body weight reflects the amount of fat.

Creep Feeding

Creep feeding is the practice of providing young, nursing lambs a source of feed that the ewes cannot access. Thus, it supplements the milk produced by the ewes and can provide valuable supplemental weight gain. This added weight gain has the most economic value for lambs managed in an intensive, early weaning production system in which lambs will be maintained in a drylot all the way to market weight. Conversely, for lambs that will be developed on pasture throughout the spring and summer, creep feeding would be of less value due to the

relative expense of this early weight gain (which may later be attained on forage). Creep feeding also is beneficial for flocks with a high number of multiple births, or flocks with ewes having limited milk production.

Young lambs may be started on creep feed as early as 10 days of age. Although significant amounts of feed are normally not consumed until 3–4 weeks of age, providing access to creep feed at an early age allows lambs to develop a habit of eating dry feed, and helps stimulate rumen development. For creep feeding to be economical, lambs must consume enough feed to increase performance. Lambs should eat a minimum of 0.5 pound of creep feed per head per day from 20 days of age to weaning.

The creep ration need not be expensive or complex. The principle behind creep feeding is to stimulate lambs to eat and therefore promote weight gain. Therefore, highly palatable feeds must be provided. At a young age, lambs prefer feeds that are finely ground and have a small particle size. Feedstuffs high in palatability for young lambs include soybean meal, ground corn, sweet feeds, and alfalfa hay. These feeds should be replaced daily to keep them fresh. A simple mixture of 80–85% ground or cracked corn and 15–20% soybean meal, with free-choice high quality alfalfa hay, is a very palatable early creep ration. The feed being fed to the ewes may also be included free choice in the creep feeder. Early in the creep feeding period, stimulating intake is of primary concern. These diets should be formulated to contain 20% crude protein.

As the lambs get to 4–6 weeks of age and older, coarser feeds become more palatable. Providing feeds early will enhance the lambs' acceptance of these coarser feeds. As the lamb gets older, intakes and growth rates should increase. Additionally, the proportion of the gain that is derived from dry feed versus milk increases. During this time, lambs may be gradually switched to a complete pelleted ration or a ration containing cracked corn and supplement. Over time, the ration should be changed to represent what will be fed once the lamb is weaned. Complete feeds are available commercially, which can be convenient yet expensive. Pelleted supplements to be mixed with cracked corn are generally cheaper, and are also widely available. At weaning, protein requirements of lambs drop to 15–16%. An advantage of the complete feeds and protein supplements is that they are fortified with antibiotics, vitamins, and minerals that are important for lamb health and performance. Lambs should be vaccinated with *Clostridium perfringens* C and D to prevent overeating disease 2–3 weeks prior to weaning (6–8 weeks of age), and receive a booster at weaning.

Finishing Market Lambs

Feeding programs for market lambs will vary considerably depending on the production system of the flock. *Winter-born* (December–February) lambs are normally provided creep feed, and should be weaned and placed on a high grain ration by 60 days of age to take advantage of the most efficient gains, which are made in the first 100–120 days of age. The objective is near maximum rate of gain, and consequently the use of high-growth breeds is wise. In this system lambs grow rapidly, ownership of the lambs is for a relatively short period of time, and lambs are marketed at a seasonally high price.

In this system, lamb nutrition need not be complicated or expensive. Lambs are already adapted to consuming grain in the form of creep feed. Once weaned, they can be transitioned to a lower cost growing-finishing diet that contains a high grain content.

For *late winter and spring-lambing* flocks (March–May), a substantial portion of lamb gain

may come from grass. These systems provide for cheap lamb gains as lambs graze during the spring and summer months with the ewes. Maximum rate of gain is not desired during the summer; rather, the lambs are kept in a growing mode with a moderate gain of 0.25–0.5 pound per day. Deciding when and how much to feed these lambs will be influenced primarily by targeted marketing time/weight and by available forage and lamb weight. Many producers have successfully finished spring-born lambs in the fall using grain supplements fed to lambs while on pasture. Others confine lambs to a drylot beginning in the fall and feed a high grain total mixed ration (TMR) until they are marketed.

Lambs born in the fall (September–November) have high quality pasture available early in their life, and as they become older the pasture availability normally declines. Many producers use pasture for the ewes and lambs into October, often relying on the high quality of the regrowth on hay fields. Later in the fall pasture becomes more limited, except for forages such as stockpiled fescue, the brassicas, or small grains. The fall-born lamb is often weaned at 60–90 days of age and placed in a drylot for finishing and marketed in the late winter or spring months.

Diets for finishing lambs need not be complex. Lambs need some type of "scratch factor" to stimulate rumen function, a high energy source, protein in appropriate amounts, and proper supplementation of minerals.

- *Whole-grain rations* consist of a whole grain (whole corn or barley) and a commercially available pelleted protein supplement. These protein supplements normally contain 36–40% crude protein, and are designed to be mixed with whole grain or barley for a complete ration. An added benefit to many of these supplements is that they may contain Bovatec, which aids in the prevention of **coccidiosis** and also promotes weight gain and feed efficiency. Feeding grain in the whole form provides adequate "scratch factor" as a roughage source. Typically the ration contains 85–90% grain and 10–15% supplement. Once lambs are adapted to the ration, it should be offered free choice in a self feeder for continuous access. Because lambs consume a diet that is very high in TDN, the rate of gain is very fast, sometimes exceeding 1 pound per day.

- *Mixtures of roughage and grain* in a TMR are commonly used in large commercial feedlots. Grains are processed by cracking or dry rolling. Roughages include chopped hay, alfalfa pellets, and peanut hulls. Byproduct feeds such as soyhulls and corn gluten feed provide fairly high amounts of fiber, though they don't have the same effect in the rumen that true roughages do. Although silages may be fed in such TMRs, there are limitations. During warm weather, silages spoil quickly. Lambs will sort through corn silage and refuse to eat pieces of cob and larger chunks of the stalk. Although the dry rations can be put in a self feeder, it is more common to provide fresh feed daily.

- *Pasture-based finishing programs* rely on high quality forages as a significant source of nutrients as well as a source of fiber. If the pasture is of sufficient quality and quantity, lambs can be finished on pasture only. This has been successfully done with stands of straight alfalfa, although blends of grass and legume with more than 50% legume would be satisfactory. Cost of gain is very low. Rate of gain is slower than with rations using grain.

- *Pasture plus grain* is another option. Once again, high quality pasture consisting of grass-legume mixtures kept at a young stage of growth is required. With these systems grains are used as an energy supplement to the pasture. No supplemental protein is required because the pasture provides adequate amounts. At least 1 and up to 2 or 3 pounds of grain (whole corn or barley) is fed per lamb daily. Rate of gain can be comparable to the TMR fed in a drylot.

Problems can arise with finishing lambs. Although the drylot prevents access to worm larvae, coccidiosis can be a problem. In pasture-based systems regular deworming is necessary to prevent losses to these parasites. Proper balance of Ca and P is necessary to prevent urinary calculi. This is a major concern with high grain rations, because grains are extremely low in Ca and quite high in P content. Slow adaptation to high grain rations is necessary to allow the rumen microbes to transition from a high fiber diet to a high starch diet. Moving too quickly in this transition can cause acidosis, rumen upset, polioencephalomalacia, and the potential for sudden death due to enterotoxemia.

Protein requirements for desired level of performance according to lamb weight are listed in table 5-4 (pp. 126). Feed efficiency for older lambs coming off pasture and into the drylot will normally range from 5 to 5.5 pounds of feed per pound of gain. Young, lightweight, fast growing lambs will have feed efficiencies of around 2.5, and as they approach market weight you can expect this to increase to 3.5–4.0.

Development of Replacement Ewe and Ram Lambs

Nutrition from birth to first lambing influences the lifetime productivity of the ewe. Ewe lambs should be in production by the time they are 12–14 months of age, because ewes that lamb first as yearlings rather than 2-year-olds have higher

Table 5-4. Protein concentration (% of dry matter) of rations for lambs of varying weights and performance levels.				
	Average daily gain			
Lamb wt. (lb)	**0.50 (lb)**	**0.65 (lb)**	**0.80 (lb)**	**0.95 (lb)**
40	17.3	21.3	25.4	29.4
65	12.2	15.0	17.7	20.5
90	9.7	11.8	13.9	16.0
115	9.0	9.2	10.8	12.5

Source: Morrical, D. 1991. In: D. Morrical, ed., Proceedings 13th Annual Iowa Sheep Symposium.

lifetime production. Therefore, development of replacement ewe lambs over the summer months prior to breeding affects the overall productivity of the flock. Ewe lambs should be targeted to reach 70% of their mature weight at breeding.

Winter-born ewe lambs generally have early rapid growth resulting from creep feeding and grain diets prior to forage being available. Winter-born ewe lambs that will be kept for flock replacements should be prevented from becoming excessively fat, which has been shown to reduce future milk production. Development of these winter-born ewe lambs is best accomplished through pasture grazing and additional grain supplementation as needed to enhance gains.

Early and late spring-born lambs traditionally are developed primarily through forage-based systems. Potential replacements should be identified and weaned so they may be properly grown and managed. These ewe lambs may need to receive supplemental corn or barley (0.5–1.5 pound per head per day) to achieve daily gains needed to reach target body weight prior to breeding (table 5-5). The amount of supplement

needed will vary with forage quality and availability, as well as anticipated breeding date. As forage quality and availability declines during the summer, supplemental grain feeding will become necessary if breeding dates are early. Periodic weighing of ewe lambs will assist in measuring progress toward target weight. Shearing of replacement ewes will enhance growth rates during the hot summer months. An effective deworming program is also crucial for optimum gains.

Growing and developing ram lambs can be fed similarly to market lambs in an accelerated program. Well grown ram lambs can be used for breeding soon after they reach puberty, as young as 8–10 months of age. Due to sex differences, ram lambs with high genetic potential for growth will not become excessively fat until reaching 130+ pounds. At this time, rams need to be limit-fed to avoid excess fat deposition and also to "toughen" them for the breeding season. Rations containing 12–15% crude protein should be used for growing and developing ram lambs. As with market lambs, requirements for protein decline as they get heavier. See table 5-5 for nutrient requirements.

Table 5-5. Daily nutrient requirements of developing replacement ewe and ram lambs.

	Body wt. (lb)	Wt. gain (lb)	DM Intake/ day[a] (lb)	Energy TDN (lb)	Protein (lb)	Ca (g)	P (g)	Vit. A (IU)	Vit. D (IU)	Vit. E (IU)
Ewe lambs	66	0.50	2.6	1.7	0.41	6.4	2.6	1,410	166	18
	88	0.40	3.1	2.0	0.39	5.9	2.6	1,880	222	21
	110	0.26	3.3	1.9	0.30	4.8	2.4	2,350	277	22
	132	0.22	3.3	1.9	0.30	4.5	2.5	2,820	290	22
Ram lambs	88	0.73	4.0	2.5	0.54	7.8	3.7	1,880	222	24
	132	0.70	5.3	3.4	0.58	8.4	4.2	2,820	333	26
	176	0.64	6.2	3.9	0.59	8.5	4.6	3,760	444	28

[a] To convert dry matter to an as-fed basis, divide by percent dry matter of the ration.

Source: Adapted from National Research Council. 1985. Nutrient Requirements for Sheep, 6th revised ed. Washington, D.C., table 1, pp. 45–47.

MINERALS AND VITAMINS FOR SHEEP

Proper animal nutrition means giving the animals an appropriate amount of all nutrients necessary for optimum production. This requires knowledge of the nutrients themselves, factors that affect the requirements of animals, and the feeds used to deliver those nutrients. Cost is always a consideration for profit-motivated producers. This interplay of factors can become very intricate, but it need not.

For the ewe flock, proper nutrition involves giving animals all the good quality forage they want, and supplementing that with nutrients that may be deficient. So the basics of animal nutrition are good forage management, such as proper fertilization, a mixture of grasses and legumes, maintaining forage at a nutritious stage of growth, and providing forage in adequate quantities.

Supplements are just that—sources of nutrition that are given to animals in addition to their basic ration, with the intent of increasing the intake of that critical nutrient. Thus, we can't properly supplement without knowing the requirements of the animals, or without knowing the amount of nutrition provided by the basal ration.

Table 5-6 shows the various minerals and vitamins of concern, levels found in good forage, and the requirements for these nutrients by various classes of sheep. The requirements are based on the National Research Council's *Nutrient Requirements of Sheep* (1), and the forage values are based on recent pasture samples taken in southwest Virginia.

Macrominerals

Many minerals are required in the diet of sheep. Macrominerals are required in larger amounts, with requirements expressed as a percent of the diet or as grams per head per day. Macrominerals are shown on the first six rows of table 5-6. Some of these are already present in sufficient quantity in forages, so supplementation is not needed, while others are never sufficient and must always be supplemented. Finally, there are those that are marginal, meaning that the amount in the forage and the amount needed are close, so supplementation is sometimes needed and sometimes not. The following list shows the abundance of several macrominerals in typical forage.

- adequate   potassium (K)
- deficient   sodium (Na) (when combined with chlorine [Cl] makes salt)
- marginal   calcium (Ca), magnesium (Mg), phosphorus (P), sulfur (S)

Calcium content is often adequate in forages, and legumes have higher levels than do grasses. Grains and grain crop silages have very low levels of Ca. Phosphorus is just the opposite—it is high in grains and low in forages, often because soils are low in P fertility levels. Because P is important for reproduction and growth, it is often included in minerals for the ewe flock year-round. Magnesium is often low in lush forage growing in early spring or when springlike conditions occur. A deficiency of Mg causes grass tetany, a problem in both cows and ewes.

Microminerals

Minerals needed in very small quantities are called microminerals, or trace minerals. The requirement by animals for these minerals is expressed in milligrams per head per day or in parts per million. Just as with the macrominerals, some are adequate, others are deficient, and several are marginal in typical forage.

Table 5-6. Minerals and vitamins in forage and required by sheep.

Nutrient	Good forage	Class of sheep and requirements (in diet dry matter)		
		Mature ewe		Young lamb
		Early pregnancy	Nursing twins	Fast gain
Calcium (%)	0.45	0.25	0.4	0.55
Phosphorus (%)	0.40	0.2	0.3	0.25
Potassium (%)	2.0	0.5	0.8	0.6
Magnesium (%)	0.25	0.12	0.18	0.12
Sulfur (%)	0.25	0.15	0.25	0.15
Sodium (%)	0.0005	0.10	0.15	0.10
Iron (ppm)	100	40	40	40
Copper (ppm)	8	10	10	10
Manganese (ppm)	70	40	40	40
Zinc (ppm)	30	30	30	30
Selenium (ppm)	0.15	0.3	0.3	0.3
Vit A (IU/lb DM)	50,000	1,000	1,200	500
Vit D (IU/lb DM)	500	100	100	100
Vit E (IU/lb DM)	10	7	7	7

- adequate   manganese (Mn), iron (Fe)
- deficient   selenium (Se)
- marginal   zinc (Zn), copper (Cu)

Zinc, Cu, and Se are all important in many physiological functions, including the immune response and disease-fighting ability. Our soils are often deficient in Se, making forage grown on those soils also deficient. Consequently, it is strongly recommended to include Se in mineral mixtures for sheep of all ages.

The U.S. Food and Drug Administration (FDA) oversees Se in livestock feeds, because it is a cancer-causing element at high levels. The agency has established rules for inclusion of supplemental Se and expressed those in three different ways. Those rules, indicating maximum levels of supplemented Se for sheep, are:

- 0.3 parts per million (ppm) in the total diet
- 0.7 mg per head per day
- 90 ppm in a free-choice mineral mixture.

Because Se is not stored in the body for very long, frequent intake or dosing of Se is critical. A good sheep mineral that contains at least 50 or 60 ppm Se must be available at all times.

Copper (Cu) can be toxic to sheep. Although there are important functions of Cu in the body, and thus it is a required mineral, excess amounts are concentrated in the liver rather than being excreted. Over time, this excess of Cu can destroy liver tissue, resulting in death of the animal. Our soils, and thus the forages grown on them, contain Cu levels that are close to the animals' requirements. Consequently, sheep minerals for the mid-Atlantic and northeast regions should not include any Cu. Forage levels of Cu are too low for cattle and goats, thus cattle minerals always have Cu added to them, and goat minerals should. Therefore, mineral mixtures formulated for cattle or goats can be toxic to sheep due to their high Cu concentration.

Vitamins

Sheep, with their ruminant digestive system, can make vitamins from the raw materials consumed in their diet. They do this very well with all of the B vitamins; thus, these are generally not any concern with sheep. Vitamins A and E are made from compounds found in green forage. Vitamin A can be stored in the liver for 2 or 3 months after sheep have been eating green forage for several months. Consequently, when consuming fresh pasture or well made hay, no supplemental vitamins are needed.

However, when sheep are eating forage that is old, weathered, mature, or otherwise low in vitamin A precursor, then this vitamin should be added to the mineral mixture. Other feeds that will result in inadequate vitamin A levels are corn silage, corn stalks, and straw.

Vitamin D is made from exposure to sunshine. For sheep housed indoors for more than 2–4 weeks, such as lambs being finished in confinement, vitamin D should be included in the diet.

Most commercial minerals for sheep designed for free-choice feeding will contain added vita-mins A, D, and E. When making a TMR, vitamin premixes can be added to the formulation if a free-choice mineral is not going to be fed.

Mineral Intake

Sheep do not eat the same amount of mineral throughout the year. They have a craving for salt, and consume a complete mineral to get salt. Some ingredients, such as dicalcium phosphate and especially magnesium oxide, are not very palatable; thus intake may be lower when these ingredients are included. Often grain products or artificial flavor enhancers are added to mineral mixes to encourage higher intake.

Intake is higher when consuming lush fresh forage, such as in the early spring. During the dry summer months intake is lower, as is the case when sheep are eating hay. If a water source is nearby, intake is higher than when water is a great distance away. In addition to nearby water, intake is higher if mineral feeders are located in shady areas or along paths frequently traveled by sheep.

Producers should monitor intake periodically. Put out a known amount of mineral and keep track of the number of days a group of sheep takes to consume it. Divide by the number of head to calculate the intake per head per day. This should be an average of 0.5–2 ounces per day.

Forms of Mineral Supplements

Minerals and salt products are available in loose, granular form and in block form. Because these blocks are hard enough to shed rainwater, it is sometimes difficult for sheep to get enough mineral from licking these blocks. In addition, sheep have broken their teeth on blocks. Finally, few if any complete minerals are in block form. Loose minerals must be put in a covered feeder of some type to keep rain out so they don't cake and become hard. Loose mineral mixes are recommended for sheep.

Types of Mineral Supplements

Sheep producers with forage-based feeding programs normally provide minerals in a self feeder. They normally do not mix minerals with other feeds that are fed each day, as is the case with swine, poultry, dairy, and beef feedlots. Several types of free-choice mineral mixtures are available for sheep. These are:

White salt: Some white salt contains only Na and Cl. This is not an adequate mineral supplement. White salt often also contains iodine, and is therefore called iodized salt.

Trace mineral salt (TMS): TMS is white salt with added trace minerals. No macrominerals are included. It is often colored red from the Fe compounds added. Unless specifically stated, TMS contains no added Se. TMS with added Se is considered to be the minimum acceptable mineral supplement for sheep, and only when sheep are consuming high quality pasture.

Complete mineral: This is a mixture containing salt, the macrominerals Ca and P, and trace minerals. It may or may not have added Se. It may have added Mg, but perhaps not enough to prevent grass tetany. Often the ratio of Ca to P is in the product name, such as 2:1 or 4:1. Because P is the needed item and Ca is normally adequate, a lower ratio (less Ca, more P) is more appropriate for forage-based feeding programs. A higher ratio just dilutes the P with Ca-containing ingredients.

Free-choice mixtures are sometimes medicated with feed additives. Although there is a much longer list of approved products for cattle, several helpful products are included in minerals for sheep. Probably the most helpful are those products that help combat coccidiosis, which is a gut disorder caused by a protozoan parasite.

A major problem with additives to feeds is the lack of precise dosing to the animal. Intake of the feed determines intake of the medication. The variability in intake of free-choice minerals has already been addressed. More precise dosing occurs when additives are included in a grain supplement that is hand-fed each day. Even more precision occurs when these products are included in a TMR, although few sheep producers feed their sheep in this manner.

Lambs Fed a High-Grain Diet

The rapidly growing lamb fed a high grain diet can experience many nutritionally related problems. One of these is urinary calculi, a blockage of the male urinary tract caused by the development of "stones." An unsupplemented high grain ration contains an excess of P and negligible amounts of Ca. The requirement (table 5-6, p. 129) is for Ca in higher amounts than P. This reversal of the Ca:P ratio results in a change in the pH of the urine and the development of mineral-based precipitates in the urinary tract.

One solution to this problem is to use ammonium chloride in the ration. This changes the pH of the urine back toward normal, thus preventing the stones from forming. However, the Ca:P imbalance persists. This is best fixed by feeding the lamb a mineral supplement that provides lots of Ca and little or no P. Ground limestone (feed grade) added to a complete ration at the rate of 1% of the mixture is recommended. In this way the diet will contain the recommended Ca:P ratio of at least 2:1.

Summary

High quality forages consisting of mixtures of grasses and legumes provide the basis for good sheep nutrition in the mid-Atlantic and northeast regions. These forages also provide many of the needed minerals and vitamins for sheep. However, several minerals will likely be deficient, thus mineral supplements must be offered.

These supplements should be in loose form, fed in a feeder to keep out the weather. Free-choice minerals for sheep must contain added Se, and should not have any added Cu. The basic ingredient is salt.

Special attention must be paid to the grow-finish lamb receiving a high grain ration. The imbalance in Ca:P must be rectified to reduce the incidence of urinary calculi.

Mineral supplementation need not be complicated or expensive. Intake of minerals by sheep needs to be monitored to ensure that amounts adequate to meet the needs are consumed. Excessive intake is costly and does not result in higher production.

By focusing on forage production and quality first, then providing minerals that are likely to be deficient, producers can cost effectively meet the mineral needs of their sheep.

FLOCK HEALTH CONSIDERATIONS FOR SHEEP

A few key diseases can be devastating to a sheep flock. Preventing these problems from getting out of hand is the best strategy. This requires a solid vaccination program for a few key diseases, close observation of the flock for animals with abnormalities, and prompt treatment of those that develop problems. In addition, a sound system for disposing of dead animals is necessary.

Internal Parasites

Sheep are very susceptible to internal parasitism. They become infected by consuming plants upon which the larvae of the parasites are found. The larvae develop inside the gut of the sheep, eventually reaching maturity. They produce eggs that are expelled from the sheep in the manure.

The eggs hatch in the environment, and the larvae eventually migrate to the plants where the sheep can ingest them with the pasture they consume. The life cycle repeats itself every 2–3 weeks, depending on the worm species and climate conditions. Two species often associated with sheep parasitism are *Haemonchus* and *Ostertagia*.

During harsh climate conditions the larvae cease to develop to maturity. They instead burrow into the gut wall and wait for more favorable conditions. In the mid-Atlantic and northeastern states the cessation of activity occurs in the fall. They resume development and activity in the spring months. These worms are not affected by all dewormer products.

Worms are tough. They can persist in the environment for more than a year. Therefore, rotational grazing that uses alternating periods of grazing and rest periods will not reduce the parasite load on a pasture. However, some situations have reduced loads on the pasture, including:

- regrowth of the pasture after harvesting hay,

- a new seeding of either annuals or perennials,

- regrowth of the pasture after grazing with another species (such as cattle), and

- sequential deworming done in synchrony with the life cycle of the parasite.

The use of dewormers is helpful in controlling the problem, but cannot be the only technique used. Constant exposure to the same dewormer compound can create a population of parasites that may become resistant to the product.

Sheep develop resistance to parasites with repeated exposure. Consequently, mature ewes are somewhat resistant, while the young lambs are the most susceptible. Management systems that focus on the young lambs are needed to control the damage done by worms. If you have sheep, then you have internal parasites on your farm. They cannot be ignored. Futher details on parasites and their control can be found in chapter 8 and in reference 2.

Coccidiosis

Another type of internal parasite is a protozoan that causes coccidiosis. The oocysts (eggs) of this organism are ingested by sheep, hatch and develop in the gut, and cause damage to cells in the gut wall. Symptoms include diarrhea, often associated with some blood, general unthriftiness, and reduced intake and performance.

Coccidiosis is mostly a sanitation problem. Feeders and waterers contaminated with feces contribute to spreading the organism, which often happens with sheep raised in confinement. Although coccidiosis can occur in sheep on pasture, it is less common. Sheep develop resistance with repeated exposure, thus lambs are the most susceptible.

Because the coccidia organism is a protozoan, dewormer products have no effect. Some feed additives are quite effective, including the ionophores (brand name Bovatec) and products specific for coccidia control, such as Amprolium and Decoquinate. Many feed supplements and mineral formulations for sheep contain one of these feed additives to control coccidiosis.

Foot Rot

Two different bacterial strains join forces to cause foot rot in sheep. The first of these, *Fusobacterium necrophorum,* exists on all sheep farms. It enters the foot through abrasions and causes mild inflammation and limping. However, when the second bacteria, *Dichelobacter nodosus,* is present and enters the wound, foot rot can result. *D. nodosus* causes liquification and dissolution of tissue in the foot, resulting in serious damage, swelling, bad odor, and lameness.

Eradication of *D. nodosus* is possible, because it cannot live in the environment longer than 2–3 weeks. In other words, it must inhabit the foot of a sheep or it dies. Eradication procedures include aggressive treatment of all sheep in the flock, not just those with signs of foot rot. If a pasture does not have sheep on it for 3 weeks it can be considered free of the foot rot-causing organisms.

Producers should focus on preventing foot rot from entering their flocks. Organisms enter a farm in the feet of infected sheep, or in the bedding of trucks used to transport infected sheep. All sheep coming from another location should be considered suspects for harboring the foot rot organism. Upon arrival they should be run through a foot bath containing an appropriate treatment solution, and isolated for 30 days.

For more detailed information on foot rot see reference 3.

Reproductive Diseases

Few disease situations are more devastating to the owner of a sheep flock than the reproductive diseases, of which there are just a few. The problems caused by these diseases include sterility and abortions.

Only one, epididymitis, affects the male. Rams with this disease develop scarring of the epididymus, a structure associated with the testicles. This damage prevents the flow of sperm cells through the tract, rendering the ram sterile. Once the damage is done it cannot be repaired, so the ram is useless as a breeding animal. A vaccine exists, and it should be used in high risk situations. Because young rams get the disease from older rams, these two groups should not be housed together. A breeding soundness exam conducted several weeks prior to the planned start of the breeding season should include a check for epididymitis.

Three diseases of the female cause serious problems. Symptoms of all three include late-term abortions and the delivery of weak and unthrifty lambs, sometimes accompanied by vaginal or uterine prolapse. Those diseases and their causative agents are:

- enzootic abortion in ewes, which is caused by the bacterium chlamydia. A vaccine is available.

- campylobacter (vibriosis), which is caused by the bacterium campylobacter. A vaccine is available.

- toxoplasmosis, which is caused by a protozoan. No vaccine is available.

Symptoms of each disease are often seen in ewes lambing for the first time. The infective agent can be transferred from dam to daughter at the time of birth, or from ewe to ewe via aborted tissues. In the case of toxoplasmosis, cats acquire the organism from eating infected tissue then distribute it around the barn area in their feces.

A vaccination program should be used as a preventive measure against abortion diseases. Grain should never be fed on the ground. If abortion occurs, isolate the aborting ewe and submit the fetus and associated tissues for diagnostic evaluation. Some producers have used high feeding rates of tetracycline in late pregnancy as a preventive or treatment for the remaining flock. Consult with a veterinarian for advice in dealing with these troublesome diseases.

Metabolic Diseases

A number of metabolic diseases affect sheep. These are caused by deficiencies, excesses, or imbalances of nutrients in the diet. Most of these are easily prevented by properly feeding the animals and by using good management procedures. Because they are not caused by an infective organism, vaccination is not possible. Some of these diseases and the type of sheep involved are listed below.

- Ketosis (pregnancy toxemia, lambing paralysis) occurs in ewes in late pregnancy, often in those carrying twins. It is caused by inadequate energy intake.

- Milk fever (hypocalcemia) occurs in ewes in late pregnancy or early lactation. A disruption in Ca metabolism leads to low blood Ca levels.

- Grass tetany (hypomagnesemia) occurs in ewes grazing lush pasture in early spring. The hallmark of the disease is low blood Mg levels.

- White muscle disease is caused by a deficiency of Se. It affects all ages of sheep, but especially young lambs.

- Urinary calculi (urolithiasis) occurs in male lambs fed high grain rations that have an imbalance in the ratio of Ca to P, causing a blockage in the urethra and failure to urinate.

- Copper toxicity affects all ages of sheep, but especially mature ewes. Minerals for sheep should not have added Cu. Avoid using mineral mixes formulated for cattle or goats.

Other Diseases

Lambs are especially susceptible to clostridial diseases, such as overeating disease and tetanus. Properly vaccinating ewes for these diseases in late pregnancy provides high levels of anti-bodies in the colostrum. Lambs should also be vaccinated at 6–8 weeks of age according to label instructions. These devastating diseases are easily prevented through proper vaccination procedures.

Sound information dealing with vaccination schedules and overall management practices to ensure high productivity and health can be found in reference 5.

In summary, a health management program for livestock should contain these components:

- Prevent exposure to disease-producing organisms or situations (via sanitation, isolation of new arrivals, disease eradication).

- Maintain a high level of resistance in the animals (via nutrition, vaccines, selection).

- Once it occurs, prevent the spread of disease to other animals (via observation, quarantine, diagnosis, treatment, proper disposal of dead animals and tissues).

GENETICS AND SELECTION

A breeding program is a planned management scheme designed to result in desirable genetic change for traits of economic importance in the flock. Producers are challenged with the task of making simultaneous genetic progress in the economically important traits of reproductive efficiency, maternal ability, growth performance, and end product merit. Doing so requires incorporation of proper selection within a designed breeding system.

Breeding Systems

The general breeding objective for commercial sheep producers is to optimize production within the given resources of the operation. Increasing

production in the commercial flock has been shown to be most effective through the use of a well managed crossbreeding program. Cross-breeding refers to the mating of animals from different breeds. Crossbred animals have two major advantages over straightbred animals: (i) crossbred animals exhibit **heterosis** (hybrid vigor), and (ii) crossbred animals combine the strengths of the breeds used to form the cross (breed complementarity). Heterosis refers to the superiority in performance of the crossbred animal compared to the average of the straightbred parents.

Heterosis is maximized when the breeds crossed are genetically diverse. Breeds that have been developed for different purposes and have different origins (Suffolk vs. Finnsheep) exhibit more genetic diversity than breeds that have been placed under similar selection criteria. The amount of heterosis expressed for a given trait is also related to the *heritability* of the trait. Heritability is the proportion of the measurable difference observed between animals for a given trait that is due to genetics (and can be passed to the next generation). Reproductive traits are low in heritability (0–10%), and therefore respond relatively slowly to selection pressure because a very small percentage of the difference observed between animals is due to genetic differences (a large proportion is due to environmental factors). The amount of heterosis is largest for the traits that have low heritabilities. Therefore, crossbred females are superior to straightbreds for reproductive performance due to advantages received from heterosis. Heterosis in the crossbred female is termed maternal heterosis, and is a primary advantage for using crossbreeding programs. Crossbred ewes exhibit significant advantages in fertility, prolificacy, and lamb survival compared to straightbred ewes. Traits that are moderate in their heritabilities (20–30%), such as growth rate, are also moderate in the degree of heterosis expressed. Highly heritable traits (30–50%), such as carcass traits, exhibit little heterosis.

Crossbreeding also allows the producer to take advantage of the strengths of two or more breeds to produce offspring that have acceptable levels of performance in several traits. As an example, blackface breeds such as Suffolk and Hampshire generally excel in growth rate and carcass merit, whereas whiteface breeds such as Dorset and Polypay typically have superior maternal characteristics. Combining the breed types results in offspring that have desirable growth and maternal characteristics. It is important to realize that the crossbred offspring will not be superior to both of the parent breeds for all traits.

Crossbreeding Systems

Selection of a crossbreeding system depends on several factors, including:

- the number of ewes in the flock,
- the number of available breeding pastures,
- labor and management,
- the amount and quality of feed available, and
- the production and marketing system.

Some common crossbreeding systems are discussed below.

Rotational Cross

The two-breed rotational cross is a simple and very popular form of crossbreeding. In this system, two breeds are mated and the resulting female offspring are kept as replacements and mated to one of the breeds. In following generations, females are bred to the opposite breed of their sire. For example, if Dorset and Suffolk were crossed to make 1/2 Dorset x 1/2 Suffolk females who were then bred to Dorset, the resulting lambs would be 3/4 Dorset x 1/4 Suffolk. These females would then be mated to Suffolk rams. For their entire lives, females would

be mated to the ram breed opposite their sire. This system would require a minimum of two breeding pastures, one for each breed of sire, and ewes need to be identified by breed of sire. An advantage to this system is the use of the crossbred ewe, with pounds of lamb marketed per ewe increased approximately 34% compared to a straightbred system. Over several generations, 67% of the maximum amount of heterosis is realized. Additionally, replacements may be selected from a large number of ewe lambs.

If three breeds are used in the system instead of two, pounds of lamb marketed per ewe increases approximately 43% compared to a purebred system and average heterosis over several generations attains 87% of maximum. However, three breeding pastures are necessary, and significantly more management is required with the three-breed versus two-breed rotational cross.

Terminal Sire Systems

The addition of another breed as a terminal sire to a two- or three-breed rotational cross system further enhances the production system. In this rota-terminal system, approximately 50% of the ewe flock is mated to the terminal sire breed (a different breed than used in the two-breed rotation), with the resulting offspring all marketed (no replacement females retained in the flock). The other 50% of the flock then operates as a two-breed rotation as outlined above. The two-breed rotation functions to produce all replacement females for the flock. Terminal sire breeds should be selected for growth rate and carcass merit. Older and poorer producing ewes are the best candidates for mating to the terminal sire. Younger ewes should be genetically superior due to selection and should be used to produce the replacement females. The rota-terminal system has been shown to increase pounds of lamb marketed per ewe by up to 50% compared to a purebred system. Maximum heterosis is realized in the lambs sired by the terminal breed, and all females are crossbred. The rota-terminal system requires more management in that a minimum of three breeding pastures are required. Additionally, less selection may be practiced on potential replacements, as a larger percentage of the eligible ewe lambs must be retained to maintain flock size. A viable option with the rota-terminal system is to purchase all crossbred replacement ewe lambs. This option would significantly reduce the degree of management required with the rota-terminal system, as all ewes would be mated to the same breed of sire.

Breeds

Large differences exist among breeds for several economically important traits. Breed classifications for mature size, growth, and prolificacy are presented in table 5-7. For commercial flocks, it is unlikely that any one breed can meet production goals as effectively as a combination of

breeds used in a planned mating system. Several criteria must be considered when making breed selection decisions:

- production system,
- market demands,
- quantity and quality of feedstuffs available,
- climate/environment,
- breed complementarity, and
- cost and availability of seed stock.

Breeds must be selected that contribute positively to the overall production system. Traits important for ewe breeds in crossbreeding programs include early puberty, moderate mature size, high fertility, optimum milking ability (appropriate for feed resources), longevity, and acceptable growth characteristics. Traits important in selecting a ram breed for use in crossbreeding programs include high growth rate with

Table 5-7. Classification of sheep breeds.			
Breed	**Mature size**	**Growth rate**	**Prolificacy**
Barbados Blackbelly	Small	Low	Moderate
Columbia	Large	High	Moderate
Dorset	Moderate	Moderate	Moderate
Finnsheep	Small	Low	High
Hampshire	Large	High	Moderate
Katahdin	Moderate	Moderate	Moderate
Polypay	Moderate	Moderate	High
Rambouillet	Large	High	Moderate
St. Croix	Small	Low	Moderate
Southdown	Small	Low	Moderate
Suffolk	Large	High	Moderate
Source: Sheep Production Handbook, *vol. 7. 2002. American Sheep Industry Association, p. 23.*			

acceptable mature size, lamb survivability, and carcass merit. With the proper use of a cross-breeding system, compromises between ewe and ram breed can be avoided. The popularity of Dorset x Suffolk flocks in the region is an example. Highly prolific breeds such as Finnsheep contribute most efficiently when used as 1/8 to 1/4 crossbred ewes. Although higher percentages of prolific breeds can be advantageous in some intensive systems, typically forage-based production and management systems favor limiting these breeds' genetics to 1/2 or less. Production systems that use fall lambing will use genetics that breed out of season. Breeds noted for this ability include Dorset, Polypay, Rambouillet, Finnsheep, hair breeds (e.g., Katahdin, St. Croix, Blackbelly), and crosses of these breeds. Considerable variation exists within these breeds for fall lambing potential, and selection for this trait needs to be a priority for operations that use an extended breeding season.

Ram Selection

From a genetic standpoint, ram selection is the most important decision a sheep producer makes. The vast majority of genetic improvement in the flock is the direct result of ram selection. For flocks with small numbers of ewes, the importance of an individual ram is even further exaggerated, because one ram alone accounts for a large proportion of the genetics represented in each lamb crop. Relative to other production and management decisions, ram selection is an infrequent occurrence. However, these decisions have long-term impact relative to the productivity and profitability of the sheep enterprise.

The first step in ram selection includes thoughtful determination of the role of the ram in contributing to the existing flock genetics. The breeding system used, marketing system, management level, and feed/environmental resources are important considerations for determining this role. For example, traits of importance in rams will vary greatly if the ram will be used to sire replacement females versus a ram that will be used strictly as a terminal sire. The following criteria should be considered:

- Performance record: Ideally, ram selection would include evaluation of a complete performance record on potential rams. This performance record would include adjusted records (or expected progeny differences generated through the National Sheep Improvement Program) for birth type, weights, fleece attributes, carcass merit, and dam lifetime production. Unfortunately, many times these records are not widely available. Although the heritability of condition of birth is low (single vs. twin vs. triplet), lambing percentage can be increased by selecting for multiple births over time. Of particular importance is the lifetime production of the dam, including number of lambs born per lambing and total weaning weight. Growth traits are typically expressed as weights measured at weaning (60–90 days), at 120 days, and at a year of age. Weaning weights are a function of both growth genetics of the lamb and milk production of the dam, whereas postweaning weights are primarily a function of differences in individual growth genetics. Selection for growth must be in concert with selection for appropriate mature size.

- Conformation/soundness: Visual appraisal is generally a poor method of selection for the traits just discussed. However, conformation as it relates to soundness is critically important to the function of the ram. Rams that stand and travel squarely and freely on their feet and legs are most desirable. Mouth soundness is particularly imperative, and rams exhibiting parrot mouth or monkey jaw conditions should be avoided. In most cases, muscling is assessed by visual appraisal, as is body capacity (depth of rib, spring of rib). A breeding soundness exam that includes semen evaluation should be performed.

- Source: A variety of sources are available to purchase rams. Seed stock suppliers who are able to furnish extensive performance records offer the best opportunity to make informed selection decisions. Select breeding stock from flocks with compatible goals and selection strategies relative to the intended role of the ram to be purchased. Ram testing stations exist in the region, and allow for the comparison of rams from different flocks in addition to providing performance information.

Ewe Selection

In most breeding systems, replacement ewe lambs will be generated from within the flock. Therefore, attention to maternal traits in the rams siring potential replacements is critical. The following are important considerations for selection from the pool of potential replacements:

- Performance record: Ewe lambs should be retained from highly productive dams. Identifying these dams through a record-keeping system is therefore the first step in identifying potential replacements. Dams that lamb early in the lambing season, produce multiple births, and excel in pounds of lamb weaned (reflective of milking ability) are the best candidates to produce replacements. In the absence of such records, identifying maternal potential in ewe lambs based solely on visual appraisal is difficult.

- Age: Preference should be given to ewe lambs born early in the lambing season (first 50 days). These ewe lambs are more likely to reach puberty earlier and breed and lamb early as yearlings, thus keeping

the subsequent lambing season short. Older ewe lambs are also more likely to reach target body weight by their first breeding season than young ewe lambs, and this coupled with age enhances their ability to breed as ewe lambs.

- Conformation/soundness: As previously discussed for rams, structural soundness and mouth soundness are also critical in ewe lambs. Additionally, ewes with adequate body capacity and muscling are preferred. Appropriate frame size is important as it relates to mature size. As mature size increases, so do nutritional requirements and thus carrying costs.

Production Records

Production records are important not only for selection, but also as a management tool. Basic performance records start with individual animal identification at birth. Simple records would include birth date, type of birth, and type of rearing. In many instances, individual lambs could be identified as to their dam as well as sire (or perhaps breed of sire in multiple sire breeding groups). These basic records can be very useful to the shepherd in terms of monitoring overall prolificacy of the flock, breed types and crosses within the flock, and individual reproductive performance of ewes. Additionally, the ability to identify an individual ewe and her lambs is an excellent management tool during lactation. More extensive performance records, including individual birth and weaning weights of lambs as well as postweaning growth measures, would also be advantageous to commercial flocks. Addition of these records allows for calculation of ewe productivity (total pounds of lamb weaned) and provides the opportunity for more accurate selection for growth traits. To be used properly in selection, all records need to be adjusted to a common basis. Growth measures such as weaning weight need to be adjusted for sex, type of birth/rearing, lamb age, and age of dam. These adjustment factors are readily attainable from several sources and are rather simple to apply.

Finally, collection of performance records enables the shepherd to monitor the rate of progress in the flock. By doing so, proper emphasis can be placed on individual traits with selection, and areas can be identified that may be responsive to management changes.

PREDATOR CONTROL

Coyotes and domestic dogs are the two major predators of primary concern to grazing sheep. Effective predator control starts with a solid perimeter fence. High tensile electrified fence is most effective in deterring potential predators. The fence must be electrified close to the ground, no more than 6–8 inches above the ground for the entire length of the fence, because predators are most likely to crawl under the fence. Gaps or holes underneath the fence are areas where predators may enter under the electrified wire. These areas must be filled in so that the bottom electrified wire will be effective. Temporary electric fence used on the interior of pastures will not be effective in warding off predators. Woven wire fence that is free of holes will help deter predators. One strand of electric wire placed at the bottom of the woven wire fence assists with predator control.

Even with solid perimeter fencing, predators are still a concern in many areas. Many producers use guard animals for additional control. The most common are guard dogs, llamas, and donkeys. These guard animals must be raised and cared for with the sheep and treated as working animals (not as pets). Guard dogs seem to be most effective against coyotes and dogs, although many producers have also had success with llamas and donkeys. Sheep grazing with

cow-calf pairs seem to be less prone to predator attack.

Penning the sheep in a barn or shelter at night is the surest way to avoid losses. Most sheep are lost in the dawn and dusk to coyotes, and dog strikes may occur at any time. Confinement of the sheep is not always feasible due to facility restrictions and increased labor requirement. Confinement may also hinder grazing performance during the hot summer months when sheep prefer to graze during the cool part of the day.

Other methods of predator control include repellents and various frightening devices. The control methods mentioned above, in combination with an understanding of predator behavior and tendencies, are key to an integrated predator control program.

PRODUCTION SYSTEMS

For any livestock enterprise, a key to profitability is applying production and management practices to a targeted market for the end product. For the sheep enterprise, this relates to the production of a lamb of a certain type/quality (grade) and weight, to be marketed at a particular time of year. Based on historical prices, this generally relates to a 70–120 pound lamb, sold sometime after January 1 and prior to early July. This large window allows for a variety of potentially feasible production systems, depending on the resources of the individual operation. Of course, high-value markets at other times of the year may be available to producers in specific locations and to specific buyers.

The one facet of the sheep industry that does seem to have some predictability is the seasonal pattern of the mainstream lamb market. Figure 5-1 depicts seasonal trends in Virginia's lamb market based on average prices in weekly mar-

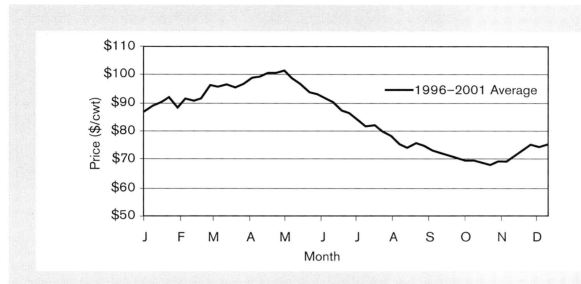

Figure 5-1. Virginia lamb prices ($/cwt.), choice and prime, YG (yield grade) 1-3, 95–125 lb.

kets in the Shenandoah Valley region. Though there may be single year variations and variation associated with special sales (particularly those targeting ethnic holidays), the highest prices paid for lambs in weekly or monthly scheduled sales have tended to be in April and May. Predictably, the lowest prices of the year have been paid during the September–November period as many lambs are marketed directly off grass when forage availability declines.

Knowledgeable producers must factor this typical seasonal price pattern into their individual production and marketing programs. This can be successfully accomplished through several systems, depending on the goals and resources of the operation. For comparison, the following production systems will be examined:

- Early winter lambing (December–February)
- Late winter lambing (February–March)
- Spring lambing (April–May)
- Fall lambing (September–November and December–January)

A number of variables, discussed below for each of the four systems, warrant consideration relative to their feasibility for an individual operation.

- Facility requirements
- Labor/management resources
- Genetics
- Forage/pasture resources
- Parasite control program
- Predator control program
- Economic returns

A sample budget for each system is provided in tables 5-8 to 5-12 (pp. 142–146). Budgets calculate returns to owner labor, management, and equity, meaning that there is no charge for labor in the budget. Within each system, it was assumed that replacement females would be retained (not purchased) and ewe lambs that did not breed were sold as market lambs. Distribution of sale weights for the lamb crop assumes that all lambs are sold on the same day. Marketing and transportation costs are included in each budget.

Table 5-8. Enterprise budget: Early winter lambing system.

100 ewes
15 percent culling rate

140	% lamb crop			100	% of lambs enter feedlot
10	% death loss			60	lbs. avg. weaning wt.
1.26	lambs raised/ewe			5.0	to 1 postwean feed conv.

Item				Unit	Price	Quantity	Total
Cash income							
Lambs	4	@	1.00	cwt[a]	$99.00	4.00	$396.00
Lambs	44	@	0.90	cwt	$102.00	39.60	$4,039.20
Lambs	50	@	0.80	cwt	$104.00	40.00	$4,160.00
Lambs	13	@	0.65	cwt	$99.00	8.45	$836.55
Cull ewes	12.0	@	1.50	cwt	$30.00	18.00	$540.00
Cull ram	0.6	@	2.00	cwt	$30.00	1.20	$36.00
Wool	6.5	#/hd.		lbs	$0.20	669.50	$133.90
					Total cash income		$10,141.65
Cash expenses							
			Feed waste				
Mixed hay			10.0%	ton	$80.00	35.21	$2,816.41
Alfalfa hay			10.0%	ton	$110.00	0.00	$0.00
Shelled corn			2.0%	bu	$2.75	626.10	$1,721.77
Soybean meal			0.0%	ton	$225.00	1.12	$251.83
Pelleted prot. supp.			0.0%	ton	$395.00	0.79	$313.83
Corn silage			5.0%	ton	$30.00	0.00	$0.00
Limestone			0.0%	ton	$60.00	0.03	$1.77
Di cal[b]			0.0%	ton	$320.00	0.00	$0.00
Feed processing				cwt	$0.55	389.48	$214.21
Salt & mineral				cwt	$21.50	19.28	$414.51
Vet & med				ewe	$3.85	100.00	$385.27
Supplies				ewe	$2.00	100.00	$200.00
Pasture	5.0	ewe/ac		acre	$18.00	20.00	$360.00
Replacement ram				head	$350.00	0.60	$210.00
Shearing				head	$2.50	103.00	$257.50
Taxes				$	--	--	$0.00
Haul sheep				head	$1.55	123.60	$191.58
Market sheep				head	$1.85	123.60	$228.66
Virginia checkoff				head	$0.50	48.00	$24.00
Bedding	80	lb/ewe		ton	$80.00	4.00	$320.00
Bldg. & fence repair				--	--	--	$200.00
Utilities				ewe	$0.90	100.00	$90.00
Machinery, non-crop				ewe	$1.78	100.00	$178.00
					Total cash expenses		$8,379.34
Annual debt payments							$0.00
Return to equity, management, & oper. labor							$1,762.31

[a] cwt-hundred weight.
[b] Di cal-dicalcium phosphate.

Table 5-9. Enterprise budget: Late winter lambing system.

100 ewes
15 percent culling rate

150	% lamb crop		100	% of lambs enter feedlot	
10	% death loss		60	lbs. avg. weaning wt.	
1.35	lambs raised/ewe		5.0	to 1 postwean feed conv.	

Item				Unit	Price	Quantity	Total
Cash income							
Lambs	66	@	1.00	cwt	$87.00	66.00	$5,742.00
Lambs	27	@	0.90	cwt	$85.00	24.30	$2,065.50
Lambs	27	@	0.80	cwt	$82.00	21.60	$1,771.20
Lambs	0	@	0.65	cwt	$75.00	0.00	$0.00
Cull ewes	12.0	@	1.50	cwt	$30.00	18.00	$540.00
Cull ram	0.6	@	2.00	cwt	$30.00	1.20	$36.00
Wool	6.5	#/hd.		lbs	$0.20	669.50	$133.90
					Total cash income		$10,288.60
Cash expenses				Feed waste			
Mixed hay			10.0%	ton	$80.00	36.40	$2,912.36
Alfalfa hay			10.0%	ton	$110.00	0.00	$0.00
Shelled corn			2.0%	bu	$2.75	492.50	$1,354.37
Soybean meal			0.0%	ton	$225.00	1.03	$230.88
Pelleted prot. supp.			0.0%	ton	$395.00	0.46	$183.65
Corn silage			5.0%	ton	$30.00	0.00	$0.00
Limestone			0.0%	ton	$60.00	0.03	$1.89
Di cal			0.0%	ton	$320.00	0.00	$0.00
Feed processing				cwt	$0.55	306.25	$168.44
Salt & mineral				cwt	$21.50	19.40	$417.05
Vet & med				ewe	$7.26	100.00	$726.04
Supplies				ewe	$2.00	100.00	$200.00
Pasture	4.0	ewe/ac		acre	$18.00	25.00	$450.00
Replacement ram				head	$350.00	0.60	$210.00
Shearing				head	$2.50	103.00	$257.50
Taxes				$	--	--	$0.00
Haul sheep				head	$1.55	132.60	$205.53
Market sheep				head	$1.85	132.60	$245.31
Virginia checkoff				head	$0.50	93.00	$46.50
Bedding	80	lb/ewe		ton	$80.00	4.00	$320.00
Bldg. & fence repair				--	--	--	$200.00
Utilities				ewe	$0.90	100.00	$90.00
Machinery, non-crop				ewe	$1.78	100.00	$178.00
					Total cash expenses		$8,397.54
Annual debt payments							$0.00
Return to equity, management, & oper. labor							$1,891.06

Table 5-10. Enterprise budget: Spring lambing system.

100 ewes
15 percent culling rate

160	% lamb crop			100	% of lambs enter feedlot
10	% death loss			80	lbs. avg. weaning wt.
1.36	lambs raised/ewe			6.0	to 1 postwean feed conv.

Item				Unit	Price	Quantity	Total
Cash income							
Lambs	7	@	1.30	cwt	$82.00	9.10	$746.20
Lambs	101	@	1.15	cwt	$85.00	116.15	$9,872.75
Lambs	14	@	0.90	cwt	$90.00	12.60	$1,134.00
Lambs	0	@	0.65	cwt	$100.00	0.00	$0.00
Cull ewes	12.0	@	1.50	cwt	$30.00	18.00	$540.00
Cull ram	0.6	@	2.00	cwt	$30.00	1.20	$36.00
Wool	6.5	#/hd.		lbs	$0.20	669.50	$133.90
					Total cash income		$12,462.85
Cash expenses			Feed waste				
Mixed hay			10.0%	ton	$80.00	32.82	$2,625.66
Alfalfa hay			10.0%	ton	$110.00	0.00	$0.00
Shelled corn			2.0%	bu	$2.75	565.15	$1,554.17
Soybean meal			0.0%	ton	$225.00	0.04	$9.70
Pelleted prot. supp.			0.0%	ton	$395.00	1.37	$542.97
Corn silage			5.0%	ton	$30.00	0.00	$0.00
Limestone			0.0%	ton	$60.00	0.00	$0.00
Di cal			0.0%	ton	$320.00	0.00	$0.00
Feed processing				cwt	$0.55	344.84	$189.66
Salt & mineral				cwt	$21.50	20.98	$450.98
Vet & med				ewe	$8.58	100.00	$857.68
Supplies				ewe	$2.00	100.00	$200.00
Pasture	3.0	ewe/ac		acre	$18.00	33.33	$600.00
Replacement ram				head	$350.00	0.60	$210.00
Shearing				head	$2.50	103.00	$257.50
Taxes				$	--	--	$0.00
Haul sheep				head	$1.55	134.60	$208.63
Market sheep				head	$1.85	134.60	$249.01
Virginia checkoff				head	$0.50	108.00	$54.00
Bedding	0	lb/ewe		ton	$80.00	0.00	$0.00
Bldg. & fence repair				--	--	--	$200.00
Utilities				ewe	$0.90	100.00	$90.00
Machinery, non-crop				ewe	$1.78	100.00	$178.00
					Total cash expenses		$8,477.97
Annual debt payments							$0.00
Return to equity, management, & oper. labor							$3,984.88

Table 5-11. Enterprise budget: Fall lambing system. (September–November).

50 ewes
15 percent culling rate

135	% lamb crop			100	% of lambs enter feedlot
10	% death loss			60	lbs. avg. weaning wt.
1.22	lambs raised/ewe			5.0	to 1 postwean feed conv.

Item				Unit	Price	Quantity	Total
Cash income							
Lambs	23	@	1.10	cwt	$100.00	25.30	$2,530.00
Lambs	21	@	0.90	cwt	$103.00	18.90	$1,946.70
Lambs	9	@	0.80	cwt	$105.00	7.20	$756.00
Lambs	0	@	0.65	cwt	$100.00	0.00	$0.00
Cull ewes	6.0	@	1.50	cwt	$30.00	9.00	$270.00
Cull ram	0.3	@	2.00	cwt	$30.00	0.60	$18.00
Wool	6.5	#/hd.		lbs	$0.20	334.75	$66.95
					Total cash income		$5,587.65
Cash expenses				Feed waste			
Mixed hay			10.0%	ton	$80.00	15.00	$1,200.33
Alfalfa hay			10.0%	ton	$110.00	0.00	$0.00
Shelled corn			2.0%	bu	$2.75	382.43	$1,051.69
Soybean meal			0.0%	ton	$225.00	0.62	$139.25
Pelleted prot. supp.			0.0%	ton	$395.00	0.59	$232.63
Corn silage			5.0%	ton	$30.00	0.00	$0.00
Limestone			0.0%	ton	$60.00	0.01	$0.85
Di cal			0.0%	ton	$320.00	0.00	$0.00
Feed processing				cwt	$0.55	238.60	$131.23
Salt & mineral				cwt	$21.50	9.61	$206.62
Vet & med				ewe	$3.82	50.00	$190.83
Supplies				ewe	$2.00	50.00	$100.00
Pasture	5.0	ewe/ac		acre	$18.00	10.00	$180.00
Replacement ram				head	$350.00	0.30	$105.00
Shearing				head	$2.50	51.50	$128.75
Taxes				$	––	––	$0.00
Haul sheep				head	$1.55	59.30	$91.92
Market sheep				head	$1.85	59.30	$109.71
Virginia checkoff				head	$0.50	44.00	$22.00
Bedding	0	lb/ewe		ton	$80.00	0.00	$0.00
Bldg. & fence repair				––	––	––	$100.00
Utilities				ewe	$0.90	50.00	$45.00
Machinery, non-crop				ewe	$1.78	50.00	$89.00
					Total cash expenses		$4,124.81
Annual debt payments							$0.00
Return to equity, management, & oper. labor							$1,462.84

Table 5-12. Enterprise budget: Fall lambing system (December–January).

50 ewes
15 percent culling rate

135	% lamb crop				100	% of lambs enter feedlot	
10	% death loss				60	lbs. avg. weaning wt.	
1.22	lambs raised/ewe				5.0	to 1 postwean feed conv.	

Item				Unit	Price	Quantity	Total
Cash income							
Lambs	2	@	1.00	cwt	$99.00	2.00	$198.00
Lambs	21	@	0.90	cwt	$102.00	18.90	$1,927.80
Lambs	21	@	0.80	cwt	$104.00	16.80	$1,747.20
Lambs	9	@	0.65	cwt	$100.00	5.85	$585.00
Cull ewes	6.0	@	1.50	cwt	$30.00	9.00	$270.00
Cull ram	0.3	@	2.00	cwt	$30.00	0.60	$18.00
Wool	6.5	#/hd.		lbs	$0.20	334.75	$66.95
					Total cash income		$4,812.95
Cash expenses				Feed waste			
Mixed hay			10.0%	ton ·	$80.00	14.31	$1,144.99
Alfalfa hay			10.0%	ton	$110.00	0.00	$0.00
Shelled corn			2.0%	bu	$2.75	306.33	$842.41
Soybean meal			0.0%	ton	$225.00	0.55	$124.18
Pelleted prot. supp.			0.0%	ton	$395.00	0.37	$146.81
Corn silage			5.0%	ton	$30.00	0.00	$0.00
Limestone			0.0%	ton	$60.00	0.01	$0.85
Di cal			0.0%	ton	$320.00	0.00	$0.00
Feed processing				cwt	$0.55	190.30	$104.67
Salt & mineral				cwt	$21.50	9.61	$206.62
Vet & med				ewe	$6.06	50.00	$302.78
Supplies				ewe	$2.00	50.00	$100.00
Pasture	5.0	ewe/ac		acre	$18.00	10.00	$180.00
Replacement ram				head	$350.00	0.30	$105.00
Shearing				head	$2.50	51.50	$128.75
Taxes				$	--	--	$0.00
Haul sheep				head	$1.55	59.30	$91.92
Market sheep				head	$1.85	59.30	$109.71
Virginia checkoff				head	$0.50	23.00	$11.50
Bedding	80	lb/ewe		ton	$80.00	2.00	$160.00
Bldg. & fence repair				--	--	--	$100.00
Utilities				ewe	$0.90	50.00	$45.00
Machinery, non-crop				ewe	$1.78	50.00	$89.00
					Total cash expenses		$3,994.18
Annual debt payments							$0.00
Return to equity, management, & oper. labor							$818.77

Early Winter Lambing System

The lambing season in this system occurs from December 15 through February 15, with an average lamb birth date of January 15. This is an intensive management system, with lambs creep fed and weaned at 60–70 days of age. Lambs are fed in a drylot on high-energy ration for maximum weight gain. The system works toward a targeted marketing date of April–May for the lamb crop to capture seasonal highs in the lamb market.

Facility requirements—Facility requirements are the highest of the four systems. An indoor lambing facility is required in most areas, and shelter is required for lactating ewes and lambs. A feeding facility is advantageous to minimize environmental challenges (mud, wind, cold/wet) to optimum weight gain postweaning.

Labor/management resources—Early winter lambing is generally a high labor system. Labor resources devoted to lambing time are critical to optimize the percent lamb crop. Expertise is needed in lambing management as well as lamb feeding.

Genetics—Breed types that optimize lamb crop percentage as well as growth are most desirable. The ewe flock likely will require a relatively high percentage of genetics with early lambing potential (Dorset cross, low percentage Finn cross). High percentage blackface ewes are likely not as desirable due to early lambing season.

Forage/pasture resources—Pasture resources are devoted entirely to the ewe flock and developing replacement ewe lambs. There are more ewes per acre than in other systems because the lambs are fed a grain-based diet rather than forage-based diet.

Parasite control program—Because parasites are generally more readily controlled in mature sheep versus lambs, a parasite control program for this system may be the easiest to design and implement relative to other systems.

Predator control program—The lamb crop will not be grazed, and therefore primarily mature ewes will be exposed to predators. Predator control will focus on ewe lamb replacements that will be developed on grass.

Economic returns—This system tends to net some of the lowest projected returns. The system will tend to gross the smallest income from lamb sales even though they are sold on the year's highest market. The percentage lamb crop tends to be less and the average lamb sale weight the lightest because the lambs are very young. Non-pasture feed costs tend to be the highest of the four lambing systems analyzed.

Late Winter Lambing System

The lambing season occurs from February 15 through March 15, with an average lambing date of March 1. Ewes and lambs are moved to pasture to use early spring growth. Lambs are provided grain on grass to optimize weight gains. The targeted marketing date for the lamb crop is June and early July, prior to the precipitous drop in seasonal prices that occurs in summer.

Facility requirements—Lambing facility requirements are similar to those needed in the early winter system, although a totally enclosed, heated facility is not necessary in most areas. Drylot capacity for lactating ewes or lambs is not required as sheep are moved to spring pasture as available.

Labor/management resources—As with the early winter system, labor must be devoted to lambing management to maximize the lamb crop. Use of a forage-based system for lamb development shifts management focus to parasite and predator control, as well as pasture/forage management.

Genetics—The late winter system is well suited to a variety of breed crosses. Most breeds are capable of reasonable reproductive performance with February–March lambing.

Forage/pasture resources—A large portion of lamb gains is derived from spring and early summer forages, emphasizing the importance of management for high-quality forage availability early in the grazing season. This system coincides with enhanced forage quality and quantity available from April to June.

Parasite control program—A strategic deworming program must be implemented, because lamb performance on forage is critical to the system. The lamb crop is marketed prior to heavy parasite infestation associated with mid-summer.

Predator control program—Predator concerns are high because lambs will be grazing in spring and early summer, which is high risk for coyote predation.

Economic returns—The net returns for this system tend to be at the lower end of the range projected. This system has the lowest nonpasture feed costs of the systems compared. The income from lamb sales is hampered by the relatively light weights at which the lambs are marketed in an effort to avoid the sharp drop in lamb prices during the summer. The system is susceptible to unpredictable sharp lamb price drops during June and July.

Spring Lambing System

The lambing season with spring lambing extends from April to May. Lambs graze with ewes through spring, summer, and into fall. Lambs are developed on a forage-based system, with minimal grain until winter. The system is geared to a targeted lamb marketing date of January through March to sell heavier lambs on a market that historically rises to more favorable prices after January 1 (avoid historically low prices in the fall).

Facility requirements—Minimal facilities are required for spring lambing. It is advisable to have a facility suited for lambing for optimum management of ewes and their lambs during the first week postlambing. Ewes and lambs are then moved directly to grass. Several paddocks and/or lots are necessary after weaning (in September), so that the ewe flock can be managed separately from market lambs.

Labor/management resources—Breeding, feeding, and lambing practices are less intensive than in other systems. The lambing season should be shortest of any system due to high fertility of ewes at breeding. Management focuses on parasite and predator control are critical to viability of the system.

Genetics—This is the most favorable system for percent lamb crop born due to high fertility. The seasonal nature of sheep reproduction favors spring lambing. Genotypes that are well adapted to a forage-based system are most desirable.

Forage/pasture resources—Spring lambing is a forage-based system, so pasture management requirements, as well as acreage per ewe, are greatest. The system is designed to take advantage of spring, summer, and fall forages. Practices such as grazing aftermath hay fields, stockpiling, and using fall/winter annuals are well suited to this system.

Parasite control program—A strategic deworming program is critical. Ewes and lambs graze during peak parasite infestation season. Lamb health and performance depend on controlling parasites. Higher veterinary charges reflect increased parasite control costs.

Predator control program—A predator control program is critical to maintain a high percentage of lamb crop marketed. Lambs are exposed to predation for a prolonged period of time.

Economic returns—This system projects to generate the highest net returns to labor and equity, primarily as a result of high income from lamb sales. The lamb crop is managed to produce heavy lambs that are marketed after the first of the year with higher prices. The system has relatively low feed costs with a major portion of those costs dedicated to finishing the lambs.

Fall Lambing System

The first fall lambing season runs from September 15 to November 15. Fall-born lambs are reared on a forage-based system with ewes until weaning at 60–70 days. Lambs are then developed in a drylot for marketing in April to capture seasonally high prices. The proportion of the flock fall lambing is likely to range from 50 to 70%. Therefore, a second lambing season from December through January is necessary to have all ewes in production during the year (ewes that did not conceive for fall lambing as well as first-time lambers). Lambs born from December to January would be managed in the same fashion as described for the early winter system. The second lambing season must be in early winter so that ewes can be weaned and exposed to rams in spring to lamb the following fall.

Facility requirements—For fall lambing covered facilities are not required in most areas; this system is similar to spring lambing from a facility standpoint. However, with the second lambing season, covered lambing facilities are a requirement in most areas.

Labor/management resources—Labor requirements are highest of all systems compared as a result of two lambing seasons per year. Additional management is required at breeding to

maximize fall lambing (via ram effect, teaser rams, estrous synchronization). The system essentially includes two flocks that need to be managed separately.

Genetics—Use of breeds that have out-of-season breeding potential is necessary.

Forage/pasture resources—Systems and management techniques that provide an abundance of high quality forage in the fall and early winter when ewes and fall-born lambs can be grazed are most advantageous. This system is well suited to stockpiled tall fescue use. An abundance of early spring forage is likely underused, as all ewes will by dry at this time. The fall lambing system permits a similar number of ewes per acre as the early winter system because lambs are not grazed extensively.

Parasite control program—The parasite control program is similar to that for the early winter system, because lambs will not be grazed during peak parasite infestation months. Parasites have typically gone into an arrested stage by October, which is when young lambs would be at risk.

Predator control program—The largest concern for predator control would be with young lambs in fall and early winter. Only mature ewes and developing replacements are grazed during spring and summer months.

Economic returns—This system projects to produce the second highest dollar returns of the systems analyzed. The budgets (tables 5-11 and 5-12, pp. 145 and 146) for this production system assume that 50% of the ewes lamb in the fall and 50% lamb in the early winter. The fall-born lambs are relatively heavy when they are marketed during the high price period of April, which helps generate positive returns to the system. The lambing rate for the winter lambing portion of the flock has been reduced slightly because all the ewe lambs will be first lambed in the winter. The combination of fall lambing and winter lambing produces some of the highest nonpasture feed costs.

The information provided in this section and in the budgets presented in tables 5-8 to 5-12 can serve as a guideline for decision making. Several factors affecting costs of production and market prices will be unique to specific areas and individual producers within that area. Spreadsheet templates of the budgets used in this chapter are available through Virginia Cooperative Extension (6). These may be used as a resource for construction of budgets specific to individual operations.

Two primary factors that influence profitability are market prices received for lambs, and the percent lamb crop marketed per ewe per year. Conception rate, lambing rate, lambing percentage (lambs born per ewe), and lamb survival rate are important variables affecting percent lamb crop marketed. These factors are favorably responsive to genetic selection and management. Prices received for lambs will be subject to yearly variations in the lamb market. Additionally, there can be large differences in prices received at various market outlets for similar lambs marketed during the same time frame (table 5-13). Management and marketing practices that positively influence market price received and percent lamb crop marketed clearly result in substantial improvements in economic returns.

Additional factors affecting returns with each enterprise relate to lamb income and feed costs. Lamb income is influenced not only by number of lambs sold and price received, but also by weight of the lambs. Lamb weight is a function of growth rate, and management must be applied to realize this growth potential. Within each system, enhanced growth performance

through genetics and improved management will improve total lamb income and therefore economic returns. Feed costs represent a large portion of total cash expenses, and will vary considerably between flocks. Strategies that reduce feed costs while still maintaining high levels of production are the most effective means of reducing cash expenses.

WOOL PRODUCTION AND MARKETING

In most farm flocks, income from wool represents a very small percentage of cash receipts (see tables 5-8 to 5-12, pp. 142 to 146). However, management and marketing strategies that maximize the value of the wool clip are warranted to offset the costs associated with the shearing. Wool value is determined by fiber diameter (grade), staple length (length of fleece), and yield (cleanliness and freedom from foreign material). Fiber diameter is associated with breed type; fine wool breeds such as Rambouillet have higher quality, more valuable fleeces than medium wool breeds such as Dorset, Suffolk, and Hampshire. Staple length is a function of both genetics and time since last shearing. Because more value is generally received with increased staple length, wool harvested from mature sheep with a year regrowth is generally more valuable than wool from young lambs. Yield or cleanliness of the fleece can be controlled through proper management, both at shearing time and throughout the production cycle. Avoiding the use of plastic baler twine is an example of a management practice that will enhance fleece quality, because contamination of the fleece from plastic twine that may be picked up through feeding round bales is a serious wool defect. Additionally, overhead hay feeders that allow chafe to fall in the fleece during feeding should be avoided. Most fleece contamination occurs from bedding and manure introduced

Table 5-13. Comparison of net returns per 100 ewes to operator labor, management, and equity with 10% variations in percent lamb crop and market lamb prices using the enterprise budgets presented previously.

	Lamb crop (%)	Market lamb price ($/cwt)		
Early winter lambing		$89.10	$99.00	$108.90
	126	$81	$886	$1,692
	140	$857	$1,762	$2,681
	154	$1,634	$2,659	$3,670
Late winter lambing		$78.30	$87.00	$95.70
	135	$124	$987	$1,850
	150	$918	$1,891	$2,865
	165	$1,712	$2,795	$3,879
Spring lambing		$76.50	$85.00	$93.50
	144	$1,841	$2,872	$3,904
	160	$2,733	$3,985	$5,058
	176	$3,626	$4,919	$6,213
Fall lambing		$90.00/$89.10[a]	$10.00/$99.00	$110.00/$108.90
	122/122[a]	$608	$1,450	$2,292
	135/135	$1,366	$2,282	$3,263
	148/148	$2,125	$3,180	$4,235

[a] Values represent lamb crop and prices for September–November-born lambs and December–January-born lambs, respectively.

at shearing time. The shearing area should be kept clean to prevent this, and caution should be exercised during wool packaging. Additionally, flocks with considerable variation in wool quality should separate wool based on potential value and package it accordingly (e.g., wool from white-faced sheep vs. black-faced sheep or mature ewes vs. lambs). Fleeces with an abundance of black fiber should be separated from other wools. Similarly, it is common practice to package belly wool seperately, and remove and

discard tags (wool severely contaminated with manure) at shearing time. Wool should be packaged in plastic or jute bags designed specifically for wool, and stored properly (kept dry, and off surfaces that may lead to moisture uptake) until marketed.

There are a variety of avenues to market wool. Wool pools are common in the eastern region. Wool pools serve primarily as marketing cooperatives, whereby wool from several producers is gathered at a central location and marketed. In some cases, the wool pools grade and sort wools from several producers and add value by assimilating large, uniform, properly packaged lots of wool that can be competitively marketed. Wools are frequently sold on a grade and yield basis (each fleece or lot is assigned a value based on its individual quality), or on a cash basis. Another option to market wool is direct to individuals, who may use the wool for spinning and crafts. Generally, large quantities of wool are difficult to market in this manner, although several modestly sized processors and mills exist in the region.

CHAPTER 6
Goat Nutrition and Management

Jean-Marie Luginbuhl and Edward B. Rayburn

NUTRIENT REQUIREMENTS

Introduction

Feeding may be the largest expense of any goat operation. Goats raised for meat or milk need high quality feed in most situations and require an optimum balance of nutrients to achieve maximum profit potential. Because of their unique physiology, goats do not fatten as cattle or sheep do. Because of their small size, their rates of weight gain or milk production are smaller, but they require feed as high in quality for similar magnitudes of production compared to body size. Body weight gains range from 0.1 to 0.8 pound per day, and milk production can exceed 16 pounds per day. Therefore, profitable goat production can be achieved only by optimizing the use of high quality forage and browse and the strategic use of energy and protein concentrate feeds. This can be achieved by developing a year-round forage program allowing for as much grazing as possible throughout the year.

Many people still believe that goats eat and do well on low quality feed. Attempting to manage and feed goats with such a belief will not lead to successful goat production, unless protein and energy supplements are provided.

Feeding Requirements

The goat is not able to digest the cell walls of plants as well as the cow because feed stays in its rumen for a shorter time. A distinction as to what is meant by "poor quality roughage" must be made to determine which animal can best utilize a particular forage. Trees and shrubs, which often represent poor quality roughage sources for cattle, because of their highly lignified stems and bitter taste, may be adequate to high in quality for goats. This is because goats selectively feed on the leaves, avoid eating the stems, don't mind the taste, have the ability to detoxify tannins, and benefit from the relatively high levels of protein and cell solubles found in the leaves of these plants. On the other hand, straw, which is of poor quality due to high cell wall and low protein content, can be used by cattle but will not provide even maintenance needs for goats because goats don't utilize the cell wall as efficiently as cattle.

In addition, goats must consume a higher quality diet than cattle because their digestive tract is smaller relative to their maintenance energy needs. Relative to their body weight, the amount of feed needed by goats is approximately twice that of cattle. When the density of high quality forage is low and the stocking rate is low, goats will still perform well because their grazing/browsing behavior allows them to select only the highest quality forage from that on offer.

Nutrients Required and Table of Nutritional Requirements

Goats require nutrients for body maintenance, growth, reproduction, pregnancy, and production of products such as meat, milk, and hair. The groups of nutrients that are essential in goat nutrition are water, energy, protein, minerals, and vitamins. The nutrient requirements of bucks, young goats, and does with a high production potential and at various stages of development and production are shown in table 6-1. Goats should be grouped according to their nutritional needs to more effectively match feed quality and supply to animal need. Weanlings, does during the last month of gestation, high-lactating does, and yearlings should be grouped and fed separately from dry does, bucks, etc., which have lower nutritional needs.

When pasture is available, animals having the highest nutritional requirements should have access to lush, leafy forage or high quality browse. In a barn feeding situation such as during the winter months, these same animals should be offered the highest quality hay available. Goats should be supplemented with a concentrate feed when either the forage that they are grazing or the hay that they are fed does not contain the necessary nutrients to cover their nutritional requirements. Total digestible nutrients (TDN) and protein requirements are shown in table 6-1. Comparing the nutrient requirements to the chemical composition of feeds shown in table 6-2 should give producers an idea of how to match needs with appropriate forages. Generally, low quality forages contain 40–55% TDN, good quality forages contain 55–70% TDN, and concentrate feeds contain 70–90% TDN.

Water

Water is the cheapest feed ingredient. Production, growth, and the general performance of the animal will be affected if insufficient water is available. Water needs vary with the stage of production, the climate, and the water content of forages. In some instances, when consuming lush and leafy forages, or when grazing forages soaked with rain water or a heavy dew, goats can get all the water they need out of the feed. However, water is almost always needed by some members of the herd, such as lactating does.

Table 6-1. Daily nutrient requirements for meat-producing goats.

| Nutrient | Young goats[a] | | Does (110 lb) | | | | Buck (80–120 lb) |
	Weanling (30 lb)	Yearling (60 lb)	Pregnant Early	Late	Lactating Avg. milk	High milk	
Dry matter (lb)	2.0	3.0	4.5	4.5	4.5	5.0	5.0
TDN (%)	68	65	60	60	60	65	60
Protein (%)	14	12	10	11	11	14	11
Calcium (%)	0.6	0.4	0.4	0.4	0.4	0.6	0.4
Phosphorus (%)	0.3	0.2	0.2	0.2	0.2	0.3	0.2

[a] Expected weight gain > 0.44 lb/day.

Sources: National Research Council. 1981. Nutrient Requirements of Goats: Angora, Dairy, and Meat Goats in Temperate and Tropical Countries. *Number 15. National Academy Press. Washington, D.C.; Pinkerton, F. 1989.* Feeding Programs for Angora Goats. *Bulletin 605. Langston University, OK.*

Table 6-2. Estimated nutrient composition of various feeds.

Plant type	TDN (%)	Crude protein (%)
Barley grain	84	14
Corn grain	86	9
Oat grain	77	13
Soybean meal	82	44
Soybean hulls, ground	75	14
Sunflower seeds	65	50
Wheat middlings	80	19
Whole cottonseed	88	22–24
Pasture, vegetative	60–76	12–24
Pasture, mature	50–60	8–10
Pasture, dead leaves	35–45	5–7
Alfalfa hay	50–63	13–20
Chicory	65	15
Cowpea leaves, vegetative	75–80	19
Soybean leaves, vegetative	68–77	18
Annual ryegrass, vegetative	72–84	17–27
Bermuda hay, 7 weeks' growth	54–58	9–11
Bermuda hay, 12 weeks' growth	47–50	7–9
Cereal rye, vegetative	73–84	18–27
Fescue hay, 6 weeks' growth	58–62	8–11
Fescue hay, 9 weeks' growth	48–53	7–9
Gamagrass	63	17
Pearl millet leaves, vegetative	73–81	17
Triticale, vegetative	74–83	18–28
Black locust, leaves	53–63	23–29
Honeysuckle, leaves and buds	70+	16+
Honeysuckle, mature	68+	12+
Green briar	16–17	73
Oak, buds and young leaves	64	18
Persimmon leaves	54	12
Hackberry, mature	40	14
Juniper leaves	64	7
Kudzu, early hay	55	14
Kudzu, leaves	65	18–25
Mimosa leaves	72	21
Mulberry leaves	72–75	17–26
Multiflora rose leaves	76	18–19
Privet leaves	16–20	72
Sumac, early vegetative	77	14
Curled dock	74	13
Pigweed leaves	75–82	15–23
Acorns, fresh	47	5

Sources: Adapted from National Research Council. 1981. Nutrient Requirements of Goats: Angora, Dairy, and Meat Goats in Temperate and Tropical Countries. *Number 15. National Academy Press. Washington, D.C. Table 2, pp. 26–48; Unpublished (North Carolina State University).*

Because it is difficult to predict water needs, goats should always have access to sufficient high quality water. Clear, flowing water from a stream is preferable to stagnant water; the latter may contain excessive levels of blue-green algae, which may be toxic. Nitrate in drinking water should also be of concern because it is becoming the predominant water problem for livestock. Safe levels in drinking water depend on the method of expression and are as follows (in parts per million): less than 100 for nitrate-nitrogen, or less than 443 for nitrate ion, or less than 607 for sodium nitrate.

Energy

Energy comes primarily from carbohydrates (sugars, starch, and fiber) and fats in the diet. Lush leafy forage and browse and tree leaves contain sufficient energy to cover the energy requirements of every goat other than high-producing dairy does (tables 6-1 and 6-2, pp. 154 and 155). Feed grains that are high in energy are whole cottonseed, corn, wheat middlings, soybean hulls, soybean meal, corn gluten feed, oats, barley, and sunflower seeds. Bacteria that are present in the goat's rumen ferment sugars, starches, fats, and fibrous carbohydrates into volatile fatty acids. These acids are absorbed from the rumen and used for energy. Fat is efficiently used for energy, but the amount that can be included in the diet is limited. Usually added fat should not represent more than 5% of a diet because it depresses ruminal fermentation. For example, if whole cottonseed (25% fat) is used as a supplement, it should not be more than 20% of the diet. Whole cottonseed also contains a good level of protein and phosphorus (P), and fed at 0.5–1.0 pound per day makes an excellent supplement to low quality forage. If the diet consumed by goats contains an excess of energy, that extra energy can be stored in the body as fat, mainly around certain internal organs.

Protein

Protein is usually the most expensive component of the goat diet. As with energy, lush leafy forage and browse and tree leaves contain sufficient protein to cover the nutrient requirements of low and moderately producing goats (tables 6-1 and 6-2). Feed grains that are high in protein are whole cottonseed, soybean meal, wheat middlings, corn gluten feed, and sunflower seeds. Protein is required both as a source of nitrogen (N) for the ruminal bacteria and to supply amino acids for protein synthesis in the animal's body. When the level of protein is low in the diet, digestion of carbohydrates in the rumen will slow and intake of feed will decrease. Inadequate levels of protein in the diet can negatively affect growth rate, milk production, reproduction, and disease resistance because insufficient amino acids are getting to the intestines to be absorbed by the body. Unlike energy, excess protein is not stored in the body of the goat; it is recycled into the rumen or it is excreted in the urine as urea. It is important for animals to have access to enough protein to cover their nutritional requirements. Protein nutritional requirements vary with developmental and physiological stages and level of production (table 6-1).

Minerals

Goats require many minerals for basic body function and optimum production. It is advisable in most situations to provide by free choice a complete goat mineral or a 50:50 mix of trace mineralized salt and either dicalcium phosphate or feed-grade limestone. Major minerals likely to be deficient in the diet are salt (sodium chloride), calcium (Ca), P, and magnesium (Mg). Trace minerals likely to be low in the diet are selenium (Se), copper (Cu), and zinc (Zn).

Forages are relatively high in Ca (legumes: more than 1.2%; grass: less than 0.5%). Calcium is low only if grain diets containing high levels of cereal grains or corn are fed to high-producing dairy does. Low quality, mature, or weathered forages will be deficient in P, especially for high- and average-lactating does. For example, bermudagrass hay harvested at 7–8 weeks' regrowth contains only 0.18% P. The ratio of Ca to P in the diet is important and should be kept about 2:1 to 4:1 (table 6-1).

Selenium is marginal to deficient in the soil of most areas of the eastern United States, and many commercial trace mineralized salts do not contain it. Trace mineralized salts that include Se should be provided to the goat herd at all times. If Se is absent, producers should encourage their local feed store to include it in commercial mixes or to order trace mineralized salts that contain Se.

Copper requirements for meat goats have not been definitively established. Recommendations for dairy goats range from 7 to 20 mg Cu/kg diet dry matter, the higher level being a precaution against interference from other minerals such as molybdenum, sulfur, and iron. Growing and adult goats are less susceptible to Cu toxicity than sheep, but their tolerance level is not well known. Young, nursing kids are generally more sensitive to Cu toxicity than mature goats, and cattle milk replacers should not be fed to nursing kids. Mineral mixes and sweet feed should contain copper carbonate or copper sulfate because these forms of Cu are better utilized by the goat than copper oxide. However, Cu levels in forages or supplements are of limited value in assessing adequacy unless concentrations of Cu antagonists such as molybdenum, sulfur, and iron are also considered.

Forages, especially low quality forages, often contain concentrations of Zn that are thought to be below recommended levels for ruminants. However, Zn requirements of goats have not been well defined, and little is known regarding factors that affect Zn availability in forages.

Vitamins

Goats need vitamins in small quantities. The vitamins most likely to be deficient in the diet are A and D. All B and K vitamins are formed sufficiently by bacteria in the rumen of the goat unless the goat is off feed or sick, and these two types of vitamins are normally not considered dietetically essential. Vitamin C is synthesized in the body tissues in adequate quantities to meet needs.

Vitamin A is not contained in forages, but carotene found in green, leafy forages is converted into vitamin A in the body. In addition, goats store vitamin A in the liver and in fat when intake exceeds requirements. Goats consuming weathered forages or forages that have undergone long-term hay storage should be fed a mineral mix containing vitamin A, or should receive vitamin A injections.

Vitamin D may become deficient in animals raised in confinement barns, especially during the winter. Animals should have frequent access to sunlight, because it causes vitamin D to be synthesized under their skin, or they should receive supplemental vitamin D. Good quality sun-cured hays are excellent sources of vitamin D. A deficiency in vitamin D results in poor Ca absorption, leading to rickets, a condition in which the bones and joints of young animals grow abnormally, especially in high-producing dairy goats.

Factors Influencing Goat Nutritional Requirements

Mature dry does, mature wethers, and bucks are examples of animals having maintenance requirements only. Additional requirements above those needed for body maintenance exist for growth, pregnancy, lactation, and hair production. As the productivity of meat goats is increased through selection and crossbreeding with goats having a higher production potential, such as the Boer goat for meat or selected Nubian bucks for high yields of high solids milk, nutritional requirements will also increase. Therefore, the more productive goats should be fed high quality feed, especially weaned kids being prepared for market, young replacement doelings, does approaching kidding, and does in early through peak lactation. Does nursing twins or triplets have greater nutritional requirements than does nursing a single kid.

Goats grazing very hilly pastures will have higher nutritional requirements than goats on level pastures of the same quality because they will expend more energy to gather feed.

In some situations where brush control in rough areas is the primary purpose of keeping goats, less productive animals or maintenance animals can be forced to consume lower quality feed. If their body condition deteriorates, these animals can then be grazed on better quality pastures or brushy areas. Once desirable body condition is achieved, the same animals can again be grazed to control brush.

BODY CONDITION SCORING

Introduction

As the breeding season approaches, producers should be concerned with the body condition of their breeding does. Goats should not be allowed to become too thin or too fat. Reproductive failure can result if does are under- or overconditioned at the time of breeding. Clinical symptoms of under- or overconditioned does include low twinning and low weaning rates, pregnancy toxemia, and dystocia.

Description of Body Condition Scoring

The term *body condition* refers to the fleshiness of an animal. We have devised a 9-point graduated scale, adapted from the beef system used in North Carolina. In this graduated scale, thin is 1–3, moderate is 4–6, and fat is 7–9 (table 6-3).

In most situations, goats should be in the range of 4–7. Scores of 1–3 indicate that goats are too thin, and scores of 8–9 are almost never seen in goats because excess fat is stored in the body cavity around the internal organs. The ideal body condition score (BCS) just before the breeding season is between a 5 and a 6 to maximize the number of kids born. Simply looking at a goat and assigning it a BCS can easily be misleading. Rather, animals should be physically handled. The easiest areas to feel and touch to determine the body condition of an animal (by running a hand over the areas and pressing down with a few fingers) are the ribs, on either side of the spine, the lumbar vertebrae, the shoulders, the pelvic bones, the hooks, the pins, the tailhead, the withers, and the thighs (figure 6-1, p. 160).

In doing so, one is able to determine the amount of fat covering the ribs. In general, does in good condition (BCS = 5 or 6) will have a fat thickness of not more than 0.05–0.08 inches over the loin and 0.03–0.05 inches over the backbone. In well conditioned goats, the backbone does not protrude and is flush with the loin. Does in good condition (BCS = 5 or 6) have a smooth look and the ribs are not very visible. The backbone and edges of the loins are felt with pressure, but they are smooth and round and feel spongy to the touch. Some to significant fat cover is felt over the eye muscle. Does in poor condition (BCS = 4 or lower) look angular, the ribs are visible, and the backbone and edges of the loins are sharp and easily felt. None to slight fat cover is felt over the eye muscle. High-producing dairy does may fall to this level during peak lactation. Ensure that such animals have adequate amount and quality forage, energy, supplements, and water to minimize weight loss and ensure

BCS	Description
	Table 6-3. Body condition scoring chart.
1	Extremely thin and weak, near death.
2	Extremely thin but not weak.
3	Very thin. All ribs visible. Spinous processes prominent and very sharp. No fat cover felt with some muscle wasting.
4	Slightly thin. Most ribs visible. Spinous processes sharp. Individual processes can be easily felt. Slight fat cover can be felt over the eye muscle.
5	Moderate. Spinous processes felt but are smooth. Some fat cover felt over eye muscle.
6	Good. Smooth look with ribs not very visible. Spinous processes smooth and round. Individual processes very smooth, felt with considerable pressure. Significant fat cover felt over eye muscle.
7	Fat. Ribs not visible, spinous process felt under firm pressure. Considerable fat felt over eye muscle.
8	Obese. Animal is very fat with spinous processes difficult to feel. Ribs cannot be felt. Animal has blocky obese appearance.
9	Extremely obese. Similar to an 8 but more exaggerated. Animal has deep patchy fat over entire body.

Source: Mueller, J. P., M. H. Poore, J.-M. Luginbuhl, and J. T. Green, Jr. 1995. Matching forages to the nutrient needs of meat goats.
HTTP://WWW.CALS.NCSU.EDU/AN_SCI/EXTENSION/ANIMAL/MEATGOAT/PDF_FACTSHEETS/MATCHINGFORAGESMG.PDF

that they regain weight as the lactation progresses.

How to Determine Body Condition

Practice makes perfect in determining body condition, so producers should use their animals to get a feel for how it's done. An easy way to start is to select a few animals that are overconditioned and some others that are thin to get a feel for extreme BCS. Then introduce a small group of animals and compare their BCS to the animals with extreme BCS. Producers should develop an eye and a touch for the condition of their animals and strive to maintain a medium amount of condition on their goats. When body condition starts to decrease, it is a sign that supplemental feed is needed or that animals should be moved to a higher quality pasture. Waiting until goats become thin to start improving their feeding regime may lead to large production losses and will increase feed costs.

Using Body Condition Scores

Pregnant does should have a BCS below 7 toward the end of pregnancy because of the risk of pregnancy toxemia (ketosis) or dystocia. In addition, a BCS of 5–6 at kidding should not drop off too quickly during lactation.

Producers should also be concerned with the body condition of the breeding bucks. If bucks are overfed and become too fat (BCS = 7 or

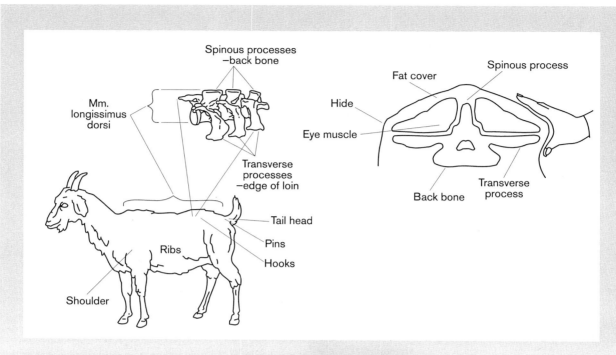

Figure 6-1. Areas to be monitored for fat cover.

Source: Mueller, J. P., M. H. Poore, J.-M. Luginbuhl, and J. T. Green, Jr. 1995. Matching forages to the nutrient needs of meat goats. HTTP://WWW.CALS.NCSU.EDU/AN_SCI/EXTENSION/ANIMAL/MEATGOAT/PDF_FACTSHEETS/MATCHINGFORAGESMG.PDF

higher), they may have no desire to breed does. Conversely, bucks that are thin (BCS = 4 or lower) at the start of the breeding season may not have sufficient stamina to breed all the does. Because of the increased activity and decreased feed intake during the breeding season, breeding bucks will most probably lose weight. Therefore, they need to be in good body condition (BCS = 6) and physical shape before the season starts.

Flushing

BCS is also used to determine whether flushing will be of benefit to breeding does. Flushing means increasing the level of feed, mostly energy, offered to breeding does starting about 1 month prior to the introduction of the bucks, to increase body weight, ovulation rate, and hopefully litter size. The increased level of energy offered to does should continue throughout the breeding season and for approximately 30–40 days after removing the bucks for adequate implantation of the fetuses in the uterus. Does in extremely good body condition (BCS = 7) will not tend to respond to flushing. On the other hand, does that are in relatively poor condition (BCS = 4 or lower) as a result of poor feed quality and supply, high worm loads, late kidding of twins or triplets, or high milk production in dairy does will respond favorably to flushing by improving their body condition.

Flushing can be accomplished by moving breeding does to a lush nutritious pasture 3–4 weeks prior to the introduction of the bucks. This cost-effective flushing method of "green flush" or "feed flush" is underused in the southeastern United States where forage is abundant. Another method is feeding 0.5 pound per day of a high energy supplement. Corn is the grain of choice for flushing; whole cottonseed is another low cost, high energy, and also high protein supple-

ment. Because the goal is to increase the intake and body weight, breeding does should be grouped according to their body condition.

GENERAL MANAGEMENT REQUIREMENTS

Nutritional Management

Nutrition of Newborn Kids

Colostrum is the first milk produced after birth. Colostrum contains a high content of antibodies (immunoglobulins), vitamin A, minerals, fat, and other sources of energy. Antibodies are proteins that help the goat kid fight diseases. The ability of kids to resist diseases is greatly affected by the timing of colostrum intake and the quantity and quality of the colostrum fed. Reports from cattle indicate that if left alone, 25% of the young do not nurse within 8 hours, and 10–25% do not get sufficient amounts of colostrum. Colostrum should be ingested or bottle-fed (in the case of weak kids) immediately after birth or as soon as kids have a suckling reflex. In cases of extremely weak kids, they should be tube-fed. The producer must be certain that all newborn kids get colostrum right after birth (within the first hour after birth, and certainly before the first 6 hours) because the percentage of antibodies found in colostrum decreases rapidly after birth. It is crucial that kids consume the antibodies in colostrum before the kids suck on dirty, pathogen-loaded parts of their mothers or stalls. In addition, the ability of the newborn kid to absorb antibodies decreases rapidly 24 hours after birth. Newborn kids should ingest 10% of their body weight in colostrum during the first 12–24 hours of life for optimum immunity. For example, a goat kid weighing 5 pounds at birth should ingest 0.5 pound of colostrum (approximately 0.5 pint) during the first 12–24 hours of life.

Summary: Body Condition

- Uses:
  - To monitor and fine-tune nutrition program
  - To head off parasite problem

- Visual evaluation is not adequate; have to touch and feel animal

- Areas to be monitored:
  - Tail head   – Ribs
  - Pins   – Hooks
  - Edge of loin   – Shoulders
  - Back bone   – Longissimus dorsi
  - Withers   – Thighs

- Scale

Thin	1–3
Moderate	4–6
Fat	7–9

- Recommendations
  - End of pregnancy   5–6
  - Start of breeding season   5–6
  - Animals should never have a BCS of 1–3 or 7–9.
  - Pregnant does should have a BCS below 7 toward the end of pregnancy because of the risk of pregnancy toxemia (ketosis) and dystocia.
  - A BCS of 5–6 at kidding should not drop off too quickly during lactation to < 4.

The extra colostrum produced by high-lactating does during the first 24 hours following kidding can be frozen for later use when needed. Ice cube trays are ideal containers: once frozen, cubed colostrum can be stored in larger containers and the trays used for another batch. Ice cubes are the perfect size for newborn kids, thus thawed colostrum is always fresh, and wastage is reduced to a minimum. Colostrum should be thawed either at room temperature or at a fairly low temperature. Colostrum should never be cooked during the thawing process, but it must be heated to control caprine arthritis encephalitis transmission.

Only first milking from healthy animals should be frozen for later feeding. The colostrum from older animals that have been on the premises for several years is typically higher in antibody content against endemic pathogens than is colostrum from first fresheners. Revaccination against enterotoxemia (overeating disease) and tetanus 4–6 weeks before the kidding date is commonly used to improve the protective value of the colostrum against these conditions.

Nutrition of Replacement Does

Doe kids needed for meat goat replacements should be grazed with their mothers during as much of the milking period as possible and not weaned early. Doe kids being raised for dairy goat replacements should be fed whole milk or milk replacer on a bottle or group feed on a nursing bucket, then weaned onto a high quality forage and grain-supplemented diet. Following weaning, doe kids should be separated from the main herd and have access to high quality forage and receive good nutrition through first kidding at 1–2 years of age, depending on the nutritional plane. Leaving doe kids with the main herd will result in undernourished does that are bred too young and too small; these animals may not reach their production potential. A yearly supply

of replacement does that are healthy, of good size, and free of internal and external parasites is essential to the success of any meat goat enterprise.

Nutrition of Breeding Bucks

Mating places a high nutritional demand on bucks. Therefore, depending on their body condition, breeding bucks should be enrolled in an increased nutritional program approximately 6 weeks before the breeding season. If bucks have been grazed on pasture or browse, concentrate supplementation must be introduced gradually to avoid risks of enterotoxemia.

Suggested Supplemental Feeding Program for Meat Goats

As a general recommendation, trace mineralized salt containing Se should be given to all goats year-round. A complete goat mineral should be offered free choice year-round in most production situations. When goats are raised on browse, abundant supply should be made available to allow goats to be very selective and to ingest a high quality diet that will meet their nutritional requirements. When forage or browse is limited or of low quality (< 10% protein), lactating does (and does in the last 30 days of gestation), developing/breeding bucks, weanlings, and yearlings should be fed 1.0 pound per day of a 16% protein mixture (77:20:2.5:0.5 ground corn:soybean meal:goat mineral:limestone). Alternatively, ground corn and soybean meal can be substituted by whole cottonseed for lactating does. Low to medium quality forage (> 10% protein) will meet requirements of dry does and non-breeding bucks. Goats can be forced to eat very low quality feed, including twigs, tree bark, etc., but producers should be aware that this practice will hurt the productivity of superior meat-producing goats and reduce body condition.

Suggested Supplemental Feeding Program for Dairy Goats

The nutrient requirement of an animal is determined by the sum of the animal's requirements for maintenance, activity, growth, pregnancy, and milk production (including milk fat content). These requirements are summarized for goats in table 6-4 (p. 164). For a lactating doe the nutrient requirement can be estimated based on the animal's weight, level of milk production, and milk fat content. Lactating does vary greatly in weight due to breed and age. Milk production varies due to the genetic ability of an animal and her size, age, and stage of lactation. Figure 6-2 presents milk production in a high-producing Nubian herd in relative terms (e.g., 1.0 is 100% of maximum production and 0.5 is 50% of maximum production) to bring goats of different ages and peak production together on the chart. In this Nubian herd goats increased in milk production with age; they had an average peak milk production of 6.8, 9.3, 11.2, and 12.2 pounds per day at 1, 2, 3, and 4 or more years of age.

Milk fat is a major determinant in the energy required to produce a pound of milk because fat contains 2.5 times more energy than does sugar or protein. Also, as milk fat increases, milk protein increases (figure 6-3, p. 165). Milk fat and protein content vary over the lactation (figures 6-4 and 6-5, pp. 165 and 166).

Estimating the Animal's Supplementation Needs

An example of calculating the nutrient requirements of a 176-pound mature doe giving 8

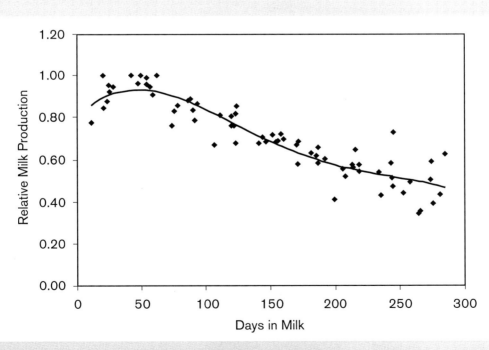

Figure 6-2. A goat's level of milk production varies over the lactation, peaking at about 60 days in milk. This chart is based on 44 lactation curves of different-aged goats from a grade Nubian herd. The slight increase in late lactation is likely due to lower producing goats being dried off before the end of a 305-day lactation.

Table 6-4. Nutrient requirements of the goat.

Body wt.	DMI[a]	TDN	CP[b]	Ca	P	Vit. A	Vit. D
			(lb)			(1,000 IU[c])	(IU)
			Maintenance[d]				
22	0.62	0.35	0.05	0.0022	0.0015	0.4	84
44	1.06	0.59	0.08	0.0022	0.0015	0.7	144
66	1.43	0.80	0.11	0.0044	0.0031	0.9	195
88	1.78	0.99	0.14	0.0044	0.0031	1.2	243
110	2.09	1.17	0.17	0.0066	0.0046	1.4	285
132	2.40	1.34	0.19	0.0066	0.0046	1.6	327
154	2.71	1.50	0.21	0.0088	0.0062	1.8	369
176	2.99	1.66	0.23	0.0088	0.0062	2.0	408
198	3.26	1.81	0.26	0.0088	0.0062	2.2	444
220	3.52	1.96	0.28	0.0110	0.0077	2.4	480

[a] DMI-dry matter intake.
[b] CP-crude protein.
[c] IU-international units.
[d] Maintenance includes stable feeding and early pregnancy. Increase this by 25% for early pregnancy and on pasture. If goats are managed under range conditions, increased activity may require that the maintenance level be increased by 50–75%.

Body wt.	DMI[a]	TDN	CP[b]	Ca	P	Vit. A	Vit. D
			(lb)			(1,000 IU[c])	(IU)
Additional requirement for late pregnancy at all weights							
	1.56	0.87	0.18	0.0044	0.0031	1.1	213
Additional requirement for growth at all weights							
Daily gain (lb/day)							
0.11	0.40	0.22	0.03	0.0022	0.0015	0.3	54
0.22	0.79	0.44	0.06	0.0022	0.0015	0.5	108
0.33	1.19	0.66	0.09	0.0044	0.0031	0.8	162
Additional requirement for milk production per pound of milk							
% Milk fat							
2.5		0.333	0.059	0.002	0.001	1.7	345
3.0		0.337	0.064	0.002	0.001	1.7	345
3.5		0.342	0.068	0.002	0.001	1.7	345
4.0		0.346	0.072	0.003	0.002	1.7	345
4.5		0.351	0.077	0.003	0.002	1.7	345
5.0		0.356	0.082	0.003	0.002	1.7	345
5.5		0.360	0.086	0.003	0.002	1.7	345
6.0		0.365	0.090	0.003	0.002	1.7	345

Source: Adapted from National Research Council. 1981. Nutrient Requirements of Goats: Angora, Dairy, and Meat Goats in Temperate and Tropical Countries. *Number 15. National Academy Press. Washington, D.C., table 2, p. 26.*

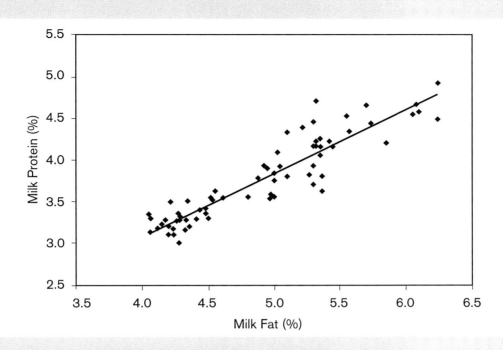

*Figure 6-3. The concentration of protein in goat milk varies with the fat content.
This chart is based on 44 lactation curves of different-aged goats from a grade Nubian herd.*

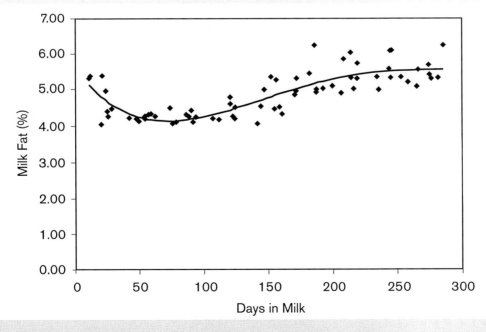

*Figure 6-4. The concentration of fat in a goat's milk varies over the lactation,
with a low at about 60 days in milk. This chart is based on 44 lactation curves
of different-aged goats from a grade Nubian herd.*

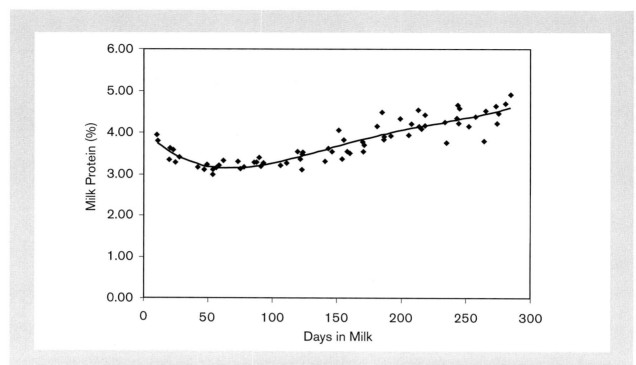

Figure 6-5. The concentration of protein in a goat's milk varies over the lactation, with a low at about 60 days in milk. This chart is based on 44 lactation curves of different-aged goats from a grade Nubian herd.

pounds of milk per day at 4% milk fat is presented in table 6-5 (p. 167). Dry matter intake increases with milk production at about one-third pound of dry matter per pound of milk produced. This differs from animal to animal, and those animals that are the highest producers are those that have higher dry matter intake capacity. At this level of milk production, grain supplements are required. If a doe producing this amount of milk were fed common forages without grain supplementation she would lose body condition as her body used its fat, protein, and minerals to make the milk. Then milk production would go down to the level that feed intake and forage nutritive content allowed. At peak milk production a high-producing doe may not be able to eat as much as her milk production requires and the doe may lose condition. It is

important that does are in good but not excessive body condition before freshening and that they be fed adequately after peak milk to regain a reasonable amount of condition.

High-producing does can consume dry matter at more than 5% of their body weight per day. When the animals are eating only hay and grain, it is relatively easy to measure the hay fed, the amount not eaten, and the grain fed to determine total feed intake. By adjusting this to a dry matter basis using forage test values, dry matter intake can be estimated (figure 6-6, p. 168).

However, when does are grazing, it is difficult to measure forage intake. We can estimate possible DMI and limit grain feeding to levels

	TDN (lb)	CP (lb)	DMI (lb)
Table 6-5. Estimating the total digestible nutrient, crude protein, and dry matter intake of a lactating doe.			
176 lb lactating doe			
Nutrient for maintenance	1.66	0.23	2.99
Nutrient for activity (maintenance + 25%)	0.42	0.06	0.75
Milk production			
4.0% milk fat (nutrient/lb milk)	0.346	0.072	0.33
lb milk	8	8	8
Nutrient for milk production	2.77[a]	0.58	2.64
Total requirement (lb)	4.84[b]	0.86	6.38
Nutrient density needed in ration (%)	75.9	13.5	

[a] 0.346 x 8
[b] 1.66 + 0.42 + 2.77

appropriate for the quality of the forage the animals eat and the level of milk produced.

If the example goat were eating 6.4 pounds DM per day, the TDN requirement could be met with about 3.5 pounds of high energy grain dry matter (85% TDN) along with 2.9 pounds of high quality hay dry matter (65% TDN). If the forage in the ration is 16% CP, the grain would need to be only about 12% CP. See chapter 2 for detail on ration balancing.

With pasture feeding, ration balancing is more difficult and producers often rely on rules of thumb and experience. To help develop and apply such rules, some principles of animal nutrition should be reviewed.

Does that are able to eat more forage DM will be capable of producing more milk at a given level of supplementation. Higher DMI may be due to larger body size, the genetic ability to consume more forage, or the animal's genetic ability to produce more milk, which is a driving force determining appetite.

Eating forage at a higher rate will maintain higher milk production. Young forage is lower in fiber and is more quickly and highly digested, allowing the animal to eat more of the feed. Therefore, young forage will allow the animal to produce more milk at a given level of supplementation than a more mature forage. Legumes are lower in neutral detergent fiber (NDF) than grasses at similar maturity, so legumes will be digested faster and be consumed at higher rates, supporting higher levels of milk production.

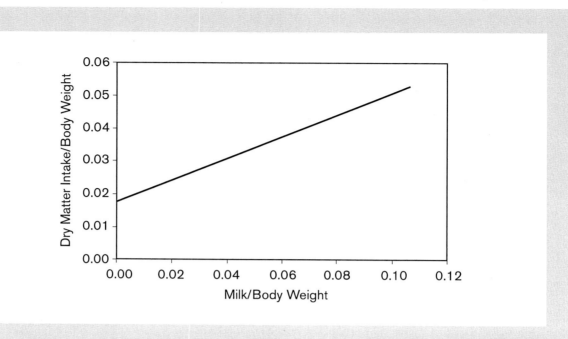

Figure 6-6. Dry matter intake of a lactating doe relative to her body size can be estimated from the animal's level of milk production relative to her body size (based on production of milk with 4% fat). Considerable variation can occur from one animal to another.

Grain that is high in energy will sustain more milk per pound of grain than grain low in energy. Corn- and soybean meal-based supplements are high in energy. Grain mixes based on oats and other fibrous feeds are lower in energy and will do less to maintain milk production but are less apt to interfere with forage digestion or to cause rumen upset if fed at excessive levels. Some of the fibrous byproduct feeds are good sources of CP and ruminally digestible energy. In general high-energy grains fed at reasonable levels are the most cost-effective supplements for dairy animals on high protein pasture or browse.

Grain that is too high in protein can increase the loss of energy from the animal's body because excess protein is converted to ammonia in the rumen and has to be converted to urea and excreted from the body. On forage high in protein, grain supplements should be high in energy and low to moderate in protein. Corn (10% CP on a DM basis) is usually the cheapest energy supplement that is high in energy and low in protein and is well suited for low to moderate levels of milk production on forages high in CP. Supplements for high-producing animals on high quality pasture seldom need to have more than 14% CP.

When animals are fed low quality forages, grain that is high in protein will often allow the animal to increase DMI. On low quality forages low CP may limit the rate at which rumen bacteria digest the forage. A small amount of high protein grain will allow the rumen bacteria to grow and multiply faster and digest the forage faster, allowing the animal to eat more of the low quality forage. In general, low quality forages are best used to feed dry animals because it is difficult and expensive to supplement these feeds for milk production.

The eating behavior of an individual animal is learned based on what its mother eats (which was learned) and experience with new foods (chapter 1). Animals that are raised in one habitat, such as a brushy hillside, will learn to browse and eat weeds. When moved to an improved pasture, they will not be as good at grazing as the animals raised on that pasture. Offspring raised on the pasture then returned to the hillside will not do as well on the browse and weeds as those that are raised there. This is a critical issue when moving goats from one grazing habitat to a new and different grazing habitat.

Rate of Grain Feeding

A simple way to feed grain is to base the amount on the level of milk production per day. A rule of thumb is to feed 0.25–0.33 pound of grain per pound of milk. This can be reduced to zero for low-producing animals. For high-producing animals this can be increased to 0.40 pound of grain per pound of milk (figure 6-7). Too little grain will result in reduced milk production. Too much grain feeding is expensive and can cause rumen acidosis and a reduction in milk fat content.

Because goats differ in size and because size determines the animal's maintenance requirement and affects how much feed the animal can eat, it is helpful to look at daily milk production as a fraction of the animal's body weight. A 120-

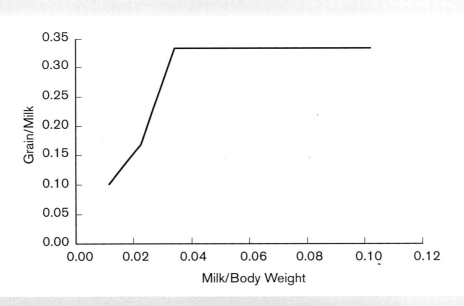

Figure 6-7. General guideline for grain feeding the lactating doe. The rate of grain feeding is a function of milk production per pound of body weight and the energy content of the ration's forage and grain. This starting guide should be limited based on forage fiber content.

pound doe making 12 pounds of milk is producing 10% of her weight per day. A 160-pound doe making 16 pounds of milk is also producing 10% of her weight. These two animals, eating the same forage at a similar rate, would need grain supplement at the same rate per pound of milk. If forage quality indicated to feed grain at 0.25 pound of grain per pound of milk, the first goat would need 3 pounds of grain and the second would need 4 pounds of grain. Figure 6-8 shows the concentration of TDN and CP needed in the total ration dry matter intake for average dairy goats producing milk containing 4% fat.

If a herd contains different breeds of goats that differ in their potential to produce milk fat and solids (such as Nubian vs. Toggenburg), an adjustment to fat-corrected milk may be justified. Milk yield can be converted to an energy

equivalent yield of milk at 4% butter fat by the following equation:

$$\text{milk lb at 4\% butter fat} = (\text{milk lb} \times \text{fat \%} \times 15) + (\text{milk lb} \times 0.4)$$

Limits to Grain Feeding

When feeding grain for high production (milk production greater than 6% body weight), grain feeding should be limited to two times the hay dry matter intake to maintain adequate fiber in the ration for proper ruminal function. On high quality hay-grain feeding may have to be kept to half of the hay intake.

High energy (low fiber) grain supplements should be limited based on the NDF content of the forage being consumed (figure 6-9). Higher quality forage has lower levels of NDF and grain

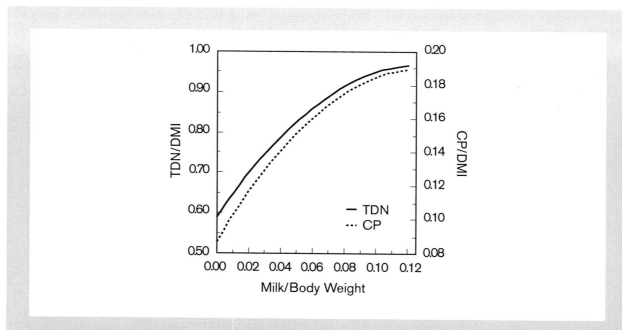

Figure 6-8. The concentration of total digestible nutrients and crude protein in a dairy goat's ration depends on milk production per pound of body weight and fat content of the milk. This diagram is based on 4% fat milk.

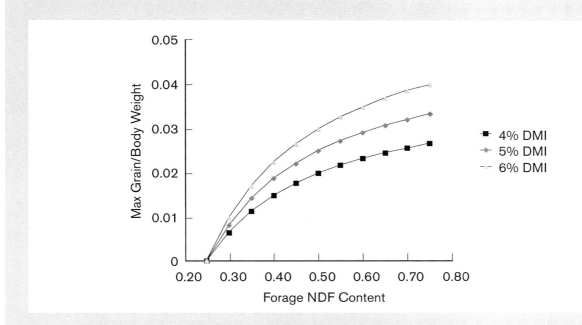

Figure 6-9. The maximum grain feeding is determined by the forage's neutral detergent fiber content, the animal's size, and the animal's dry matter intake.

feeding levels should be lower or there is the risk of rumen upset due to too little effective fiber in the ration. For example, a goat eating dry matter at 5% of her body weight can receive 2.5% of body weight as grain if the forage has 50% NDF but less than 2% of body weight as grain if the forage has 40% NDF.

On pasture or browse that is high in protein, some high-energy grain may be of value at low production levels. Most forage protein is converted to ammonia in the rumen by the bacteria digesting the forage. When protein is in excess to energy there is not enough energy available to the ruminal bacteria to convert this ammonia back into bacterial protein. When this is the case the ammonia goes across the rumen wall into the bloodstream and has to be converted to urea for excretion in the urine. This requires energy. Some high-energy grains, such as corn,

fed under these conditions can supply energy to the bacteria, enabling them to use some of this otherwise excess protein. If animals are already overly fat, this could be of no value and even detrimental. If the animals are low in body condition, it will be of value.

The grain to milk ratio shown in figure 6-7 (p.169) should be adjusted up when forage quality is lower and down when forage quality is higher than the 70% TDN forage for which this chart was developed. The maximum rate of grain feeding should also be changed based on forage quality as shown in figure 6-9.

High-producing does at peak lactation often make more milk than the ration provides nutrients. The doe's body then takes the needed nutrients from her fat, muscle, and bones to make up the difference. This is normal but cannot be too great or go on for too long before milk production goes down or the animal has metabolic disorders such as ketosis (energy deficiency), milk fever (hypocalcemia or Ca deficiency) or grass tetany (hypomagnesia or Mg deficiency).

In early lactation many dairy producers will "lead feed" supplements. This means that they will feed grain supplements for a higher level of milk production than what the animal is making that day, knowing that milk production is increasing daily.

Common Nutritional Diseases

Pregnancy Disease or Ketosis

During late pregnancy, nutritional requirements are as high as they are during lactation, especially if the pregnant doe is carrying more than one fetus. Not only are extra nutrients needed by the developing fetuses, but the fetuses also crowd the abdominal cavity and reduce ruminal volume. As a result, adequate amounts of feed cannot be consumed. Because of this, does fed a poor quality and bulky diet (especially if they are fat) can develop ketosis and die due to inadequate energy intake. Grain and protein meal and to a lesser extent whole cottonseed are preferred feeds to overcome this problem.

Inadequate nutrition during late pregnancy will also result in small, weak kids at birth, and high early death losses, especially with twins or triplets. When forage or browse is low in quality, (40–55% TDN; 10% protein or less), meat goat does in late pregnancy and early lactation should be provided with about 1 pound per day of a 16% protein concentrate. Dairy does in early lactation should be fed according to milk production.

Urinary Calculi or Urinary Stones

In goats, clinical obstruction of the urinary tract is most frequently seen in young, castrated males; the calculi are usually composed of cal-

cium phosphate salts. Castrated goats raised for goat shows, goats kept as pets, and overfed bucks are at high risk for developing the condition due primarily to the feeding of excessive cereal grain in the diet. If the diet contains too much P relative to Ca, supplemental Ca from feed-grade limestone is one way to maintain the Ca:P ratio between 2:1 and 4:1. As a preventive measure, 10 grams of ammonium chloride can be fed per head per day or ammonium chloride can be added at 2% of the concentrate ration. Ammonium sulfate is sometimes used in place of ammonium chloride at a rate of 0.6–0.7% of the total ration.

Grass Tetany

Grass tetany is associated with low levels of Mg in the blood and can occur when goats in early lactation are grazing lush, leafy small grain, annual ryegrass, or well managed grass pastures. Grass tetany is more likely to occur on soils low in P but high in N and potassium (K) because this combination tends to inhibit Mg uptake. Spring fertilization of pastures with N or K will increase the risk of animals having grass tetany. Under those conditions, it is advisable to provide a mineral mix that contains 5–10% Mg to counterbalance the excess K.

Gastrointestinal Parasite Management

Introduction

Gastrointestinal parasites are a major source of economic loss in meat goat and sheep production systems throughout the United States. As small ruminants, goats and sheep share the same gastrointestinal parasites. Losses result from reduced growth rate, weight loss, a marked decrease in milk production, illness, and death. The most important of the gastrointestinal parasites of goats include roundworms and coccidia (also see chapter 8).

Effective control of these two groups of parasites makes a significant contribution to the health and well-being of goats. Eradication of these parasites is impossible, but the simple presence of a parasite in an animal does not indicate disease. An animal will show the symptoms of disease only when parasite loads become excessive, when an animal's natural immunity to disease becomes suppressed, or in an animal with a low BCS.

Roundworms

Of the family of gastrointestinal roundworms, the really important ones are the barber pole worm (*Haemonchus contortus*) and the brown stomach worm (*Ostertagia circumcincta*). Other species can and occasionally do cause economic losses to goat producers, but they are of lesser significance. The barber pole worm is the most important and most common blood-sucking gastrointestinal parasite of goats. One thousand barber pole worm larvae can suck up to 2.5–3.0 ounces of blood per day, resulting in anemia. Anemia can be detected as paleness in the mucous membrane around the eyes, inside the mouth, or inside the edge of the rectum or vagina. As plasma protein is lost, edema can occur in the subcutaneous tissue. This frequently is detected as swelling under the jaw—the reason the condition is often called "bottlejaw"—or low on the abdomen. The barber pole worm usually does not cause diarrhea, unlike some other gastrointestinal parasites such as the brown stomach worm.

Adult female barber pole larvae have a tremendous egg-laying potential (5,000–10,000 eggs or more per day). Eggs are passed in the feces and contaminate the environment. Eggs hatch and pass through three larval stages, the third being infective for the next host when ingested. The successful development of these stages outside the host depends on the climate (see chapter 8). Barber pole eggs and larvae require warm,

moist conditions for continued development. Barber pole larvae can also undergo a process called arrested development, in which they lie dormant in the abomasum (the true stomach of ruminants) following infection and don't become adults until several months later. This is an important adaptation for keeping the worm around through cold winters when eggs and larvae don't survive well on pasture. As a result, we see an increase in gastrointestinal parasite transmission from spring to fall.

Anthelmintics

Anthelmintics are the traditional method of treatment against gastrointestinal roundworm parasites. However, goats are not a major target of research by the pharmaceutical industry. One unfortunate consequence of the lack of pharmaceutical interest and information is a considerable amount of "off license" or "extra label" use of products in goats. Animals are often treated at dose rates recommended for sheep with little regard to whether these dose rates are appropriate. Off license or extra label drug use requires consultation with a veterinarian. The use of inappropriate dose rates for anthelmintics is particularly worrisome because *underdosing* is probably the most important factor influencing the development of anthelmintic resistance by gastrointestinal parasites (see chapter 8). There is a considerable body of evidence on the existence of physiological and pharmacological differences between sheep and goats that supports the view that anthelmintic treatments should be administered at higher dose rates in goats than sheep. Furthermore, the problem of resistance of barber pole worms and other gastrointestinal parasites to anthelmintics is a great concern for goat producers. Unlike sheep, goats have a relatively poor ability to increase an effective immune response against gastrointestinal **nematodes**. In addition, immunity to parasites declines during the **periparturient** period and during periods of illness or malnutrition. Surveys have shown that adult goats generally have the same level of infection as kids.

The major classes of broad spectrum anthelmintics are: benzimidazoles, imidothiazoles, tetrahydropyrimidines, and macrocyclic lactones. The only U. S. Food and Drug Administration (FDA)-approved anthelmintics for goats are fenbendazole (trade names Panacur or Safeguard) from the benzimidazole class and morantel tartrate (Rumatel) from the tetrahydropyrimidine class. Because of the scarcity of products labeled for goats, the FDA has undergone a minor species–minor use effort to increase the number of anthelmintics labeled for that species.

Benzimidazoles. Thiabendazole or TBZ, a member of the benzimidazole family of drugs, was one of the most frequently used dewormers because of its safety and effectiveness against numerous species of intestinal parasites. Other benzimidazoles include oxfendazole (Synanthic), albendazole (Valbazen), and fenbendazole (Panacur, Safeguard). Synanthic and Valbazen are teratogenic (may cause birth defects) at 22.5 milligrams per kilogram of body weight and should not be administered to lactating dairy animals or to animals in the first trimester of pregnancy. Panacur 10% suspension and Safeguard 10% suspension are the only presently marketed benzimidazoles labeled for goats that can be used to treat lactating goats without incurring milk withdrawal. They are administered orally at a dose of 5.1 milligrams per kilogram of body weight. Goats metabolize this group of drugs differently and therefore require approximately two times the normal sheep dose. Therefore, if administered to lactating goats at a dose higher than prescribed, consultation with a veterinarian is necessary. Widespread resistance to benzimidazoles has been found in barber pole worms, and once they become resistant to one anthelmintic of the benzimidazole class, barber pole worms become resistant to all of them.

Imidothiazoles. Imidothiazoles are not labeled for goats. Goats require a higher dose than sheep for products of the imidothiazole class to be effective (1.5 times the sheep dose rate is recommended for goats). However, the safety margin for imidothiazoles (Levasole, Tramisol) is lower than for other anthelmintics and side effects such as salivation may be seen, particularly when the injectable form is used. Imidothiazoles should not be administered to lactating dairy animals or female dairy animals of breeding age. Resistance to imidothiazoles does not seem to be as widespread as with benzimidazoles.

Tetrahydropyrimidines. Morantel tartrate (Rumatel) is available as a medicated premix and is labeled for goats. Research indicates that morantel use does not result in drug residues in milk and is effective at 9.9 milligrams per kilogram of body weight for goats. In addition, morantel is safe to use in pregnant goats.

Macrocyclic lactones. Avermectins and milbecycin are members of the macrocyclic lactone class of anthelmintics. Macrocyclic lactones are not labeled for goats. Ivermectin (Ivomec) has a high therapeutic index due to its superior potency against parasites. Ivomec is available in injectable, pour-on, and oral forms. In goats, the oral form appears to be more effective against gastrointestinal trichostrongyles. The withdrawal time for oral treatment is also shorter. However, it has been reported that when the injectable form of ivermectin was used for goats, it was more effective at keeping fecal eggs per gram at a constant low level than when an oral drench of Ivomec was given. Ivomec should not be administered to lactating dairy animals or female dairy animals of breeding age. Other avermectins of the macrocyclic lactone class are doramectin (Dectomax) and eprinomectin (Ivomec Eprinex). The active ingredient of milbecycin is moxidectin (Cydectin). Dectomax should not be used to treat female dairy animals 20 months of age or older. Ivomec Eprinex and Cydectin have no milk withdrawal for dairy animals. As a consequence of the macrocyclic lactone having lethal effects on many arthropods, including "good" and "bad" dung-breeding insects, there is growing concern about interference with the natural process of biodegradation and the resulting environmental impact of this class of drug.

Drug Resistance

Populations of the barber pole worm have developed different degrees of resistance to all available pharmaceutical dewormers, ranging from low to complete resistance. The highest resistance has been observed with Ivomec®, Valbazen®, SafeGuard® and Panacur®, and low to moderate resistance has been observed with Levasol® and Tramisol®. Resistance to Cydectin® is prevalent and increasing on many farms.

Cydectin® should not be used on farms unless a limited number of animals are treated at one time. If Cydectin® is used on all animals at once, development of resistance will be accelerated. Resistance has developed because past recommendations did not consider refugia, which is the proportion of a population of worms that are sensitive to dewormers or in "refuge" from a dewormer. When treating all animals in a herd as has been practiced in the past, only resistant worms survive. If these animals are moved to a "clean" pasture that has not been exposed to goats for four to six months or longer or was hayed, only resistant worms can develop in that pasture. However, if only animals in need are treated, and they then go back to a "dirty" pasture with a low to moderate level of infectivity, as is now currently recommended, the resistant worms can breed with sensitive worms and maintain a worm population that should still respond to dewormers. In other words, the population of worms in refugia provides a pool of genes to dilute the resistant genes. **This is the most important component of maintenance of a population of worms that will remain susceptible to dewormers.** Past recommendations included deworming ewes over winter. We now know that this leads to survival of resistant worms and in the spring an outbreak of more resistant barber pole worms can occur. Current recommendations include selective treatment of only animals in need according to the FAMACHA system. Untreated animals will harbor sensitive worms.

FAMACHA

FAMACHA is a selective treatment whereby the producer decides which animals to deworm according to a procedure described below. FAMACHA was developed by a group of veterinarians and scientists in South Africa and was validated in the southern U.S. by members of the Southern Consortium for Small Ruminant Parasite Control (SCSRPC; www.scsrpc. org). A complete description of FAMACHA can be found on the web site. FAMACHA is a tool used by producers that consists of examining the color of the mucous membrane of the lower eyelid, matching the color to that of a chart that ranges from red or healthy to almost white or anemic (figure 8–3, p. 213). The lighter the color, the more anemic an animal is.

Anemia occurs as a result of the adult worms removing more blood than the animal can replace. There may be other causes of anemia (coccidiosis, liver flukes, lice, ticks, fleas, Cu poisoning, P deficiency, kale poisoning) so the producer must be aware of the health and nutrition status of the herd. Animals with red color can be left untreated, whereas paler scores indicate that an animal should be treated. Research indicates that 20% of the herd carries 80% of the worms. Or in other words, 20% of the animals consistently are more susceptible to infection with the barber pole worms, carry the worms, and disseminate the eggs in the pasture. Identification of these animals is possible partly through the use of FAMACHA and good records, and these animals can be culled or removed. It is possible to develop a more resistant group of animals that need less frequent treatment for parasites.

Recommendations

Worm burdens will increase throughout the growing season and ideally animals will be examined regularly using FAMACHA. Lactating dairy goats and other animals that are handled regularly through a chute can easily be monitored by checking the mucous color of their lower eyelids. FAMACHA examination should occur more frequently on weaned kids and also in does in late gestation and early lactation. It is important to examine late pregnant does because the immune system becomes depressed around the time of kidding, which leaves the animal more susceptible to parasites. If selective treatment takes place before kidding, make sure that the dewormer is safe for pregnant animals. Also, watch for signs of an infection such as bottle jaw or animals that lag behind.

Different anthelmintics and different formulations of the same anthelmintic (oral vs. injectable vs. pour-on; sheep vs. cattle) may have different meat and milk withdrawal times, and it is important to consult a veterinarian prior to their use. Pour-on anthelmintics labeled for cattle do not seem to be effective if used as pour-on in goats. Restrictions described earlier in this chapter concerning the use of anthelmintics should be followed.

Coccidia

Coccidiosis is caused by single-celled protozoan parasites called coccidia that reside in the intestines of goats (also see chapter 8). All adult goats carry coccidia in their intestines, and kids ingest infectious oocysts from feed or pasture contaminated with manure excreted by adult goats. Coccidia are very host specific. Therefore, the species of coccidia that infect goats infect goats only. Coccidia found in birds, cattle, dogs, and rabbits will *not* infect goats. The coccidia of sheep, however, may be responsible for some infection in goats and therefore should be regarded as suspect.

The presence of coccidia eggs in the feces of normal goats does not indicate an infective situation. In general, if the animals do not show any clinical signs, the infection is not significant. Adults will have immunity to the parasite that is fairly effective in preventing disease, but not infection. The disease is almost always going to occur in young animals, and kids less than 5 months of age are more susceptible. Kids will become infected early on from the environment. However, the stress of weaning may depress their immune system enough to allow the coccidia to get the upper hand and cause the disease. Goats that survive through a disease outbreak are usually immune to future problems. Coccidiosis is best prevented by maintaining a sanitary environment (see chapter 8).

Symptoms of Coccidiosis

The symptoms of coccidiosis are divided into two categories: subclinical and clinical. Subclinical cases result in a decrease in feed intake and weight gain, and are difficult to detect because of the absence of diarrhea. Clinical cases can vary from mild cases with some loss of appetite, decrease in weight gain, and slight, short-lived diarrhea to severe cases involving great amounts of dark, bloody, foul-smelling diarrhea, fluid feces containing mucous and blood, persistent straining in attempt to pass feces, loss of weight, rough hair coat, dehydration, and in some cases death within 24 hours. The primary pathology associated with coccidiosis involves intestinal cell destruction. Scarring and rupture of the cilia of the lining of the intestines following treatment or recovery may result in permanently unthrifty and stunted animals because of an impaired ability to absorb digested food. The only two FDA-approved coccidiostats for goats are decoquinate (Decox) and monensin (Rumensin).

Forages for Meat Goats

Introduction

Goats offer an opportunity to effectively convert pasture forage to animal products such as milk, meat, and fiber that are marketable and in demand by a growing segment of the U.S. population. In addition, goats selectively graze unwanted vegetation in pastures and forests, thus providing biological control that will reduce dependence on certain herbicides.

Goats consume only the most nutritious parts of a wide range of grasses, legumes, and browse plants. Browse plants include brambles, shrubs, trees, and vines with woody stems. The quality of feed on offer will depend on many things, but it is usually most directly related to the age or stage of growth at the time of grazing. The nutrient composition for several common feed types found on many farms is shown in table 6-2, p. 155.

Grazing Behavior and Grazing Time

Goats are very active foragers, able to cover a wide area in search of scarce plant materials. Their small mouths and split upper lips enable them to pick small leaves, flowers, fruits, and other plant parts, thus choosing only the most nutritious available feed.

The ability to utilize browse species, which often have thorns and an upright growth habit with small leaves tucked among woody stems, is a unique characteristic of the goat compared to heavier, less agile ruminants. Goats have been observed to stand on their hind legs and stretch up to browse tree leaves or throw their bodies against saplings to bring the tops within reach.

The feeding strategy of goats appears to be to select grasses when the protein content and digestibility are high, but to switch to browse when its overall nutritive value may be higher. This ability is best used under conditions in which there is a broad range in the digestibility of the available feeds, giving an advantage to an animal that is able to select highly digestible parts and reject those materials that are low in quality.

Grazing goats have been observed to:

- prefer browsing over grazing pastures,
- prefer foraging on rough and steep land over flat, smooth land,
- graze along fence lines before grazing the center of a pasture,
- graze the top of pasture canopy fairly uniformly before grazing close to the soil level, and
- select grass over clover.

Because of their inquisitive nature and tolerance of "bitter" or high tannin material, goats may eat unpalatable weeds and wild shrubs that may be poisonous to other livestock species. The absence or the severity of poisoning is related to the quantity of material consumed, the portion and age of the plant eaten, the season of the year, the age and size of the animal, and other factors. Several ornamental plants that are grown outdoors or indoors are highly toxic. For example, goats should not have access to, or be fed clippings of, yew, azalea, delphinium, dicentra, foxglove, ground ivy, hellebore, larkspur, lantana, lily-of-the-valley, oleander, rhododendron, spider lily, or yellow jessamine.

In a pasture situation goats are "top down" grazers, meaning that they graze the top of pasture canopy fairly uniformly before grazing close to the soil level. This behavior results in uniform grazing and favors a first grazer-last grazer system. This might consist of using a high milking goat herd or weanlings as the first group, cattle plus sheep next, and horses as the last group. This management is most appropriate with lactating does or growing kids as the first group.

Goats naturally seek shelter when it is available. Goats seem to be less tolerant of wet cold conditions than sheep and cattle because of a thinner subcutaneous fat layer. A wet goat can easily become sick. Therefore, it is advisable to provide artificial shelters, such as open sheds. Nevertheless, goats with a BCS of 6 and higher will be more tolerant of wet cold conditions.

Some livestock producers confine their animals at night for protection from straying, predation, and adverse environmental conditions. However, confinement means that grazing time is reduced and that the animals spend more time in unsanitary lots or pens. Reduced grazing time due to confinement at night is more of an issue during the hot and humid summer months, because animals may not forage effficiently during the hottest periods of the day. If animals must be confined at night, allowing the animals to graze during the cooler parts of the day would increase production.

Grazing Management for Meat Goats

Grazing of forage generally provides the least expensive way of supplying nutrients to animals. Therefore, it is advantageous to develop a year-round forage program that allows for as much grazing as possible every month of the year. However, good pasture management involves much more than simply turning the animals to pasture. The principles of controlled grazing of goats or sheep are similar to those used for cattle. The primary goal is to have control of the animal's grazing pattern so that one can dictate the degree and the frequency of defoliation. To obtain efficient animal production over a number of years, the needs of the plants as well as the needs of the animals must be considered. The development of a successful forage management plan entails:

- Adjusting the number of animals grazing a certain area (**stocking density**) of pasture because some forage must be left at the end of the grazing period to maintain adequate plant production. Otherwise, overuse will weaken the plants and regrowth will be slower. Adjusting the stocking rate requires experience because forage growth is not uniform throughout the year or from year to year. It varies with differing environmental conditions such as rates of precipitation and fluctuations in temperature.

- Harvesting ungrazed forages as hay or silage at an immature stage of growth when forage growth is more rapid than it can be grazed. This will provide high quality feed when grazing is not available. Cross-electrified fencing will keep animals concentrated on small areas while excess growth accumulates on other paddocks. Under those circumstances, consider use of short-duration rotational grazing through a series of paddocks, or strip grazing a rapidly growing pasture using a movable electrified fence to allow animals access only to enough forage to carry them for 1 day.

- In more southern areas, overseeding bermuda pastures with legumes, ryegrass, cereal grains, or brassicas to extend the grazing season and to provide some high quality feed during the winter and spring.

- Restricting the use of high quality forage, when in short supply, for the supplementation of other low quality pastures, hay, or

Table 6-6. Estimated stocking rates or feed needs for goats, sheep, and cattle on pasture.			
Pasture type	Goats	Sheep	Cattle
Head[a]			
Good quality pasture system	6–8	5–6	1
Good brush-browse system	9–11	6–7	1
Head/acre			
Wheat/alfalfa system	10–12	8–9	1.5
Alfalfa pasture, Oklahoma	12–15	10–11	1.9

[a] Number of animals to consume similar amount of feed.

Source: Luginbuhl, J-M., J. T. Green, Jr., J. P. Mueller, and M. H. Poore. 1995. Grazing habits and forage needs for meat goats and sheep. Chapter 20. In: D. S. Chamblee (ed.), Production and Utilization of Pastures and Forages in North Carolina. *pp. 105-112. North Carolina Agricultural Research Service Technical Bulletin No. 305.*

silage. This can be achieved by letting goats graze high quality forage a few hours at the end of each day, or by grazing the limited high quality supply every other day.

Clearing Land with Meat Goats

When the aim is to kill or reduce the amount of unwanted vegetation, greater severity and frequency of grazing is necessary. Goats will actively select major weeds at particular stages of growth. As a rule, effective control of unwanted vegetation can be achieved in 2–3 years. It is important to consider goats' feeding strategies before deciding to use them to clear land. Because they are browsing animals, goats stunt tree growth and prevent the regeneration of forests and thus should be managed carefully in areas where forests are desired. Goats could be very useful, however, in areas where regrowth of brush and trees is not desirable.

Mixed Grazing and Stocking Rates

The differences in feeding behavior among cattle, sheep, and goats uniquely fit each species to the utilization of different feeds available on a farm. These differences should be considered in determining the best animal species to use a particular feed resource. Feeding behavior is also important in determining whether single or multiple species will best use available plant materials. Most studies indicate that greater production and better pasture use are achieved when sheep and cattle or sheep, cattle, and goats are grazed together as opposed to grazing only sheep or goats or cattle alone. This is especially true where a diverse plant population exists and brush is encroaching.

Under mixed grazing conditions (more than one ruminant species grazing in the same paddock) on fescue/orchardgrass-clover, where the forage supply is low and the nutritive value is high, goats and sheep may be at a disadvantage. Under these conditions, the animal with the largest mouth (i.e., cow or horse) has an advantage because it can grasp more material per unit of time. In addition, goats' food intake declines rapidly and may stop if the pasture is soiled or trampled, even with an ample amount of pasture remaining.

Generally one cow eats about the same amount of feed as six to eight goats (table 6-6, p. 179). Because of the complementary grazing habits, the differential preferences, and the wide variation in vegetation within most pastures, one to two goats could be grazed with every beef cow without adversely affecting the feed supply of the beef herd. The selective grazing habits of goats in combination with cattle would eventually produce pastures that would be more productive, of higher quality, and have few weed problems.

ANIMAL SELECTION, BREEDING, AND GENETICS

Production Traits

Introduction

Four key traits to be considered for genetic improvement in goats used primarily for meat production include: (i) adaptability to environmental and production conditions, (ii) reproductive rate, (iii) growth rate, and (iv) carcass characteristics. Of these four production traits, only carcass characteristics are not readily measurable on the farm. For dairy goats milk quantity and quality (fat, protein, and solids nonfat) are important breeding goals to add in place of carcass characteristics, along with udder type, feet and legs, etc. Milk yield is easily measured in small dairy goat herds; milk quality data can be obtained by participating in Dairy Herd Improvement Association owner sampling programs.

Adaptability

Adaptability is the most important of all the production traits. The profitability of any goat enterprise may be greatly diminished if the production environment impairs the animals' ability to survive and reproduce. Goats have proven to be perhaps the most adaptable of all the domesticated livestock. Indeed, goats survive worldwide in a wide range of environmental conditions. However, when taken out of one environment and placed into another, domesticated livestock of any species may not always realize their production potential (see chapter 1). Therefore, we might expect Spanish goats to perform differently in the Carolinas and Virginia than they do on the arid Edwards Plateau of Texas. Similarly, Boer goats might perform differently in South Africa than they do in North America. In addition, various breeds exhibit different degrees of adaptability. For example, we might expect Spanish goats to be inherently better adapted to extensive browsing conditions than Myotonic goats.

Adaptability is low in heritability because natural selection has already reduced the genetic variability. Therefore, adaptability will respond slowly to selection. Chapter 1 discusses the learned feeding behavior of animals and how it relates to animal adaptation to new habitats.

Reproductive Rate

In animals kept primarily for meat production, reproductive rate is the single most important factor contributing to the efficiency of production. Reproductive traits of interest in a meat goat enterprise are conception rate, kidding rate, and ability to breed out of season.

In general, goats have a high reproductive rate, and conception rate is not usually a problem. Several studies have demonstrated that although twins and triplets have lower birth and weaning weights and slower growth rates, they produce more total weight of kid per doe per year. Therefore, prolificacy, defined as the number of kids born per doe, is an important reproductive trait. Goats that have evolved in the temperate zones of the world tend to be seasonal breeders, with females coming into estrus in the fall and anestrus in late spring and summer. This breeding pattern does not always coincide with the optimal marketing period of weaned kids. On the other hand, goats from tropical regions are nonseasonal breeders and kid all year-round. Therefore, incorporating this trait of nonseasonality into a meat goat enterprise would be advantageous.

Intersex or pseudo-hermaphrodite or hypoplasia of sex organs is a reproductive problem that has received considerable attention because this condition is associated with the absence of horns. The mating of two **polled** goats will result in a percentage of intersex, sterile animals. Linked to the polled gene is a dominant gene for intersexuality that is manifested only in the homozygous polled female. Female intersexes are genetically female but externally can range from an apparently normal female to male in appearance. Some animals have an enlarged clitoris and are obviously abnormal at birth, but others may reach maturity before being detected.

Growth Rate

Growth rate can be effectively divided into two periods: preweaning average daily gain and postweaning average daily gain. A high preweaning average daily gain not only reflects the genetic potential of the growing animal, but also the mothering ability of the doe, her milk yield, or the nutritional management of replacement kids. In some production systems, kids are sold at weaning and therefore preweaning average daily gain is an important production trait to consider. In other production systems

kids are sold as yearlings or as older animals and postweaning average daily gain becomes an important production factor. For dairy doe kids the optimum growth rate allows freshening of yearlings and development of optimum body size so that animals can have the high forage intake needed to maintain higher levels of milk production.

Carcass Characteristics

Carcass characteristics of interest are dressing percentage, anatomical distribution of muscle, and the ratios of lean:fat:bone. Generally, the dressing percentage of goats is around 45%. As an animal grows, the percentage of fat in the carcass tends to increase, the percentage of bone tends to decrease, and the percentage of lean muscle stays about the same. The portions of the carcass with the largest muscle mass are the leg and shoulder. However, as a percentage, these portions tend to decrease as the animal grows.

Conformation and General Appearance

With the exception of the Boer goat, meat goat breeds are lacking in some aspects of performance or have not yet been tested in our production systems. Using a set of scales and good record keeping, meat goat producers can readily collect the information needed for the selection of animals possessing the economically important traits described while keeping carcass characteristics in mind.

Breeding

Introduction

Breeding is a very important aspect of any goat operation. Preparing the breeding does and buck(s) for the breeding season could have a large influence on the outcome and the profitability of the operation. Important factors will affect breeding indirectly, such as body condition (see earlier section in this chapter), the grouping of animals, deworming, trimming feet, using the "buck effect" to synchronize does, and vaccination.

Grouping Animals

Goats are very social animals and should be grouped together several weeks before the breeding season so that the pecking order of the animals may be established. Forming groups just prior to the breeding season will disrupt the animals' pecking order. The fighting that will ensue to establish a new pecking order within the newly formed groups will be a source of stress and will influence reproductive performance.

Deworming

Deworming breeding does and the buck(s) before the start of the breeding season is an important management technique. If flushing is planned, it is advisable to deworm prior to flushing. Wormy does will not increase their body condition during the flushing period; therefore, flushing may not increase the ovulation rate. In addition, wormy does will not breed well, may not breed at all, or may conceive and later abort.

Trimming Feet

Feet and legs should be examined closely for sores, overgrown hooves, and sources of strange smells that could be associated with infections or foot rot. Start trimming the feet of animals several weeks before the breeding season to make sure that they will be in top shape during that period of increased activity. The buck in particular will cover a lot of territory. A lame buck will cover does only sporadically, or might give up altogether. Similarly, limping does may not let bucks breed them.

The "Buck Effect"

Segregating does from bucks is crucial in the development of sound breeding programs. The best approach to separate does from bucks is to develop a secure buck pasture. The buck pasture should be far from the breeding doe herd, otherwise bucks will attempt to go through fences to breed does in estrus.

In goats, estrus can be induced with the strategic exposure of anestrus does to intact males. This response depends on the depth of seasonal anestrus and is associated with a first ovulation in 2–3 days after the introduction of the buck. The first ovulation is usually silent and of low fertility. The second ovulation 5 days later is accompanied by a fertile estrus. The response to the buck effect is influenced by the sexual aggressiveness of the buck, the intensity of the stimulation, and the body condition of the does. Immediate contact results in a greater response than fence-line contact or intermittent contact. The pheromones responsible for inducing estrus are present in buck hair, but not in urine, and are not associated with buck odor during the breeding season.

Bucks should not be given access to lactating dairy does other than at the breeding event. The scent of bucks in rut will rub off on the doe, making it difficult to produce milk that is not "goaty" in flavor. This must be prevented or it will be diffficult to market milk or cheese made from such milk.

Vaccination

Although some producers have had no health problems when not implementing a vaccination program, it is recommended that goats be vaccinated against overeating disease (enterotoxemia) and tetanus. For twice a year vaccination, breeding does should be vaccinated before the start of the breeding season and 4–6 weeks before kidding. If vaccinated once a year, it is preferable to vaccinate does prior to kidding because some immunity will be passed on to the newborns. The choice of vaccines is the following:

- *Clostridium perfringens types C and D + tetanus toxoid* in one vaccine, against over-eating disease and tetanus. This vaccine is labeled for goats.

- *Multivalent clostridial vaccine (8-way vaccine).* One example of a multivalent clostridial vaccine, labeled for sheep, is Covexin8, which is more reactive and may cause a higher incidence of adverse reaction at the injection site. Covexin8 may be used in herds that have had problems with blackleg and malignant edema (gas gangrene). Although blackleg and malignant edema are common and costly infections in sheep and cattle, they are uncommon in goats.

Is the Buck Ready for Breeding?

Bucks may be easily overlooked, but don't assume that they are reproductively sound. A buck that was sound one year may not be the next. The results of using a reproductively unsound buck will be reduced kidding rates and profits. It is a good idea to watch bucks for normal urination and also for signs of sexual behavior as the breeding season approaches. For a more thorough breeding evaluation, sit the buck on its rump. With the back of its head resting on your thigh, examine the testes. They should be roughly the same size, fairly firm to the touch and devoid of lumps. The presence of testicular abnormalities could indicate that the buck is unsound for breeding. Next, examine the sheath (also called the prepuce) and the penis if you can make it protrude. It requires some experience to push the prepuce down to reveal the penis. The penis should be checked for sores, and the pizzle (the thin wormlike process at the end of the penis) should not be hard anywhere. The presence of hard, small lumps could be an indication of urinary stones (a condition also called urinary calculi). A buck suspected of reproductive problems, whether in the testes or any part of the penis, should be examined by a veterinarian before it is allowed to breed does.

Breeding Season

Although goats are considered seasonal breeders and in our region the breeding season generally extends from September to February, many exceptions occur. Among dairy breeds (e.g., Alpine, LaMancha, Nubian, Oberhasli, Saanen, Toggenburg), some does have the ability to breed out of season and as early as July if housed or grazed with a buck. Meat-type goats such as the Pygmy and the Myotonic apparently have the ability to breed out of season. The same appears to be true for the Boer breed. Factors playing an important role in the ability of goats to breed out of season include plane of nutrition, body condition, stimulus from a buck, and day length.

For successful breeding, does and bucks should be joined for 40–45 days, which is the length of time necessary for does to complete two estrous cycles. A ratio of 20–30 does per buck is recommended for best breeding results.

Heat Detection

Does in heat become vocal, and some bleat very loudly as if in pain. Constant tail wagging from side to side is another sign of heat. In addition, the vulva will appear slightly swollen and reddened, and the area around the tail may look wet and dirty because of vaginal discharge. Other signs of heat include decreased appetite and an increased frequency of urination. Does in heat also are easily identified if a mature and smelly buck is nearby. They will pace restlessly along their enclosure for a way to get to the buck or stand close to the fence. Finally, a doe in heat may mount another doe as if she were a buck or let another doe mount her.

In spite of all these signs, it is still sometimes possible to miss heat. In general, people who have the most trouble detecting estrus usually have only one or two goats. In some instances, it may be very useful to run a teaser (vasectomized) buck with the does to detect estrus. A vasectomized buck is rendered infertile through surgery that cuts the tubes carrying the sperm from the testes to the penis. However, his libido and interest in mating still remain. An intersex animal exhibiting female genitalia with an enlarged clitoris but demonstrating male mating behavior can also be used to detect estrus. Goats (bucks, intersex females) used to detect estrus can be fitted with a harness containing a crayon that will mark the females in heat when they are mounted. If the herd is checked twice a day, marked females can then be separated and mated to the appropriate stud male.

Estrous Cycles

During the breeding season, goats come into heat or estrus approximately every 18–22 days. A transitional period occurs at the beginning and end of the breeding season during which short heat cycles without ovulation have been documented. Short estrous cycles of less than 12 days and very often of 5–7 days may occur, especially in young does. Mature does that have shortened estrous cycles in the middle of the breeding season should be considered abnormal.

The duration of estrus varies from 12 to as long as 48 hours. Within that duration standing heat (the period during which the doe stands firmly when a buck attempts to mount) lasts approximately 24 hours. On occasion, some does may find the buck sexually unattractive and will not stand to be bred. Ovulation usually occurs 12–36 hours after the onset of standing heat. At the beginning of estrus, the vaginal discharge is clear and colorless. It becomes progressively whiter and more opaque toward the end of standing heat.

Puberty, Breeding, and Body Size

Does reach puberty and may be ready to breed at 7–10 months of age. However, does should not be bred until they reach 60–75% of their expected mature weight, otherwise their growth may be stunted. Therefore, in deciding when to breed does, producers should consider their age and size, but also when they were bred last, and their body condition. Season should also be considered because kids born during the hot spring or summer months do not thrive; they experience more health problems than kids born during cooler times of the year. Meat goats can be bred every 8 months. However, such frequent breeding requires excellent management, good nutrition, and breeds that effectively breed out of season. In addition, environmental conditions during the summer months will increase death losses of kids and decrease growth rate. Breeding once a year will result in increased litter size per breeding and over the lifetime of the doe, give the doe more time to nurse kids when they grow the fastest, and allow the doe time to rest and replenish its body condition for the next breeding season.

Gestation Length

The average gestation period is 150 days, with a range of 146–155 days. Usually, older does carry and give birth to more kids than does giving birth for the first time. Parturition signs are like those observed in most mammalian species such as enlargement of the vulva and of the abdomen, udder swelling, and relaxation of pelvic ligaments. In addition, does about to give birth will become restless, paw the ground, and repeatedly lie down and then stand up. They will discharge mucous and may move away from the rest of the herd into a secluded corner or even into some underbrush or a creek bed, which is dangerous for the survival of the newborn.

Goat Genetics

Introduction

Goats of any breed or crossbreed are eventually sacrificed for human consumption. With the exception of the South African Boer goat imported via New Zealand in early 1993, there are no true meat goat breeds in the United States. However, there are a few breeds that stand out as more suitable for meat production. These breeds are the Myotonic, Kiko, Nubian, Pygmy, and Spanish goats.

Boer Goat

The Boer goat of South Africa owes its name to the Dutch word "boer," meaning farmer. The origin of Boer goats is vague and probably rooted in indigenous goats kept by Hottentot and migrating Bantu tribes, with a possible infusion of Indian and European bloodlines. The present-day improved Boer goat emerged in the 20th century when South African farmers started breeding for a meat-type goat with good conformation, high growth rate and fertility, short white hair, and red markings on the head and neck. The South African Boer Goat Breeders' Association was founded in 1959 to establish breed standards for the emerging breed. Since 1970 the Boer goat has been incorporated in the South African National Mutton Sheep and Goat Performance and Progeny Testing Program, which makes the Boer goat the only known goat breed routinely involved in performance and progeny tests for meat production. There are approximately 5,000,000 Boer goats in Africa, of which 1,600,000 are of the improved type.

New Zealand and Australian companies have imported the Boer goat into their respective countries to help improve their own meat goat industries. In April 1993, the quarantine restrictions for the New Zealand Boer goats expired, and animals became available for importation into the United States. The Australian Boer goats were released in October 1995. In June 1993, the North American Boer Goat Association was founded, breed standards were established, and a registry of animals was begun. According to New Zealand researchers, the plane of nutrition plays a greater role than the light-dark cycle in stimulating Boer goats to breed out of season.

Spanish Goat

The Spanish goat came originally from Spain via Mexico to the United States. It is now a meat-type goat found primarily on or around the Edwards Plateau of central Texas. The Spanish goat has the ability to breed out of season and is an excellent range animal because of its small udder and teats. In addition, Spanish goats are usually characterized as being very hardy and able to survive and thrive under adverse agroclimatic conditions with only limited management inputs. Within the general group of "Spanish goat" there are those that are purely Spanish, whereas others represent an amalgam of all genotypes introduced to the area. There have been obvious infusions of dairy and Angora blood in many Spanish herds, but no organized attempt has ever been made to use them for milk or mohair production.

The terms "wood" (Florida), "brush" or "briar" (North Carolina, South Carolina), "hill" (Virginia), and "scrub" (Midwest, Pennsylvania) goat tend to be used in the Southeast and elsewhere to refer to Spanish goats. Until recently, these goats were kept mainly for clearing brush and other undesirable plant species from pasture lands. Presently, they are also being used to reduce the undergrowth in hardwood forests and other timberland areas to provide buffer zones around rural communities and newly established development projects as viable protection against forest fires during periods of summer drought. In addition, they also provide an environmentally friendly alternative to herbicides under power-line rights-of-way.

In recent years, the escalating demand for goat meat and the expanding interest in cashmere production have focused attention on the Spanish goat. Current estimates of the Spanish goat population are around 500,000 head. Several Spanish goat producers in Texas have been intensively selecting for increased meat production for the past several years. From information obtained from these producers, these "selected" Spanish goats appear to greatly outperform the ordinary Spanish goat used primarily for pasture maintenance.

Myotonic Goat

The Myotonic goat has several aliases, including "Tennessee stiff-leg," "Tennessee wooden-leg," "nervous goat," "fall-down goat," and "fainting goat." The Myotonic goat is a very meaty and muscular animal. This goat breeds out of season, and in many herds it is usual for does to kid twice a year. The number of kids varies from one to four.

The Myotonic goat suffers from a recessive trait called myotonia. When frightened, it experiences extreme muscle stiffness, causing extension of the hind limbs and neck. In this startled state, if unbalanced, the animal will topple over like a statue or will stand immobile until the attack, usually lasting only 10–20 seconds, passes. According to a Texas neurologist, this type of involuntary isometric muscle contraction could build a more tender muscle than a muscle developed by strenuous use.

Little is known about the earliest history of this breed except that in the early 1880s a man appeared in Marshall County, Tennessee, with a cow and three does and a buck of a unique strain. These four goats suffered from myotonic

spells and were purchased by a Dr. Mayberry, who propagated the breed. The population of Myotonic goats is informally estimated to be around 3,000–5,000 head, with herds found primarily in Tennessee and Texas.

Nubian Goat

The Nubian goat, also called Anglo-Nubian, is considered a dual-purpose goat breed used for milk and meat production. This breed was developed in England and is a composite of dairy goat breeds from India, Europe, and Africa. Brought into the United States at the beginning of the 20th century, the Nubian has become the most popular U.S. dairy goat breed, with more than 100,000 registered breeding stock.

Alpine, Oberhasli, Saanen, and Toggenburg Goats

Alpine, Oberhasli, Saanen, and Toggenburg are popular dairy goat breeds originating in the alpine regions of Europe. They sometimes produce more milk than some Nubian does, but it often contains a lower amount of milk fat and solids nonfat.

LaMancha Goat

LaMancha is a breed developed in California from Spanish Murciana origin and Swiss and Nubian crossings. LaMancha are known for excellent adaptability and good winter production. They are also producing fleshier kids than the Swiss breeds, but are not milking as much. They have no external ear or only a rudimentary ear due to a dominant gene.

Pygmy and Nigerian Dwarf Goats

The Pygmy and the Nigerian Dwarf are miniature goats of West African origin, but they are separate and distinct breeds. The Pygmy goat was originally called the Cameroon Dwarf goat. It is a heavily muscled and short-legged goat.

Cameroon goats found their way to the Caribbean and North America as a byproduct of the slave trade in the 18th century. Cameroon goats were also exported from Africa to zoos in Sweden and Germany where they were on display as exotic animals. From there they made their way to England, Canada, and the United States. In West Africa, the Pygmy is used almost exclusively for meat production. The pygmy is well adapted to humid climates, it usually breeds all year-round, and twinning is frequent. In the United States, the Pygmy has so far been raised mainly as a pet and as a show animal, and more than 30,000 animals are currently registered with the National Pygmy Goat Association.

The Nigerian Dwarf is similar in conformation to that of the larger dairy goat breeds. Nigerian

Dwarf goats provide a surprising amount of milk (3–4 pounds per day) for their size and are prolific year-round breeders. Nigerian dwarf goats are registerable in three registries: American Goat Society, International Dairy Goat Registry, and Canadian Goat Society. Only 3,500 animals are registered in the United States.

Kiko Goat

The Kiko is a meat breed that originated from large dairy males crossed with New Zealand feral stock and then backcrossed to dairy males over two decades of intensive selection. They were then selected for twinning, growth rate, and constitution. The Kiko is thought to be a vigorous, hardy, large-framed, and early-maturing animal that doesn't need pampering.

CHAPTER 7

Horse Nutrition and Management

William J. Bamka, Daniel Kluchinski, and Jeremy W. Singer

INTRODUCTION

Horses are used in a variety of activities; there are more than 5.32 million horses in the United States (35). Many of these horses are owned and managed for profit, and a significant number are kept for recreation and sport. Regardless of the use, proper nutrition is essential for maximizing animal growth and productivity, and pastures play an important role in feeding and exercising horses. Although nutritional needs vary considerably among horses, depending on breed, age, weight, and activity level, forages can and should be a primary component of the equine diet.

The horse and its relatives are nonruminant herbivores or natural grass eaters. A horse's digestive tract differs considerably from that of a cow. The cow has a rumen in the front of the digestive tract where much of the digestion and synthesis of B vitamins and amino acids occurs. The horse has a small, simple stomach and a large cecum and colon located between the small and large intestines. Most digestion and absorption takes place forward of the cecum, which is similar to other simple-stomach animals. The cecum and colon in the horse generally serve a similar function as the rumen in the cow. Synthesis of B vitamins and amino acids occurs in the cecum and colon of the horse. The location of the cecum near the end of the horse's digestive tract likely reduces its contribution to digestive efficiency. Feed passes through the digestive tract of the horse faster than through the digestive tract of ruminants; the faster feed passage rate contributes to the lower digestion efficiency in horses. The capacity of the digestive tract of the horse is smaller when compared to ruminants, and the equine digestive system functions best at two-thirds capacity. The equine stomach is actually designed for near constant intake of small quantities of feed rather than large quantities at one time. Therefore, smaller, more frequent meals are more desirable. This makes grazing ideal for the horse.

To fully take advantage of the pasture resource, producers must practice sound pasture and grazing management. Well managed pasture during the growing season can reduce horse feed costs and completely replace all supplementation, with the exception of water and salt, for mature, idle, and recuperating horses, as well as those in the early stages of gestation. In fact, good quality pasture can provide the maintenance needs of most mature horses (22). Pasture can provide roughage at a cost lower than that of purchased grains. Studies have shown that the annual costs of horse care can range from $500 to $3,500, with 50% of the expenses related to feed (17). Well managed pastures furnish horses with high quality, nutritious feed at a relatively low cost and help to maintain healthy animals by allowing exercise and access to sunshine and fresh air. Other positive health benefits of pasturing include reduced incidence of colic and laminitis (founder) (22).

The northeastern United States has tremendous pasture potential that horse managers can harness. This potential stems from the adaptation of numerous forage species, a favorable climate, and the length of the growing season. However, many pasture managers do not use sound grazing and forage management practices to reach full pasture production potential, or are constrained by the amount of pasture acreage available. Staff of the U.S. Department of Agriculture's National Animal Health Monitoring System (NAHMS) collected data from a representative sample of equine operations in 28 states in 1998 (33). The NAHMS report indicated that more than 27% of operations that pastured equids for 3 or more months did not rely on pasture to provide at least 90% of the roughage in the horses' diet.

Unlike other livestock, horses have never been selected for feed efficiency or uniformity; therefore, the equine manager must maintain awareness of each animal's individual needs. An understanding of pasture management principles and nutritional requirements of horses is essential, and the two must be dovetailed so both resources are maintained and optimally used. This chapter will focus on the methods to determine horse nutritional needs and effective pasture and grazing management techniques to supply quality forage and nutrients.

NUTRITIONAL REQUIREMENTS OF HORSES

Important to any feeding regime is a general knowledge and understanding of the nutritional needs of horses, the nutritional status of individual horses, and the feed and forage resources available to meet the nutritional requirements of the horses. The following information is presented to provide an overview of basic animal and horse nutrition. A more complete review of animal nutrition can be found in chapter 2. Although that chapter focuses primarily on ruminant nutrition, much of the information presented is applicable to horses.

Basic nutritional requirements that support body functions include carbohydrates and fats (energy), protein, vitamins, minerals, and water. Carbohydrates, which are abundant in plant materials, and fats, which are found in feed, provide the fuel necessary for physical activity, growth, milk production, and cell repair. Protein is needed for growth, muscle development, reproduction, lactation, and tissue repair, as well as skin and hair development. When energy is low in the diet, protein can be converted to energy (by ketosis), but this happens only rarely in horses. Vitamins perform a number of functions in the body, such as acting as catalysts for metabolism.

Nutritional requirements of horses can be grouped into two components: maintenance requirements and activity requirements. Both of these requirements must be satisfied if a horse is to maintain its body weight and condition. The major nutrient requirements for different classes of horses are provided in table 7-1, (p. 190). This table, prepared by the National Research Council, estimates daily nutrient requirements of average horses over different physical conditions and activity levels. This information can be used as a guideline to establish approximate needs; however, an analysis of each horse's body condition score is needed to more fully determine animal-specific dietary needs. Once dietary needs are established, a feed ration can be developed. The process of developing a feed ration is explained in detail in chapter 2.

BODY CONDITION SCORING

Having your horse in the best condition is important for health and performance. Although producers may group horses together in similar production and weight classes to assess energy sources, levels, and feed utilization, routine assessment of each horse's body condition is necessary because horses in similar production

Table 7-1. Nutrient concentration in total diets for horses and ponies (90% dry matter basis).

Stage of production	Digestible energy[a] (Meal/lb)	Crude protein (%)	Calcium (%)	Phosphorus (%)	Vitamin A (IU/lb)[b]
Mature horses					
Maintenance	0.80	7.2	0.21	0.15	750
Stallions	1.00	8.6	0.26	0.19	1,080
Pregnant mares					
9 months	0.90	8.9	0.39	0.29	1,510
10 months	0.90	9.0	0.39	0.30	1,490
11 months	1.00	9.5	0.41	0.31	1,490
Lactating mares					
Foaling to 3 mos.	1.10	12.0	0.47	0.30	1,130
3 months to weaning	1.05	10.0	0.33	0.20	1,240
Working horses					
Light work[c]	1.05	8.8	0.27	0.19	1,100
Moderate work[d]	1.10	9.4	0.28	0.22	970
Intense work[e]	1.20	10.3	0.31	0.23	800
Growing horses					
Weaning, 4 mos.	1.25	13.1	0.62	0.34	650
Weaning, 6 mos.					
Moderate growth	1.25	13.0	0.50	0.28	760
Rapid growth	1.25	13.1	0.55	0.30	670
Yearling,12 mos.					
Moderate growth	1.15	11.3	0.39	0.21	890
Rapid growth	1.15	11.3	0.40	0.22	790
Long yearling, 18 mos.					
Not in training	1.05	10.1	0.31	0.17	930
In training	1.10	10.8	0.32	0.18	740
2-year-old, 24 mos.					
Not in training	1.00	9.4	0.28	0.15	1,080
In training	1.10	10.1	0.31	0.17	840

[a] Values assume a concentrate feed containing 3.3 Mcal/kg and hay containing 2.0 Mcal/kg of dry matter. Mcal = megacalorie (1 million calories).
[b] IU-international units.
[c] Examples are horses used in Western and English pleasure, bridle path hack, equitation, etc.
[d] Examples are horses used in ranch work, roping, cutting, barrel racing, jumping, etc.
[e] Examples are race training, polo, etc.

Source: Reprinted with permission from the National Academies Press, Copyright 1981, National Academy of Sciences.

and weight classes will vary in their nutrient needs. Body condition scoring provides a quick, reliable method to evaluate whether each horse is in proper condition. Based on the assessment, animals in abnormally high or low body condition may need to be separated further to help ensure that the individual needs of each animal are met through adjustments to the feeding program.

Scoring Method

Body condition scores are numbers used to suggest the horse's relative fatness or thinness. Most often a scoring range from 1 to 9 is used, with a score of 1 being extremely thin and 9 being extremely fat. Accurate assessment of a horse's fat cover allows for visual appraisal of

Score	Condition
Table 7-2. Body condition scoring for horses.	
1	Extremely thin. The horse is emaciated. The backbone, ribs, hip bones, and tailhead are all prominent. The neck is hollow and the bones of the shoulders, withers, and neck are easily discerned. Individual vertebrae are clearly seen and easily palpated. No fat can be palpated.
2	Very thin. The backbone is prominent, ribs, tailhead, and pelvic bones stand out. Bone structures of the neck, withers, and shoulders are evident. Individual vertebrae can be seen and are easily palpated.
3	Thin. The backbone is prominent but fat covers to the midpoint. A slight layer of fat can be felt over the ribs, the tailhead is evident, but individual vertebrae cannot be seen. Pin bones cannot be seen, but withers, shoulders, and neck are emphasized.
4	Moderately thin. A negative crease along its back. The outline of the ribs can be seen. Fat is palpable around the tailhead. Point of hip not evident. Withers, neck, and shoulders are not obviously thin.
5	Moderate. Back is level. Ribs can be felt but not easily seen. Fat around tailhead feels spongy. Withers are rounded, and shoulder and neck blend smoothly into body.
6	Moderately fleshy. A slight crease in along the back. Fat on the tailhead feels soft. Fat over the ribs is spongy. Small deposits of fat along the withers, behind the shoulders, and along the neck.
7	Fleshy. A crease is seen down the back. Ribs may be felt, but fat between ribs is obvious. Fat on the tailhead is soft. Noticeable fat along neck, behind shoulders, and withers.
8	Fat. Crease down the back is prominent. Ribs difficult to feel due to fat in between. Wither area is filled with fat and very soft fat over tailhead. The space between the shoulders is filled in and flush, and there is fat along the inner buttocks.
9	Extremely fat. The crease down the back is very prominent. Fat is in patches over the rib area, with bulging fat over tailhead, withers, neck, and shoulders. Fat along inner buttocks may rub together and flank is filled in flush.

Source: Adapted from Henneke, D.R. 1983. Relationship between condition score, physical measurements and body fat percentages in mares. Equine Veterinary Journal 15 (4):371–372. Used with permission.

the horse's energy status. Horses in a positive energy balance (too much intake of carbohydrates or fats) will store energy as fat. Body fat is reduced when the ration does not provide sufficient nutrients to maintain energy balance. Body condition scoring is accomplished by scoring the neck visually and then feeling for fat cover. This is also done for the withers, loin, tailhead, ribs, and shoulder areas (figure 7-1). The body condition score is then compared to a scoring table to determine the condition of the horse. Table 7-2, p. 191 explains the body condition scoring scale for horses (13).

After a body condition score is determined, one should use table 7-3 as a guide to compare the individual score with the desired condition score based on activity. Then adjustments in the animal's feeding should be made to raise or

Table 7-3. Desired body condition scores based on activity.	
Class of horse	Desired condition score
Dressage horse	6–8
Endurance horse	4–5
Eventing	4–5
Hunters	5–7
Open mares	4–6
Polo and polo crosse	4–5
Ponies on spring pasture	7–8
Pregnant mares	7–8
Quarter horses	6–8
Ranch horses	4–5
Show hacks	6–8
Show jumpers	5–7
Stallions (breeding)	5–7
Stallions (off season)	4–6
Standardbred racing horses	4–6
Thoroughbred racehorses	5–7

Source: Kohnke, J. 1992. Feeding and Nutrition, The Making of a Champion. Birubi Pacific. Rouse Hill, NSW, Australia. Used with permission.

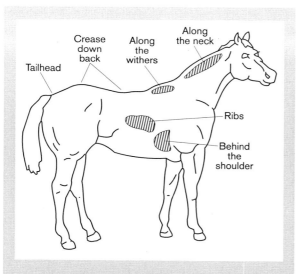

Figure 7-1.
Body condition evaluation areas.

Source: Adapted from Henneke, D.R. 1983. Relationship between condition score, physical measurements and body fat percentages in mares. Equine Veterinary Journal 15 (4):371–372. Used with permission.

lower the body condition score to the appropriate level.

CONTRIBUTION OF PASTURE TO THE FEED RATION

Horse nutrition and pasture and grazing management are often approached as separate and unrelated tasks, but they should be considered hand in hand, because they are components of the overall animal production system. Deciding what fraction of the total dietary requirements can be realistically provided by pasture requires multiple pieces of information from the various components of the animal production system. Collecting this information can be a complicated task because this information may be very site- or operation-specific, and may often change due to the many factors involved. However, collection and use of this information can be invaluable to managing both the crop and animal components of the animal production system. To that end, the following sections of this chapter outline six key factors that must be known to best determine a pasture's capacity to supply the forage and nutritional needs of a horse (figure 7-2). These factors include pasture yield, length of the pasture season, pasture forage quality, horse energy demands, rate of forage intake, and pasture carrying capacity. Discussion of pasture and grazing management practices to optimize pasture and animal productivity will follow.

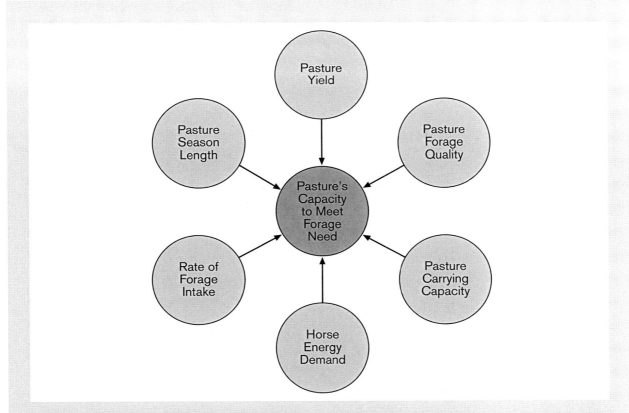

Figure 7-2. Factors that determine a pasture's capability to meet forage needs.

Pasture Yield and Pasture Season Length

Knowledge of species' growth rates, yield, and length of the growing season, either through experiential or theoretical data, can be used to determine seasonal availability of forage or the need to supply hay during growth slump periods. Dense sods such as Kentucky bluegrass will yield up to 2 tons of dry matter per acre, but production is high only in early summer and fall. Tall species such as orchardgrass and smooth bromegrass may yield twice that. When a legume is incorporated with grasses, 6 tons of dry matter per acre per year can be produced (18). A well managed grass pasture will out-produce a grass-legume mixture in spring but will produce less than mixtures during the remainder of the grazing season. Tables 1 and 2 in chapter 4 (Perennial Warm-Season Grasses) of the NRAES book *Forage Utilization for Pasture-Based Livestock Production* (NRAES–173; see HTTP://WWW.NRAES.ORG for more information) provide information on cool-season grasses, grass-legume, and warm-season grass pasture production (yield) and growth distribution in the Northeast over the year. Those tables also provide data on the annual number of animal grazing days per acre; this information can be used to supplement site-specific information when that is available.

Regrowth after grazing is another important factor that can affect growth rate and availability of forage. Regrowth rates vary by species and variety, and depend on grazing pressure, weather, and management practices. Pasture grasses and legumes have different abilities to recover from grazing. Species that have growing points underground tolerate frequent grazing better than those with growing points aboveground. Kentucky bluegrass always maintains storage and growing points underground, while timothy and smooth bromegrass have growing points that are

aboveground during stem elongation. If the latter species are grazed while their growing points are aboveground, it will damage and eventually kill them. White clover tolerates frequent grazing, but has a low tolerance to drought. Tall fescue is fairly tolerant of frequent grazing and can withstand trampling. Newly seeded pasture should not be grazed until a majority of the plants are 6–8 inches tall and then it should be grazed no lower than 2–3 inches. This will allow for adequate root growth to withstand pulling and plant reserves to recuperate from grazing.

Pasture Forage Quality

Forage analysis to determine forage quality (total digestible nutrients, protein, fiber) can be invaluable in ration balancing and determining nutrient cost basis comparisons. Published information on nutrient composition of forages can also be used. Table 7-4 provides average nutrient levels of various pasture species; however, the nutritional and mineral content of the forages from individual pastures will vary based on soil fertility, pH, species composition, stage of growth, climatic conditions, and other factors. Properly sampled, a forage analysis provides the most accurate information on forage quality. County agricultural agents can provide information on forage testing laboratories and methodologies.

Horse Energy Demands

Awareness of horse energy demands and consumption patterns is essential in determining the use of pasture, hay, concentrates, or combinations of the three in meeting animal needs. The energy requirements based on activity must be known to develop a feed ration or pasturing schedule. For example, mature horses performing minimal to no work can be maintained on high quality forages without grain supplementation. Horses that may require supplementation on pasture include fast-growing foals and weanlings,

Table 7-4. Nutrient composition of forage.

Forage species and maturity	Dry matter (%)	Digestible energy[a] (M cal/lb)	Crude protein (%)	Calcium[a] (%)	Phosphorus[a] (%)
Kentucky bluegrass					
Early vegetative	31	0.95	17.4	0.50	0.44
Mature	42	0.81	11.6	––	––
Orchardgrass					
Early vegetative	24	1.04	12.8	0.25	0.39
Mature	28	0.92	10.1	0.23	0.17
K31 tall fescue					
Early vegetative	31	1.01	15.0	0.51	0.37
Ladino clover					
Early vegetative	19.3	2.50	25.8	1.27	0.35
Red clover					
Early vegetative	24.0	3.19	22.3	1.71	0.26

[a] Based on 100% dry matter.

Source: Adapted from National Research Council. 1989. Nutritional Requirements of Horses, *5th ed. National Academy Press, Washington, DC. Table 6-1A, pp. 49–67.*

exercising horses, gestating and lactating brood mares, and stallions in breeding programs. These horse classes would have to consume unrealistic quantities of forage to meet nutritional needs, and therefore require supplementation with concentrates.

The determination of forage needs versus supply is based on the type, age, size (table 7-5), and activity (table 7-6, p. 196) of the horse. For example, to maintain a 1,000-pound horse without supplementation, a pasture must produce approximately 450 pounds of forage per acre per month. If the forage production is not adequate, due to the time of year or rate of regrowth, supplementation is required.

Table 7-5. Daily horse feed requirements as a percentage of body weight.

Age of animal	% of body weight
Nursing foal	3.5–4
Weanling	2.5–3
Lactating mare	3–3.5
Mature	2–2.5

Source: Wagoner, D.M., ed. 1977. Veterinary Treatments and Medications for Horsemen. Equine Research, Dallas, TX, p. 72.

Table 7-6. Horse energy requirements at various work levels.		
Work	**Definition**	**Kcal[a] required per hour**
Light	Slow jog, lope	2,000
Medium	Fast jog	6,000
Heavy	Gallop, jumping	11,000
Severe	Polo, speed work	19,000

[a] Kcal = kilocalorie (1,000 calories).

Source: Wagoner, D.M., ed. 1977. Veterinary Treatments and Medications for Horsemen. *Equine Research, Dallas, TX, p. 72.*

Forage Intake Rate

Consumption patterns and intake are horse-specific, and are best determined through observation. It is generally assumed that dry matter intake on pasture is similar to intake on a complete hay diet. Under normal conditions, mature horses with ample access to forage graze for approximately 14 hours per 24-hour period, primarily in daylight, but with the longest periods during early morning, late afternoon, evening, and midnight (21, 33). Typically mature horses will consume 2–2.5% of body weight in feed each day, which should include at least 1% by body weight of long roughage from hay or pasture (22). Forage consumption may also change due to health and environmental conditions. For example, decreases in grazing may occur when horses are irritated by flies, lack companionship, and during severe weather. However, decreased grazing time may also be an indication of a prolific, high quality pasture because nutritional needs are met more efficiently (21). Therefore, observational data are essential in determining pasture consumption rate.

Pasture Carrying Capacity

The terms "carrying capacity" and "stocking rate" are similar but have slightly different meanings. Carrying capacity is a measure of the number of animals that can be placed on a pasture for a season to achieve a targeted level of animal performance or economic production without causing deterioration of the pasture resource. Pasture carrying capacity depends on many variables in the animal production system. Stocking rate is the number of animals stocked per acre of grazing land in a management unit for a defined period, without reference to the condition of the pasture resource. Research on horse pastures in New Jersey (31) has indicated that most are overstocked. As part of a detailed study of 40 pastures in New Jersey, Singer found stocking rates ranging from 0.18 to 2.3 acres per horse (31). The median reported value was 0.61 acre per horse.

In addition to the number of horses, variation in the grazing behavior among individuals or herds remains an important factor in manipulating stocking rate in relation to daily intake and species selection. Horses are selective grazers that may choose only a few of many available species based on palatability. Although the horse owner may manage at an ideal production level, the animals make their own choices, resulting in mature, weedy growth in some areas and bare ground in others. As stocking rate increases, forage availability decreases, causing horses to consume plants and plant parts they would ordinarily leave, resulting in more balanced pasture utilization. Circumstance may require changes in the precalculated stocking rate to keep up with pasture growth. It is therefore important to develop a grazing management plan, and then the stocking density of each paddock must be established. Formulas exist involving animal units, dry matter intake based on percentage of body weight, and the production potential of various forage species. However, there are also many incalculable factors involved, so that experience and observation play important roles in determining proper stocking rates.

PASTURE MANAGEMENT AND GRAZING MANAGEMENT— A BALANCING ACT

Proper pasture management and grazing management are essential for optimum pasture production and utilization. Horse pastures differ in several respects from cattle pastures, although many of the principles for establishing and maintaining forages for cattle pastures also apply to horse pastures. These practices include pasture plant species selection, field preparation for planting, soil testing, fertility and pH management, clipping, weed control, and manure management.

These pasture management practices must be integrated with grazing management practices, because the effects of the animal on the pasture and the pasture on the animal are interrelated. Due to grazing behavior, horse pastures often require more intense management than cattle pastures. Horses are selective grazers, which

affects the productivity of a pasture. Generally, horses prefer to eat young, immature plants and will graze some areas down to bare ground. In other areas horses will allow plants to reach maturity, thereby lessening palatability and reducing forage quality. Clipping plants off close to the ground can also inhibit plant regrowth. Additionally, horses will not graze areas where they defecate. This type of grazing behavior is referred to as spot or pattern grazing. The net effect is a pasture containing overgrazed and undergrazed areas of forage; these short and tall forage areas are often termed "lawns" and "roughs." Equine pasture managers are also challenged because horses are more destructive to pastures than cattle. Horses tear the sod, wear paths along fences, and punch holes in the sod during wet periods. Overgrazed or poorly managed pastures supply little or no feed, can allow for the introduction of weeds and poisonous plants, and possibly serve as a source of internal parasites, such as *Strongylus* (blood worm). Muddy and wet pastures can also increase risk of injury from disease or unsafe footing, while dry barren pastures may create a problem with airborne or inhaled dust and particulate matter.

The following two sections of the chapter discuss pasture management practices and grazing management practices for horses.

Pasture Management

It is important to establish and maintain a strong stand of forages regardless of the grazing system used. Once established, a pasture must be properly managed to remain a viable resource in the animal production system. A pasture should be monitored and managed to prevent overgrazing. A question often faced by producers is whether maintaining an older established stand is more economical than renovating or reestablishing a pasture. Comprehensive coverage of pasture establishment and renovation can be found in chapter 7 (Establishing Forage Stands) of the

NRAES book *Forage Production for Pasture-Based Livestock Production* (NRAES–172; see HTTP://WWW.NRAES.ORG for more information). The following sections provide important information on horse pasture management.

Forage Species Selection

Horse owners are continuously confronted with pasture management decisions that ultimately affect the productivity, persistence, and appearance of their pastures. An understanding of pasture species, growth habits, and specific growing conditions is required for proper species selection and management. First and foremost, an assessment of soil drainage, intended grazing pressure, and production goals is necessary. This information will aid in making decisions about the types of grasses and legumes that can be planted for optimum pasture production.

Realize that species selection is not the only factor influencing production success, but is the foundation upon which productive pastures are based. Finally, the way in which the pasture is used (continuously grazed, rotationally grazed, or exercise lot) and the grazing habits and foot traffic of horses must be considered to determine management needs and production limits.

Grasses

Grasses are the mainstay of horse pastures. Table 7-7 lists recommended grass species adapted to growing conditions in the northeast United States. Cool-season grass species, rather than warm-season grasses, are most commonly grown in the region; these include Kentucky bluegrass, perennial ryegrass, orchardgrass, smooth bromegrass, reed canarygrass, timothy, and tall fescue. Cool-season grasses thrive in

Table 7-7. Selected properties of forage grass species recommended for the northeastern United States.

| Grass | Seedling vigor[a] | Tolerance to soil limitation | | | Winter hardiness | Tolerance to frequent grazing[b] |
		Drought	Wet	Low pH		
Kentucky bluegrass	M[c]	L/M	M	M	H	H
Orchardgrass	H	M	M	M	M	M
Perennial ryegrass	H	L	M	M	L	M
Reed canary grass	L	H	H	H	H	M
Smooth bromegrass	H	H	M	M	H	L
Tall fescue	H	M	M	H	M	M
Timothy	M	L	L	M	H	L

[a] The higher the vigor, the more rapid the germination and establishment of the species.
[b] Frequent grazing refers to any grazing system that provides the recommended 3-week rest period between grazing events.
[c] H-high, M-medium, L-low.

temperate climates, with the majority of their growth occurring in the early spring and late fall when temperatures are cooler. Productivity of cool-season grasses decreases during hot summer weather. Warm-season grasses usually have a tropical origin and are most productive in the hot summer months; however, they often are less hardy and can winter kill. Cool-season grasses also vary in cold hardiness. Susceptible species often winter kill when exposed to below normal winter temperatures. Table 7-7 provides information on seedling vigor, tolerance of the species to droughty or wet soil conditions, pH (soil acidity), winter hardiness, and tolerance to frequent grazing.

Although commonly recommended throughout the northeastern United States, timothy and smooth bromegrass are probably not good choices for horse pastures unless a regular schedule of pasture rotation is practiced, because they do not tolerate frequent grazing. If the amount of pasture is limited, stocking densities are high, or rotational grazing is not practiced, species that tolerate frequent grazing are essential. Tolerance to frequent grazing is most often the critical criterion for horse pasture managers when selecting species for pasture establishment or renovation. Consequently, it is not surprising that Kentucky bluegrass and common white clover are the two most abundant species in horse pastures. A common seeding mixture in the northeastern United States contains Kentucky bluegrass, endophyte-free tall fescue, and white clover. This mix performs well, unless site-specific pasture conditions limit the use of these species. Kentucky bluegrass and white clover both tolerate frequent grazing but are sensitive to dry soil conditions. Tall fescue tolerates dry conditions better than Kentucky bluegrass or white clover. Reed canarygrass tolerates wet soils, but even the new low-alkaloid varieties are not as palatable as other pasture species. Orchardgrass tolerates frequent grazing better

than smooth bromegrass and timothy, is high yielding, and establishes quickly, so it is usually recommended with Kentucky bluegrass and white clover. Perennial ryegrass can be substituted for orchardgrass, but because it is not as winter hardy, reseeding often may be necessary in more northern locations.

Tall fescue can be a useful grass because it is tough forage that persists despite drought or heavy grazing pressure, although it may not be as palatable as other grasses. Tall fescue can also be infested with a naturally occurring endophyte fungus (*Acremonium coenophialum*) that grows as a parasite or symbiont within the plant. Though tall fescue production levels are greatest in the spring, quality factors such as palatability, digestibility, and nutrient concentration are higher in the fall compared to other species. Tall fescue maintains higher quality in the fall than most grasses. Foliage remains green through early winter and deteriorates slowly. In winter, tall fescue dry matter digestibility is 10% greater compared to orchardgrass-white clover mixes.

Legumes

Legumes are a family of plants that include alfalfa, birdsfoot trefoil, and clovers. Table 7-8 (p. 200) lists recommended legume species for the Northeast, and provides information on seedling vigor, tolerance to soil moisture, and low pH. Information about potential for frost heaving, persistence, and tolerance to frequent grazing is also listed.

Legumes provide a good source of protein and calcium, and also add nitrogen to the pasture through nitrogen fixation. Common white clover and Ladino white clover are the most popular legumes for horse pastures. Usually, the clover component in pasture seed mixes is low, and clover seed in the soil may also germinate and contribute to pasture productivity. Although

		Tolerance to soil limitations			Frost heaving potential	Persistence	Tolerance to frequent grazing[b]
Legume	Seedling vigor[a]	Drought	Wet	Low pH			
Alfalfa	M[c]	H	L	L	M	H	L
Birdsfoot trefoil	L	M	H	H	L	M	M
Red clover	H	L	M	M	M	L	M
White clover	M	L	H	H	L	H	H

Table 7-8. Selected characteristics of legume species recommended for the northeast United States.

[a] The higher the vigor, the more rapid the germination and establishment of the species.
[b] Frequent grazing refers to any grazing system that provides the recommended 3-week rest period between grazing events.
[c] H-high, M-medium, L-low.

alfalfa is commonly recommended, it is seldom found in pasture mixes. Red clover is also commonly recommended, but horse managers sometimes avoid it because it may cause slobbers or excessive drooling. This condition is caused by an alkaloid (slaframine) found in red clover infected with the fungus *Rhizoctonia leguminicola*. Birdsfoot trefoil is another recommended legume, but it is not commonly included in pasture mixes. It tolerates wet soil conditions better than alfalfa or red clover, but is slower to establish and does not tolerate frequent grazing as well as the white clovers.

Fertilization and pH

Pasture fertilization and pH management should be based on soil tests. Soil testing is best conducted by private and university agricultural analytical testing laboratories. These analyses will not only provide an assessment of current soil pH and fertility levels, but also recommend amounts of limestone and fertilizer needed. Soil testing is relatively inexpensive and an excellent investment. Without such information, pasture management is based on guesswork, and the practices employed may not meet the needs of pasture plants to optimize growth and productivity, or may lead to overapplication of nutrients. On farms where stocking densities are high and manure is distributed by dragging, soil testing is even more essential to ensure that environmentally sound nutrient management is practiced. It is estimated that 85% of the phosphorus and 50–98% of the potassium from forage is recycled in urine and feces, and may accumulate in areas where horses defecate and subsequently refuse regrowth (2).

Fertilizer application when required based on soil testing should be done during the growing season. Fertilizer applied at intervals throughout the growing season has a greater effect on pasture regrowth and fall root reserves than a single annual application. Legumes require more phosphorus and potash than grasses. Grass and grass-legume mixes with less than 30% legume respond to nitrogen fertilization. Because of its high mobility in soil, one-third of the total annual nitrogen recommendation should be applied in the spring, with two or three subsequent applications. Potassium and phosphorus applied in the fall will foster winter survival of plants.

Lime can be applied any time of year to established pastures, except on frozen ground. Lime should be topdressed or incorporated with fertilizer during establishment to maintain pasture soil pH between 6.4 and 6.8 to optimize plant nutrient utilization. For information on soil testing and interpretation of fertilizer and lime recommendations, consult with local extension personnel.

Clipping

Clipping or mowing of pastures is recommended season-long to minimize weed growth and encroachment, control undesirable species, reduce weed seedhead production and propagation, and minimize potential eye irritation of horses during grazing. The practice will also promote new growth when horses are grazing unevenly or when cool-season grasses mature in the summer and shade out legumes. Pastures should not be clipped lower than 3 inches from the ground. Shorter clipping reduces leaf surface area and photosynthesis and weakens the plants because root reserves must be used for new shoot growth. Clipping should cease in early autumn so that adequate growth is left to allow for winter grazing. In general, horse pastures require more clipping than do pastures for other livestock, because horses graze more selectively. However, frequent clipping may indicate the need to adjust next year's stocking rate.

Manure Management/Dragging

Manure is more often considered a waste product and a handling nuisance. With an average

daily production rate of 45 pounds per 1,000-pound horse, proper manure management is essential. Manure from stalls and barn areas can be hauled away off site, composted, or spread on cropland. Manure left in the field by grazing horses is a resource that may provide a source of nutrients for pastures. The typical nutrient composition of horse manure is 12 pounds of nitrogen (N), 5 pounds of phosphorus (as P_2O_5) and 9 pounds of potash (K_2O) per ton. In addition, other macro- and micronutrients present can contribute to soil fertility and plant nutrition. As discussed above, the use of manure as a nutrient source should be part of the overall nutrient management plan developed using soil testing and manure analysis.

Dragging pastures will help distribute manure, eliminate grazing avoidance areas, and increase utilization of the pasture. Horses defecate in particular patterns based on gender, and many publications maintain that they will refuse to graze near fecal piles because of the odor or other sensory cues to avoid parasites (7, 16, 26, 38). Studies have also demonstrated the lasting negative effects of fecal piles on balanced grazing. When eight fresh manure piles were placed on a field and removed one at a time, horses rejected all eight areas regardless of whether the individual piles were removed 24 hours or 2 weeks after placement (3). A 1995 Rutgers University study of equine grazing behavior found that grazing time was equal for lawn, bare, and rough areas, and that urination and defecation were not limited to a particular area with stocking rates of one horse per 0.46 acre during the day (23).

Many publications claim that dragging will desiccate and destroy parasite eggs or larvae, but this claim is not research-based (5, 9, 10). Other evidence suggests that no parasitic species relevant to equine health can be destroyed by desiccation because they employ strong survival mechanisms (Sukhdeo, 2000). The most damaging horse parasites (*Strongylus*) are usually acquired in the infective larval stage during grazing, primarily in the roughs. The parasites mature inside the horse and can cause severe tissue damage while migrating through the body to lay eggs, which are later excreted in manure. Clipping and dragging are recommended as soon as horses are rotated or when necessary on continually grazed pastures. The best way to avoid parasite-related problems is to remove manure daily, but this involves extreme labor costs. An Ohio State University study (9) concluded that removing manure only twice a week reduced parasite populations 18-fold.

Weed Control

Chemical or mechanical weed control methods may be necessary to remove undesirable weed species in an established pasture or prior to pasture renovation. In general, the best time to apply herbicides is in early spring when active weed growth ensues, during the spring or early summer, or late summer and early fall. Few herbicides can be used midseason where a clover is a desirable component of the pasture, because they may kill the clover as well as weed species. Grazing restrictions may require the removal of animals from the pasture for a period of time after applications are made. Information about grazing restrictions can be found on the pesticide label.

Mowing or power-shredding is an alternative to herbicide use. Perennial weeds should be clipped before the flower bud stage and annual weeds before seed formation. Grazing can have positive and negative effects on weeds. Continuous grazing often allows prostrate broadleaf weeds such as dandelion (*Taraxacum officinale*) and plantain (*Plantago* sp.) to proliferate, but rotated horses under high stocking rates often consume these weeds. A 1990 Australian study (11) recommended that horses grazing weedy areas should be restricted from rotation for 10

days before moving to weed-free pastures so that ingested viable seeds are eliminated in the feces elsewhere.

Grazing Management

Good grazing management is essential for farm operations that depend on pasture as a key source of feed. Grazing management affects the species composition and long-term survival of pasture plants, the forage quality of pasture plants, and overall pasture yield. A poor grazing management plan can result in an economic loss to operations that rely on pastures for forage. The overall goal of the grazing management plan is to achieve even grazing of all forage within the pasture. The spot grazing behavior of horses makes this a difficult goal to achieve depending on the type of grazing system used. Continuous and rotational grazing are the two most common grazing systems used by equine operators.

Continuous Grazing

Continuous grazing is the most common grazing system in the United States. With continuous grazing, animals remain on a grazing unit throughout the season. This system offers the benefit of lower capital inputs due to a reduced need for fencing and watering stations. Overgrazing in this system can result in a plant community of less desirable species over time. When horses graze without restriction they first select the most palatable forage. Repeated grazing of these plants over time without allowing for root recovery and leaf regrowth will result in plant death. The system is generally inefficient, because many areas are spot-grazed and possibly overused while others are untouched or avoided. If a continuous grazing system is used, prostrate species such as Kentucky bluegrass and white clover are ideal. This type of growth habit protects the growing point of the plant from being damaged.

Use of continuous grazing does not necessarily mean that the stocking rate is held constant. If too many animals are pastured continuously on the same field, overgrazing may be minimized by leaving horses on pasture for only a few hours a day or by removing them from pasture to another area where they can be supplemented with hay for at least 4 hours daily. Supplementing horses with extra hay and grain while they are on pasture will not prevent overgrazing. However, in many cases, continuous grazing leaves underused areas that contain manure and become unpalatable, thereby reducing pasture efficiency or the utilization of the resource. Due to the traffic around gates and waterers, "sacrifice areas" will develop due to trampling and defecation. Sacrifice areas can be managed in continuous systems by frequently relocating water and feed units. Some horse managers seed these sacrifice areas with tall fescue because it tolerates excessive trampling better than other species and the quality of dietary forage is not interrupted. If plant species for grazing are damaged by physical exercise by horses that are turned out from stalls, they should be exercised elsewhere before grazing.

In a continuous grazing system, increasing pasture efficiency is possible by following grazing horses with cattle, sheep, or goats, or grazing concurrently. These livestock are less selective and will consume less palatable, mature species and weeds refused by horses. This method of removing plant material reduces the need to clip and prevents pastures from undergoing plant population shifts that may require reseeding or weed management (5, 9, 15). Cattle and horses are not susceptible to the same parasites, so grazing both animals on the same pasture can potentially decrease the overall incidence of parasites because they die after consumption by unaffected species.

Rotational Grazing

Rotational grazing entails subdividing a single pasture into two or more smaller grazing units. Many horse farms can benefit from some type of controlled rotational grazing system. Horses are moved from one grazing unit to another for short periods of time. This system utilizes a temporary overstocking of the grazing unit. The temporary overstocking promotes greater forage use efficiency. Other benefits of rotational grazing of horses in equal-sized paddocks include improved yields, parasite control, and opportunities for fertilization.

Key to the success of the rotational grazing system is determining the optimum time to move horses to another grazing unit. The amount of time necessary to permit forage regrowth varies with plant species, stocking rate, time of year, and rainfall. Generally, a rotational grazing plan is based on resting grazing units for periods of 2–4 weeks. Spot-grazing and overgrazing can sometimes be reduced or eliminated by dividing larger pastures into smaller ones. Providing rest periods from grazing allows pastures to recuperate, thereby enabling increased forage production and possibly higher stocking rates. Rest periods can be provided by rotating pastures or removing horses from pastures for a portion of the day.

Implementing a rotational grazing system will require an investment in fencing. Strip grazing within a paddock using portable electric fencing also uses pasture more efficiently. Horses placed behind temporary fencing must be adapted to it. Horses can become acclimated to temporary fencing by first using it in larger pastures prior to smaller ones.

It should also be noted that the use of small grazing units for more horses may not always be feasible. This is primarily due to the potential for aggressive behavior among horses. Research has shown that mares placed in continuously grazed pastures lost weight and either maintained or lost a half point in body condition score compared to those rotationally grazed on smaller pastures (19). Results from animal gain studies using yearlings are inconsistent in comparing continuous and rotational systems (19, 37).

In rotational systems, the first paddocks grazed regrow while others are grazed, and are available again in 3–4 weeks, allowing greater stocking rates than in a continuous grazing system, where up to 4 acres per horse are recommended (28). Any number of paddocks can be used in rotation, but many forage species can tolerate no more than 1 week of grazing followed by 3 weeks of rest, or 2 weeks of grazing and 1 month of rest (8, 16, 26). Therefore, a minimum of three paddocks are necessary, barring problems with drainage, drought, fertility, and weed invasion. Ultimately, carrying capacity will depend on the class of horse, soil type, fertility, rainfall, drainage, and species composition. The actual grazing area in each paddock should be approximately equal after subtracting the area occupied by trees, brush, and bare ground from the total acreage. Because horses are capable of great physical damage to pastures, it may be several seasons after establishment before a pasture can handle the pressure of maximum stocking.

When implementing a rotational grazing system, pasture managers should remember that horses are more sensitive than other livestock to changes in feed, especially the change from dry hay and grain to pasture, or from low to high quality pasture. Moving horses to a new paddock may stimulate a "greedy" response that can result in a 20% increase in forage intake even if the horses were just moved from a suitable area (26). Under such circumstances, colic, laminitis, and other digestive problems may occur. Therefore, to acclimate the horses from a low or mod-

erately productive continuous grazing system to a moderately to highly productive rotational grazing system, slow pulsed changes to a rich vegetative pasture should occur by limiting grazing time or intake mechanically with a muzzle. Depending on the frequency of rotation, the increased intake may significantly affect gain of growing horses or those needing an increase in body condition. To transition horses gradually to a change from feed to forage, Heusner (17) recommended the following procedure:

- Feed all the hay a horse will eat before grazing.

- Graze on lush pasture 30 minutes in morning and evening.

- Increase grazing time to 1 hour each in morning and evening the second day.

- The third day increase grazing to 2 hours each in morning and evening.

- On day four, repeat day three and make a judgment call. Horses should reach fill in 2 hours. If they continue to eat after 2 hours, then you may want to continue with 2 hours in the morning and 2 hours in the evening for several days.

Horse owners must skillfully integrate pasturing and pasture management to reduce or eliminate such negative occurrences.

Grazing management under any system requires year-round management, because each season provides a different challenge to pasture and grazing management. Spring growth often provides too much energy and protein for mature horses, which risk putting on excess weight or developing laminitis. High spring growth rates may require that pastures be cut for hay or stocking density be increased if the current stocking densities are too low to maintain pasture species in vegetative growth stages. Hot, dry summer conditions will reduce the growth of cool-season grasses, and pasture production diminishes. However, cooler fall weather will bring an increase in production. Extending the grazing season is desirable and can be accomplished by stockpiling and eliminating all grazing and clipping by late summer to allow for growth that can be grazed during late fall and winter. In a rotational system, this is easily accomplished, but in continuous pastures, this may not be feasible. Even these fields must be rested to allow pasture species to accumulate energy reserves and recover for winter or the continuously grazed plants will weaken and be more susceptible to winter injury. It is essential, however, that a pasture seeded during the spring of the current growing season should not be grazed during the seeding year unless it is a dry site and well managed. Late summer seedings should not be winter grazed. Established pastures should not be grazed during wet weather to prevent physical damage to plants and soils, but rather, horses should be stabled or removed to a high-traffic area or barn-side paddock.

Sacrifice Areas

Sacrifice areas are separate parcels of land where the main goal is exercise and not grazing. A sacrifice area is a small enclosure or paddock area that provides space during times when pastures are easily damaged, such as during wet soil conditions, winter, and following renovation. When land area is limited, a sacrifice area can be of value during the winter months because pastures cannot survive continuous grazing and trampling during this season. The use of a sacrifice area can result in increased pasture productivity on remaining pastures. It should be located on well-drained soils away from waterways. Vegetation will likely be sparse to nonexistent, as the area will be subjected to significant wear and tear. Consider locating your sacrifice area so that vegetated areas surround it; these will serve

as a filtration system to reduce sediment and nutrient removal from the sacrifice area.

Fencing

Fencing is a critical component of all grazing systems. Portable electric fencing provides the most efficient and economical way to create temporary paddocks for rotation. Wide, colored poly tape is inexpensive, but flags may be necessary on the fence to enable horses to see it clearly, even when the animals are experienced with this type of fence. A single strand of wire should be strung at a height of 33 inches, double strands at 20 inches and 36 inches, and triple strands at 16, 28, and 40 inches (21). Jordan et al. reported that horses adapt easily to portable fencing and will respect two strands of 14-gauge wire if voltage exceeds 2,000 volts, even at stocking densities as high as four animals per 0.1 acre. However, horses cannot be expected to remain in their areas if only a single strand of wire is used and the current is low or disconnected. Small paddocks may select against weaker horses if they are forced to compete for food. Grazing groups containing particularly dominant or subservient horses should be carefully monitored for changes in body condition. Comprehensive coverage of all aspects of fencing for pasture systems is provided in chapter 9 (Tools for Management of Pasture-Based Livestock Production) of the NRAES book *Forage Utilization for Pasture-Based Livestock Production* (NRAES–173; see HTTP://WWW.NRAES.ORG for more information).

Water, Minerals, and Shelter

Pastures should contain a clean, reliable source of water, mineral salt blocks, and shelter from the sun and inclement weather. Typical consumption of water by an adult horse is 0.4 gallon per 100 pounds per day (1). The problem of stationary watering systems, feeders, and shelter is solved in the rotational system by creating a common area that runs the full length of each adjacent paddock and is accessible to the one in use by an open gate, while the others are closed. This common area can also be situated in the center of surrounding square or radiating paddocks. Plenty of space must be provided to allow normal competition for water as well as any supplemented hay and grain. A common area will be subjected to heavy traffic and maintaining vegetation may be difficult.

Horse managers must remember that most forages are deficient in sodium and vary in vitamin and mineral content, in part due to the available minerals derived from the soil. Sodium chloride (salt), the mineral needed in the greatest amount in the horse's diet, can be supplied by mixing it with feed, fed in the loose form, or supplemented via trace mineral salt blocks (36). However, mineral blocks may not contain calcium and phosphorus, which must be supplemented separately if adequate amounts are not being provided through grain or forage. Mineral blocks are often placed near the water supply; this can contribute to the development of an excessive use area. Placing minerals away from water or other congregation areas can redistribute animal impact and avoid overuse. Detailed information regarding the water and shelter component of pasture systems can be found in chapter 9 (Tools for Management of Pasture-Based Livestock Production) of the NRAES book *Forage Utilization for Pasture-Based Livestock Production* (NRAES–173; see HTTP://WWW.NARES.ORG for more information).

Plant-Related Health Problems

Tall Fescue/Endophytes

Many health problems of horses on pasture are plant-related. As previously discussed, one potential problem is tall fescue and endophytes. The tall fescue endophyte *Acremonium coenophialum* produces ergot alkaloids responsible for hormonal interference in broodmares, resulting in abortions, foaling difficulties, and milk production problems. The endophyte's toxins may cause reproductive and other disorders in horses grazed on infected fescue. Mares affected by the toxin may have a variety of problems, particularly in late pregnancy. Ingestion of the toxin may result in lack of udder development, prolonged gestation, lack of colostrum, and decreased milk production. Additionally, mares grazing infected fescue are less likely to become pregnant (6). For other classes of horses, there have been no reported health problems when grazing tall fescue. Data collected in 1998 by the USDA's National Animal Health Monitoring System from equine operations in 28 states revealed that 61.6% of samples collected nationally from pastures tested positive for endophyte (34). Endophyte was found in 56.3% of pastures in the Northeast. Older stands of tall fescue may be naturally infested and can be tested for infection rate. The threshold of tolerance for infected tall fescue in a pasture is not precisely known. Small amounts of tall fescue are common in most fields and should not be an automatic cause for alarm. Once significant infection of a pasture is determined, managers can avoid grazing pregnant mares there, graze infected pastures in concert with other feed sources to dilute the toxin, or destroy and reseed infected pastures with endophyte-free varieties of tall fescue. If endophyte-infected fescue is present in pastures, remove mares from the pasture during the last 60–90 days of gestation and feed a fescue-free diet. Alert your veterinarian if there is a possibility that mares have been exposed to endophyte-infected fescue during the latter stages of gestation.

When purchasing seed, be sure to read the seed label, because many varieties of endophyte-enhanced tall fescue and perennial ryegrass are available for use in home lawns to enhance pest resistance; use of these varieties in pastures should be avoided. When establishing or renovating pastures that are to include tall fescue, an endophyte-free forage variety of tall fescue should be used. Recently, researchers have inserted a nontoxic endophyte (Max Q) into tall fescue varieties to improve stand persistence and animal performance benefits. The nontoxic endophyte has been evaluated in grazing trials conducted in the Southeast and Midwest with beef cattle, where results have shown elimination of fescue toxicosis problems. Extensive testing is currently being conducted to evaluate the health benefits and safety on horses.

Ryegrass/Staggers

Ryegrass staggers is a disorder associated with perennial ryegrass. Ryegrass staggers occurs when plants are under environmental stress and when pastures are grazed severely. The endophyte fungus *Acremonium lolli* produces neurotoxins in the grass. Ryegrass staggers can affect sheep, cattle, and horses. Early symptoms are characterized by a difficulty in flexing the legs, which causes an unusual gait. In severe cases animals may have difficulty walking and may fall repeatedly. Endophyte-free seed should be planted when establishing or renovating pastures.

Alsike Clover/Photosensitivity

For horse pastures, alsike clover (*Trifolium hybridium*) should be avoided in seeding mixtures. Alsike clover is found in some general forage mixes because of its tolerance to wet soil conditions. However, it contains unidentified compounds that cause photosensitivity in susceptible light-colored horses and alsike clover poisoning. The incidence of both increases with wet growing conditions and when alsike clover comprises the majority of the forage being consumed. The photosensitivity is also known as "dew poisoning" because it occurs most frequently when the pasture is wet and the skin of the animal is moist. The condition is mostly evident in thinly haired and white skinned areas around the lips, nose, mouth, and feet. Symptoms include reddening, dry necrosis of the skin, or edema and discharge. Alsike clover poisoning can cause liver failure, neurological problems, and death.

Red Clover/Slobbers

Red clover is often avoided in horse pastures because it can cause the slobbers, which is excessive drooling caused by an alkaloid (slaframine) found in infected red clover. Although it is an undesirable condition, it does not harm the horse.

Sorghum/Prussic Acid

Sorghum species, including sorghum, sudangrass, sorghum-sudangrass hybrids, and johnsongrass, should not be used for horse pasture. A compound known as prussic acid, hydrogen cyanide, or hydrocyanic acid is found in frost- or drought-stressed sorghum and other related plants. In healthy plants, the compound is a component of the chemical dhurrin, located in plant leaves. Under normal conditions, the plant material is not toxic. However, under stress such as drought or frosting, the hydrogen cyanide is released from the dhurrin and cyanide concentrations increase in plant tissue. If livestock ingest drought- or frost-stressed sorghum species, poisoning can occur. As with any toxicant, the response is related to the concentration of the toxin, the amount ingested, and the condition of the animal. When eaten, cyanide is absorbed in the bloodstream and ultimately prevents hemoglobin from transferring oxygen to cells in the body. The result is asphyxiation. Death occurs quickly, and symptoms are usually observed too late to provide any treatment.

Poisonous Plants

Poisonous plants can affect horses in many ways, including death, chronic illness, reproductive abnormalities, nervous system disorders, and decreased weight gain. Poisonous plants are typically invader species that germinate in pastures when poor grazing or pasture management persists. During periods of drought or overgrazing, often there is a lack of good quality forage in pastures, which may lead horses to investigate poisonous plants in a pasture or within reach beyond the pasture or paddock fencing. Several different chemical compounds capable of poisoning can be found in a variety of plants. The chemicals range from the alkaloids found in the nightshade family to the glycosides found in wild cherry and sudangrass. It is beyond the scope of this chapter to discuss all plant species that could be toxic to horses. Toxic species common to the Northeast include white snakeroot (*Eupatorium rugosum*), nightshade (*Solanum* spp.), bracken fern (*Pteridium aquilinum*), milkweed (*Asclepias syriaca*), jimsonweed (*Datura stramonium*), and yew (*Taxus* sp.), and hardwood species such as black locust (*Robinia pseudoacacia*), red maple (*Acer rubrum*), black walnut (*Juglans nigra*), and oak (*Quercus* spp.).

Negative effects will occur from consuming high levels of these plants; therefore, horse pasture managers should walk pastures to scout for large populations or pockets of poisonous weed species and remove any found. The best defense against poisonous plants is to promote productive stands of desirable grasses and legume species through a sound pasture management program. Additional information regarding poisonous plants can be found in chapter 6 (Invertebrate Pests, Weeds, and Diseases of Forage-Livestock Systems) of the NRAES book *Forage Production for Pasture-Based Livestock Production* (NRAES–172; see HTTP://WWW.NRAES.ORG for more information).

SUMMARY

If the equine operator wishes to use pasture as a key source of feed, it is important to establish and maintain a strong and vigorous stand of forage. The agronomic practices for establishment of forage species in the pasture are the same as those used in beef cattle, dairy, or sheep operations. Pasture management practices such as weed control, fertilization, and liming will also be similar to those employed by other livestock operations. Equine pasture managers are often challenged by horses' tendency to be more destructive to pastures than other livestock. Also the spot-grazing behavior of horses often requires a more intense grazing management program. If the equine operator is to obtain maximum production of quality forage, it is essential to integrate both pasture management and grazing management. The equine pasture manager should be familiar with the factors that affect the capability of pasture to meet the overall forage needs of the production system.

CHAPTER 8

Parasite Control: Basic Biology and Control Strategies for Pasture-Based Systems

William P. Shulaw

INTRODUCTION

Parasite control is best viewed as an integral component of a comprehensive herd health management plan. However, husbandry differences among farms make a basic knowledge of parasite biology and life cycles important to the livestock producer who wishes to make the most efficient use of his or her resources. It is also important for producers to become knowledgeable about regional differences in the relative importance of specific parasites and the environmental factors that may affect their life cycles. The costs and benefits of various strategies for control must be considered in light of producer resources, expectations, and desired results.

INTERNAL PARASITES

Roundworms or Nematodes

Possibly the most important internal parasite in pasture-based management systems is the roundworm. For cattle, sheep, and goats, the most important of these worms live in the gastrointestinal tract, where they cause tissue damage or feed on blood (23, 30). A discussion of the basic biology and management of these parasites in sheep, a species in which parasite control is most critical, provides the background for control programs in goats and cattle because the basic biology of the parasites is similar. Differences between internal parasite control in sheep and cattle will be discussed later.

Managing internal parasites in sheep is really a function of *pasture management* for most of the year. Delays in management or failure to recognize the necessity of managing the pasture for the control of internal parasites can lead to mid- to late-summer situations where the pasture is extremely dangerous for the sheep. At that point, even frequent deworming may fail to completely control the harmful effects.

Basic Biology

Pastures upon which sheep were grazed the previous season will have varying numbers of surviving, overwintered worm larvae. This number will depend upon the intensity of grazing the previous season, the class of sheep grazed (mature animals vs. lambs) and their level of infection, the amount of heat and dryness during the previous grazing season, and the nature of the winter weather and snow cover in temperate regions. When sheep consume these larvae in the spring, the larvae become egg-laying adults in a few days, and the eggs result in a new generation of larvae on the pasture that can infect grazing lambs and ewes. Larvae migrate up the forage in films of moisture; however, the majority of them will be in the bottom two to three inches. As the moisture film evaporates, the larvae tend to move back down the plant to shaded areas. Those that are exposed to severe dry conditions may die; however, most worm species produce sufficient eggs to ensure that enough survive to maintain pasture infectivity. The entire cycle from eggs to eggs can take as few as 21 days under ideal conditions of moisture and warm temperatures (1, 5) (figure 8–1, p. 210). Continuous worm production cycles magnify the pasture larval burden and may result in

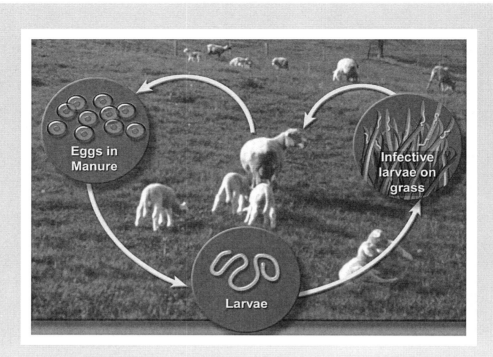

Figure 8–1. The typical life cycle of the roundworm lasts 21 days.

very dangerous pastures by mid- to late summer (figure 8–2). At that time as much as **95% of the total** *farm* **worm burden is on the** *pasture.* Although extreme heat and dryness will kill larvae on pasture, they can survive reasonably well with just a little forage for shade. During periods of drought, the fecal pellets provide enough moisture for larval development from the egg, and larvae can survive if the fecal material is protected from extreme drying (30).

Another major source of parasite larvae on pastures in early spring is the eggs passed in the feces of lactating ewes (5). During late gestation and early lactation, a ewe's immune system is typically somewhat suppressed. This allows larvae acquired the previous fall, which have been living in the tissues of the stomach wall in a state of arrested development (called hypo-

biosis), to resume development to egg-laying adults. This phenomenon is called the "periparturient rise" in fecal egg counts (FEC). The periparturient rise lasts six to eight weeks, and it ensures that the pastures will become contaminated with a new generation of worm larvae. In fact, because the larvae of the blood-feeding *Haemonchus contortus* are generally not as tolerant to cold as some of the other worm species, this period of arrested development and subsequent development to the egg-laying adult stage is an important survival mechanism for this species in the more temperate regions of the world.

Sheep do acquire some amount of resistance to internal parasites with age and exposure to them. However, lambs are essentially non-immune until they have been exposed during grazing. Low-level exposure is necessary for resistance to develop, but producers must avoid high-level

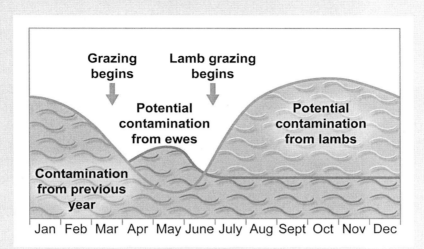

Figure 8–2. If unchecked by strategic control methods, pasture larval contamination from both lambs and ewes can reach extreme levels by mid- to late summer.

exposures that result in loss of performance or clinical disease. Lambs exposed during a grazing season have usually developed some resistance by the time they are 6–8 months old. Until resistance has developed, worm infections can develop more easily in lambs, and worm egg output in lambs is generally much higher than in ewes. Nevertheless, resistance is never complete and can be overcome if the exposure is high enough or if the animal's resistance is weakened by poor nutrition, disease, or other stress.

Controlling parasite populations then centers around using selective or targeted treatments and pasture management strategies that keep larval contamination on pastures low and which minimize the use of dewormers to avoid selection for drug resistance in the parasites (3, 26, 27, 28).

Drug Resistance

Drug resistance of internal parasites in sheep and goats has become a serious problem in the major sheep- and goat-rearing areas of the world (12, 26, 27, 28, 29). Drug resistance to all the

chemical classes of anthelmintics (dewormers) has also been described in flocks in the United States, and a general understanding of how it develops is crucial to understanding modern approaches to control of parasitism.

Drug resistance develops as a genetic trait of the worm just as some production traits are genetically controlled in the sheep. Unfortunately, once it is present in the flock, it will usually be permanent. If resistance develops to all chemical classes of available dewormers, it may be very difficult to graze sheep on that property. Resistance develops when the farm's worms are exposed frequently to the same drug, and this is perhaps the most important cause of drug resistance. Although parasitologists disagree about whether drug classes should be rotated annually, continuous use of the same product usually results in resistance with enough time. Another important cause of drug resistance is *underdosing*. Underestimating the sheep's weight, incorrect calculations of the dose to be given, incorrect dilution of products that must be mixed, and improperly calibrated equipment are common causes of underdosing. This exposes worms to less than lethal doses of the drug and increases the selection pressure in favor of resistance (7).

Once resistance is present to a specific dewormer, other drugs in that chemical class may also be less effective; this is known as side resistance. There are presently only three drug classes licensed by the U.S. Food and Drug Administration (FDA) for use in sheep in the United States. These products and classes include Ivomec Sheep Drench and Cydectin drench (the avermectin/milbemycin class); Tramisol and Levasole drench and oblets and Prohibit Soluble Drench (levamisole hydrochloride), and Valbazen (albendazole–the benzimidazole class). Many shepherds have used fenbendazole (Panacur, Safeguard) or thiabendazole (Omnizole or Thibenzole) in the past. If resistance to one of them is present, resistance or partial resistance to albendazole may already be present. All available dewormers in the United States today are members of one of these three classes of drugs. It is important for shepherds to know whether each class is effective in their flock before finding out otherwise during an episode of clinical parasitism. Unfortunately, it is unlikely that new classes of dewormers will become available for grazing sheep, goats, or cattle in the near future (7, 12).

Selective Treatment and Pasture Management Strategies

Elimination of all roundworm parasites in pasture-based systems is not feasible at the present time. For ruminant grazing systems to remain sustainable, we must use parasite control strategies that do not place total reliance on chemical dewormers, and they must incorporate a knowledge of the parasite life cycle such that a small level of parasite infection exists but that severe, production-robbing burdens and animal deaths do not result. In light of the current con-

cerns of drug-resistant worms, the concept that parasitologists refer to as "refugia" (meaning "in refuge") needs to be understood. Refugia simply refers to that portion of the farm's worm population that doesn't get exposed to dewormers or is exposed very infrequently (25, 26). These worms may be larvae existing on pasture, larvae in arrested development in the sheep's stomach, or worms in sheep that remain untreated with dewormers. If new infections, or reinfections of treated sheep, occur from the population of worms in refuge, selection for drug resistance doesn't occur, or occurs very slowly (26, 27, 28).

One approach to maintaining a "refugia" on the pasture is the concept of selective or targeted treatment. This approach is based on our understanding that in most livestock populations, worms are not evenly distributed across all the animals in a group. Usually, only about 20–30% of the animals harbor about 70–80% of the flock's worm burden. If we could determine which animals make up this 20–30% and effectively deworm them, we could accomplish our goal of treating the animals that most need treatment and which are contributing the majority of pasture larvae contamination, and at the same time, maintain a residual population of worms that haven't been selected for resistance by exposure to a dewormer (26).

One such approach that has been shown to be very useful in many parts of the US and other countries is the FAMACHA system (25) (figure 8–3). This system was developed in South Africa in response to a severe problem with drug-resistant *Haemonchus contortus*. This worm is a voracious blood feeder that can create severe anemia in sheep and goats, and it is a very important parasite in the US. The FAMACHA system uses a patented card that allows farmers to "score" their sheep (or goats) on a 1–5 scale based on a comparison of the color of the inside of the lower eyelid with the colors on the card. These colors range from shades of red to nearly white and have been shown to have good

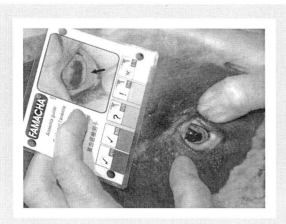

Figure 8-3. The FAMACHA scoring system allows selective deworming of animals with the heaviest worm burdens and which are most in need of deworming. Because it estimates a level of anemia, it is only useful where the blood-feeding worm Haemonchus contortus is the major worm of concern. Selective deworming, as opposed to deworming all animals in the group, reduces selection for drug-resistant worms.

Source: The Ohio State University

correlation with the level of red blood cells in the body. Animals that score 1 or 2 have rather normal levels of red blood cells. Animals that score 4 or 5 are dangerously anemic, and animals that score 3 are somewhat anemic. Lambs and lactating ewes that score 3, 4, or 5 need deworming while those that score 1 or 2 usually do not need treatment. Regular examination of the animals at times when risk of infection and disease are highest during the grazing season allows treatment of animals that are in most need of it while avoiding selection for-drug resistant worms in the animals that are not treated. Because *Haemonchus contortus* is such an important worm species in much of the US, the FAMACHA system can be useful for many producers; however, it is not applicable to the control of other species of worms that do not cause

anemia (13). More information on the FAMACHA system is available at www.scsrpc.org.

Other selective treatment strategies to maintain a population of unselected worms include treating only thin animals or animals with poor body condition scores; treating only lactating females, nursing twins, or triplets; and leaving an arbitrary 10–20% of animals in a group untreated when the group is moved to a new pasture where there are few worm larvae.

One practice that used to be recommended, but which is now considered to be very dangerous with respect to selecting drug-resistant worms, is the "treat-and-move strategy." This technique is especially useful for young growing lambs, and its effectiveness was demonstrated by research in the early 1980s. It involves treating all the lambs in a group and then moving them to a "safe" or "clean" pasture, which is defined as one with no, or very low numbers of, worm larvae on it. This can be a hayfield that has been harvested and allowed to regrow, a pasture that has had cattle or horses on it earlier in the grazing season, or one that hasn't yet been grazed by sheep. Lambs with very low worm burdens that are placed on pastures that have no worm larvae on them will remain relatively uninfected for several weeks to several months. This is an ideal situation for the lambs and for the shepherd, but unfortunately, we now know that it can be a powerful force for selecting drug-resistant worms.

It works like this: No dewormer is truly 100% effective and some worms survive treatment. In addition, we now know that genes for drug resistance exist in the important worm species in virtually all domestic sheep populations across the world. The proportions of worms carrying these resistance genes vary from flock to flock; but they are there and we can select for them. Treating all the animals in a group and then moving them to a safe pasture allows the survivors of treatment to enjoy a reproductive advantage. In most cases, it is likely these survivors will be the ones carrying the resistance genes. Their progeny will then develop on the new pasture with little or no competition from worms that do not have the resistance genes. Depending on the season and weather, the immune status of the sheep, the stocking density, and length of time the new pasture is grazed, the resistant worms in those animals can build to significant numbers and create a pasture capable of making considerable change in the gene pool of the farm's total worm population.

Because moving animals with low worm burdens to pastures relatively free of worm larvae can help maximize production and animal health, this is still an attractive strategy. However, the current concerns of drug-resistant worms and the likelihood that no new chemicals may be readily available in the future require that we modify this approach to maintain a small population of unselected worms. The FAMACHA system as described earlier could be used prior to a move to a clean pasture to identify a portion of animals to deworm.

Another approach is the so-called "delay the move after the dose" strategy. This allows the treated animals to become lightly reinfected before going to the clean pasture. This helps ensure that contamination of the new pasture will occur with larvae from worms that have not had drug selection pressure put upon them. This may be especially useful if signs of parasitism, such as anemia or bottle jaw, have already appeared in lambs before moving to the clean pasture. The number of days to graze the infected pasture before moving depends on how heavily infected it is and the relative susceptibility of the animals (lambs versus less susceptible non-lactating ewes), but in general 4–7 days of grazing offers a useful compromise. It is impor-

tant to note that deworming with moxidectin will not allow this strategy to work because of its persistent activity in killing incoming larvae from pasture, which can be as long as 35 days (15).

A variation of this strategy is to "move then dose." This means grazing the new pasture a few days before deworming to allow some contamination to occur. Less information is available to recommend the length of time to graze before treatment, and it will depend on the level of egg shedding when the animals are moved. Animals with severe parasitism caused by *Haemonchus* can be shedding tremendous numbers of eggs, so the pasture can become contaminated relatively quickly, and they may suffer more stress from the move. In most cases it should be safe to wait a week before treatment if the animals are apparently healthy (15).

Another pasture management strategy that is useful to both sheep, goat, and cattle producers is called "alternate species grazing." This grazing strategy takes advantage of the relative specificity of worms for their normal host (2). Spring grazing of last year's sheep pastures with another species, such as cattle, followed by sheep grazing beginning again in late June or early July, allows use of the contaminated pasture in the spring. This pasture will become largely uncontaminated for the sheep in June or July as a result of die-off of remaining overwintered larvae. Likewise, sheep may graze pastures contaminated by sheep parasites in the spring without treatment until their move to safe pastures in May or early June. The pasture grazed by the sheep can then be safely used by an alternate species, such as cattle or horses, following removal of the sheep. Goats are not the same species as sheep, but do share the same parasites. Therefore, goats should not be considered as an alternate species in this strategy. Llamas should also be considered susceptible to sheep parasites. Grazing cattle and sheep together at the same time can enhance forage utilization, but generally will not provide the level of protection from parasite exposure that alternating the pastures will. A combination of approaches with selective use of dewormers offers the best hope for effective parasite control over the long term (3, 26, 27, 28, 29).

Shepherds frequently ask whether rotational grazing practices lower the risk of parasitism because of the misconception that frequent moving of the sheep moves them away from the worm larvae. Worm larvae on pasture can survive several months, even if the weather is hot and dry, if they have some shelter in the forage. When the sheep are moved back to the paddock in typical pasture rotation systems, the larvae are usually there waiting for them. Because stocking density on rotationally grazed pastures is usually higher than is used in conventional grazing systems, pasture contamination is likely to actually be greater. Age and immune status of the grazing sheep, mixed species grazing, weather, season, and presence of drug resistance (partial or complete) all complicate the decision-making process. Monitoring fecal egg output can give the grazer insight into the pasture's infectivity just as watching forage growth helps plan the harvest.

Determining Worm Resistance to Dewormers
Although it is essential for sheep producers to develop complementary strategies for sustainable parasite control, most will need to use chemical dewormers, at least occasionally or for selected animals, for the foreseeable future. When dewormers are used, it is crucial that they actually work with a high degree of effectiveness if the control strategy is to be successful.

There are currently only two ways to determine if the dewormer you wish to use is effective.

The first is the fecal egg count reduction test (FECRT) (4). This approach estimates the ability of a drug to reduce egg counts in feces compared to a control group. This method requires 15–20 animals per group for each dewormer tested, including the untreated control group. Therefore, in small flocks perhaps only one chemical class per grazing season can be tested. A quantitative egg counting method, like the McMaster method, must be used, and the fecal egg count at the time of treatment must average at least 200–300 eggs per gram of feces for it to be valid (figure 8–4, p. 216). It is important that the 15–20 animal group size is used because of the wide variation in egg counts typically seen across a group of animals. Fecal samples are collected 12–14 days after treatment unless ivermectin or moxidectin is being evaluated where 15–16 days is more appropriate. The FECRT can be performed by many veterinarians, and the equipment needed is not difficult to obtain or expensive. If a dewormer is still highly effective on a farm, we expect that the egg count reduction in the treated group will be 95% compared to the untreated control group.

The main drawback to the FECRT is that by the time you can detect developing resistance to a dewormer, the proportion of resistant worms in the total worm population is relatively high, and continued use of the product in traditional ways may result in a rapid increase in the resistant proportion to the point where the drug is virtually useless. If that point has not already been reached, it will take very selective and careful continued use to maintain a practical level of effectiveness.

A second way to detect resistance to dewormers is something called the larval development assay. In this assay, multiple drug classes can be evaluated at one time, and their effectiveness is estimated by determining how readily worm eggs develop to infective larvae in the presence of a

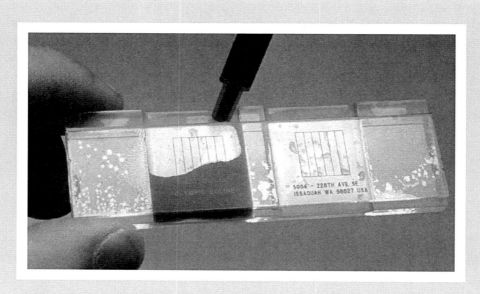

Figure 8-4. The McMaster slide and counting technique is one method of quantifying egg output.

Source: Courtesy Rupert Herd, The Ohio State University.

series of increasing concentrations of dewormer (14). This is usually done in a plastic plate with multiple small cavities containing nutrients for larval development and the dewormer classes to be tested. A significant advantage of the technique for the producer is that a single composite sample of fecal material from only 10–15 representative animals can be sent to the laboratory for testing. Small flock owners can get information about all three chemical classes of dewormers with one set of samples. Results are available in about two weeks.

The larval development assay can detect developing resistance in the worm population at an earlier stage than can the FECRT (14). This can give a producer a "heads up" that a dewormer must be used very carefully if he or she expects to be able to continue using it effectively. It can also be used as a monitoring tool to detect changes in resistance patterns over time. As

with the FECRT, the average egg count for the sample sent for the assay must be high enough for the laboratory to harvest enough eggs to put in the plastic plate. For the most accurate results, samples should not be sent from animals that have been recently treated with a dewormer.

Presently, the only larval development assay available in the United States is the DrenchRite® assay which is conducted in Dr. Ray Kaplan's laboratory at the University of Georgia. The assay requires considerable time and technical expertise and therefore, must be scheduled in advance. Samples cannot be stored and must be collected and promptly shipped by overnight courier; however, this is not difficult. There are specific instructions for collecting and packaging the sample, and they are also easy to do. The contact for arranging a DrenchRite® assay is Ms. Sue Howell, Department of Infectious Diseases, Room 2212, College of Veterinary Medicine,

University of Georgia, Athens, GA 30602; voice: (706)-542-0742. Additional information on the assay can be found at HTTP://WWW.SCSRPC.ORG/ under the "Smart Drenching" link.

Currently, parasitologists recommend testing for dewormer effectiveness about every two years. Testing will involve some cost, but if you are one of the unfortunate producers whose options have become very limited, it may help avert a costly disaster or a season of very poor performance. It may signal a need for you to make some major management changes in your sheep or goat operation.

Collecting and Handling Fecal Samples

Collecting fecal samples is relatively easy. We find that using resealable plastic bags works very well. Turn the bag inside out over your hand. You can then insert a finger into the rectum of the sheep with your palm facing upward. Gently stimulate the rectum by massage and pull fecal material back into your palm with your curved finger. When you have collected enough, usually about a tablespoonful, remove your finger, turn the bag right side out, expel as much air as you can, and seal it. Mark the animal's identification number on the bag with a permanent marker.

Samples should be placed in a Styrofoam cooler or other suitable container and kept at about 38–40° F until they can be examined by your veterinarian or diagnostic laboratory. Egg counting should be done within 2 or 3 days, and certainly within a week, to avoid egg losses and inaccurate counts.

Special Considerations for Goats and Cattle

Goats may have different forage preferences than sheep if they have a choice. However, in pasture systems the development of a parasite burden for goats closely parallels that of sheep. Strategies for control are similar to that of sheep,

and resistance of internal parasites of goats to available dewormers has been documented in the United States. Goats differ from sheep in that they tend to metabolize the available dewormers more efficiently. Generally speaking, this requires an upward adjustment in the dose of dewormer given to goats. Because there are few dewormers approved by the FDA for use in goats in the United States, use of products other than exactly as they are labeled requires consultation with a veterinarian.

Although calves, like lambs and kids, are born with no immunity to parasites, exposure to them on pasture during their first grazing season allows them to develop an immune response. Most calves coming off their first season of grazing with their mothers have acquired a worm burden. Deworming them before they enter the feedlot or before they are weaned and moved to new pastures or drylots for overwintering is almost always a cost-effective strategy. By the end of the second grazing season, cattle generally have developed a rather strong immunity to gastrointestinal nematodes. Mature, well-nourished cows seldom suffer clinical parasitism as mature ewes sometimes do. Immunity to internal parasites in cows can wane, however, during periods of high stress or when they are severely undernourished. Egg counts (expressed as eggs per gram of feces) of mature beef cattle tend to be very low, perhaps as low as an average of 5 eggs per gram of feces over the year. In addition, in spite of a relatively large volume of fecal output, pasture larval counts tend to build much more slowly than in ewe/lamb grazing systems. Egg counts in adult cows approaching 100 eggs per gram of feces is usually a sign of a relaxation of their immunity because of poor nutrition or other disease (11).

There is controversy as to the value of deworming adult beef cows (20). Several studies have shown an improvement in average calf weaning weights or reproductive performance of the cow when cows are dewormed in the spring and summer. However, these results do not seem to be repeatable across all regions of the country or across different management systems. Similarly, some reports have indicated improved performance in calves and cows when cows were dewormed in the fall or in both the fall and spring. Likewise, not all reports show significant positive results. The cost-effectiveness of various approaches has not been thoroughly evaluated. However, many factors come into play when attempting to address the benefits of adult beef cow deworming, and conclusive recommendations that fit most pasture systems are not likely to be forthcoming (20).

Similar to the situation with adult beef cows, there is no conclusive evidence to indicate that deworming adult lactating dairy cows will routinely be beneficial (21). Several older studies conducted in other countries have shown potential benefit, but in many, the results are not directly applicable to typical North American systems. It is important to note that older studies were performed with dewormers administered in the dry period because no products were available for lactating cows. Many of these studies have been reviewed by Bowman (1).

Recently, two studies conducted in North America have been published describing the production effects of deworming lactating dairy cows with eprinomectin (no milk withholding) (16, 22). This work was conducted in herds located in Quebec and Prince Edward Island, Canada. The study population was 942 cows from 28 herds in those two provinces. Minimum pasture exposure for these herds was access to a grass-covered exercise area, but 71% of adult animals received at least some of their nutrient needs from pasture. Cows were dewormed once at calving with eprinomectin. This work indicated that treated cows produced an additional 0.94 kg

of milk per day when compared to nontreated controls. Analysis of the reproductive records of a subset of 20 of these herds indicated that treated cows had fewer services per conception provided that the interval from calving to first service was less than 90 days. A tendency for a reduction in the calving-to-conception interval for treated cows ($P = 0.06$) was also observed. This work suggests that deworming lactating dairy cows at calving with a long-acting product such as eprinomectin may be economically beneficial in the northern temperate zones of the United States if cows receive at least some of their nutrition from pastures where they may be exposed to worm larvae. Additional studies from other areas of North America will be helpful to describe the potential benefit of deworming at calving. It is probably safe to say that, if possible, producers should avoid grazing lactating cows on pastures likely to be heavily contaminated by less-resistant heifers and calves. Alternative use in the spring (until mid- to late June) of pastures grazed in the previous fall would likely reduce exposure of cows to infective over-wintered larvae. Adult dairy cows maintained in confinement with no access to pastures are unlikely to benefit from deworming.

Parasitologists generally agree that control programs for grazing yearlings, stockers, and replacement dairy and beef heifers are warranted (6, 19). Parasitism in these animals seldom develops to the point of severe death losses such as is frequently seen in pastured lambs; however, the animals may grow more slowly and fail to reach their full potential for growth or milk production. This is likely due to reduced appetite and forage consumption as well as the direct effects of the parasites. Similar to approaches in sheep, reducing the potential for larval buildup on pasture during the grazing season by prophylactic treatments in the spring is more appropriate than treatment after significant pasture contamination has already occurred. As early as 1980, it was shown in Ohio that two dewormings, 3 and 6 weeks after turnout in early April, resulted in significantly improved growth on permanent pastures and improved future milk production in dairy heifers (9, 10). Similar results were later observed for beef heifers and steers. In other regions a little farther south, three treatments at 3-week intervals were needed. With the advent of cattle dewormers with more prolonged activity, the spacing can be timed to fit the known duration of activity for the dewormer. In the northern regions of the United States, turnout to grass may occur later in the spring, and the number and timing of the treatments will need to be adjusted. A bolus releasing small amounts of ivermectin over a period of several weeks became available in the mid-1990s. This approach offered a significant convenience in reducing the labor needed for treatment and was effective in improving the performance of yearling cattle on infected pasture. This bolus is not available as of this writing. Alternate species grazing strategies can also be used in an overall parasite control program for young cattle.

A word of caution concerning control of roundworm parasites in cattle is in order here. Completely preventing infection, or lowering it to very low levels, is not desirable or necessary. The cost-effectiveness of doing this is, of course, a major consideration. However, it is important to remember that some degree of exposure to worm larvae is necessary for development of a normal immune response in these younger cattle. In addition, as in sheep, resistance to dewormers has been reported for internal parasites in cattle, although this has not yet been a major a problem in the United States. Parasitologists warn that although this problem has not become as serious as in sheep and goats, the potential should not be ignored. The goal of our internal parasite control strategies in cattle, goats, and sheep should be to minimize the use of dewormers while maintaining optimal performance and profit.

Tapeworms

Unlike the roundworms, which are very small, tapeworms of cattle, sheep, and goats are relatively large, and their segments can occasionally be seen in the stools, especially in sheep. Adult tapeworms are relatively harmless to the animal, and most loss is caused when a particular species, *Thysanosoma actinoides*, only common in some localized regions of our western states, invades the bile duct of sheep and causes condemnation of the liver at slaughter. Very large numbers of tapeworms in lambs have sometimes been associated with poor performance, but this is poorly documented.

The life cycle of the common tapeworms (i.e., *Moniezia* spp.) of cattle, sheep, and goats differs from that of the roundworms discussed earlier. The segments passed in the feces are really part of the worm itself and are filled with eggs. The eggs are released after the segment dries out and breaks open, and then they are scattered. Before grazing ruminants can become infected, free-living mites that normally live in the pasture must eat the eggs. When the grazing animal incidentally consumes the mite, the now immature tapeworm form (a cysticercoid) is released in the intestine, and it develops into an adult tapeworm. Development in the mite is necessary for the tapeworm egg to develop into an adult.

Control of tapeworms in cattle and sheep is seldom necessary. Most animals develop an immune response that clears the majority of tapeworms from their intestines. Strategies to disrupt the surface humus layer of the pasture to interrupt the mite life cycle have been suggested, but there is little experimental evidence to support this. Control usually centers on treatment of the animals when large numbers of segments are seen. Currently, only fenbendazole and albendazole are approved for the treatment of tapeworms in ruminants.

Protozoa

Although a number of protozoa parasitize domestic livestock, two diseases seen in pasture-based systems, coccidiosis and toxoplasmosis, are relatively common and important to the livestock owner.

Coccidiosis

Protozoa in the *Eimeria* and *Isospora* genera produce coccidiosis. Several species of *Eimeria* occur in cattle, sheep, and goats, and those species are host-specific, meaning that the species seen in cattle do not cause disease in sheep. In addition, only a few species of coccidia are considered important disease producers. Coccidia are very common and virtually all herds and flocks are infected.

The life cycle begins with oocysts passed in the feces by an infected animal. After a short period (as little as 1 or 2 days) of development, called sporulation, the oocysts are infectious. When a susceptible animal ingests an oocyst, *sporozoites* emerge from the oocyst and penetrate cells lining the intestine. After several more cycles of multiplication and penetration of cells, oocysts form and are passed in the feces.

A single ingested oocyst has the potential to produce hundreds of thousands of new oocysts. Destruction of the cells lining the intestine by the developing immature forms of the parasite leads to the diarrhea, often bloody, and dehydration, which are the cardinal signs of coccidiosis. It is important to understand that a single ingested oocyst is multiplied exponentially within the intestine, and the destruction of cells is proportional to the number of oocysts consumed. Similarly, clinical disease is a function of the number of oocysts ingested. With some exceptions, young animals are the primary victims of disease. After the initial episode of infection and oocyst shedding, most animals develop a protective immunity but are not totally resistant to new infections. Some animals will continue as healthy shedders of small numbers of oocysts for years. This serves to provide a continuous low level of contamination of the environment with oocysts. Oocysts are resistant to destruction and can survive several months in the environment, especially if it is moist and shaded. Drying and sunlight reduce the numbers of viable oocysts.

Coccidiosis generally develops when animals are exposed to high levels of oocysts or when they are exposed to lower levels of contamination but are weakened by stress or poor nutrition. Buildup of oocysts in the environment, especially when animals are housed, leads to exposure sufficient to cause disease. In pasture-based systems there may be enough dilution of the oocysts to prevent significant buildup. Infections will still occur but without apparent disease. However, damp areas around waterers, creep feeders, and buildings, and where animals congregate for shade can provide an environment sufficient to produce disease.

Prevention of coccidiosis is best accomplished by maintaining a sanitary environment. In herds with a history of problems or where preventing a buildup of oocysts in the environment is difficult, the use of drugs in the feed to reduce the level of cycling within the animal can prevent clinical disease. These drugs are called coccidiostats and do not prevent infection entirely. They do allow the animal to develop immunity while reducing the risk of disease. Decoquinate, lasalocid, and monensin are approved by the FDA for use in cattle; decoquinate and lasalocid are approved for sheep; and decoquinate and monensin are approved for goats. For these drugs to be effective, the animals must consume the correct amount daily. This can be a challenge in pasture-based systems.

Toxoplasmosis

The protozoan *Toxoplasma gondii* causes toxoplasmosis. The cat is the normal host for this parasite, and the cat is the only animal that excretes oocysts in the feces. The cycle goes to completion, with formation of oocysts, only in the cat. Unlike coccidiosis, however, oocysts may infect most warm-blooded animals. The resulting developmental stages in animal tissues (no longer termed oocysts) are also infective. In nature, the cycle involves cats shedding oocysts that reinfect other cats and small rodents. Young cats are infected by ingesting oocysts from the contaminated environment or by ingesting infected rodents. After a period of oocyst shedding lasting about two weeks, the cat becomes relatively immune and subsequently sheds few oocysts. People may become infected by accidentally ingesting oocysts from areas contaminated by cats (e.g., litter boxes, sandboxes), exposure to aborting ewes and does and their discharges, or by consuming inadequately cooked meat infected by the developing forms. Food preparation areas may also become contaminated by raw meat, and can serve as a source of contamination of other foods such as fruits and vegetables. People who have impaired immune systems may develop very serious illness. Pregnant women may abort or deliver babies that are deformed or mentally challenged.

Cattle are not usually affected by toxoplasmosis, but sheep and goats are very susceptible. If a nonimmune, pregnant ewe or doe ingests oocysts, the fetus may become infected and an abortion results. Toxoplasmosis is one of the most common causes of abortion in sheep in the United States. Once sheep and goats have been infected, they develop enough immunity that they will not abort again if exposed. Likewise, if they become infected while not pregnant, the infection goes unnoticed. No drugs are currently approved for effective treatment or prevention of toxoplasmosis. Preventing infected cat feces from con-

taminating the environment and food supply is the only reliable prevention strategy. On pasture, ingestion of oocysts is possible, but the greatest risk to sheep and goats is exposure to stored feeds, such as grain and hay supplies, that have been contaminated by cat feces. Contamination of hay storage, barn, and barnyard areas with oocysts also provides an opportunity for exposure.

If it is possible to remove cats from the environment and keep them from returning, control can be accomplished. If keeping domestic or feral cats out of the environment is not possible or is difficult, incoming new cats that eat the rodents present on the contaminated premise may become infected. An alternative strategy is to maintain a healthy, neutered, resident cat colony to reduce the risk from stray cats. A rodent control program should be developed to reduce the overall risk of infection, and it should include feeding the cats to reduce their need to hunt.

EXTERNAL PARASITES

External parasites cause economic damage to livestock by reducing condition or productivity, causing physical damage to animals and their by-products, and transmitting disease between animals or between animals and humans. Numerous species are involved, but most fall into the classes of flies, lice, mites, and ticks. Many of these have relatively complicated life cycles or significant mobility, and it can be difficult to develop control strategies for parts of the cycle that do not involve the animal host. This section will discuss some aspects of the biology and control of the more economically important external parasites in grazing livestock.

Flies

Heel Flies or Cattle Grubs

The cattle grub is really the larval stage of the heel fly or warble fly. Two species of this fly are found in the United States, and cattle and bison are the principal species affected. Occasionally they are found in goats, horses, and rarely, humans. Heel flies do not bite, but lay their eggs on hair shafts of the legs and lower parts of the animal's body during the months of March to July or August, depending on the weather and region of the country. Young grub larvae hatch from the egg and penetrate the skin near a hair follicle. They migrate through the subcutaneous tissues and over the connective tissues of muscles for the next 7–8 months, eventually forming a jellylike cyst under the skin of the back. One species of the grub has a predilection for migration around the tissues of the esophagus, and the other species spends 1–2 months inside the spinal canal before reaching the back during the late fall or winter. The grubs create small breathing holes in the skin and then feed and grow for about 6 weeks. Subsequently, they enlarge the hole and drop out to the ground, where they burrow under surface debris and pupate. They emerge in the spring as flies to begin the cycle again (18).

Fly activity tends to make the animals nervous and decreases normal grazing activity. Larval migration and activity causes considerable damage to the high value areas of the carcass over the back and reduces the value of the hide. Severe illness in an animal is rare. Control of adult heel flies is not feasible at this time. Systemic insecticides that kill the migrating larvae are used to break the cycle and to prevent damage to the carcass and hide. Currently available compounds provide excellent control and could make it possible to eradicate the cattle grub if there was a national will to do so. Timing of the treatment is important. Fly activity in the northern states stops with the first frost. Treating too early can allow reinfection. Treatment in the late summer or fall kills migrating larvae, but killing the parasites when they are located near the esophagus (gullet) or spinal canal can cause a reaction that leads to salivation and bloating or

paralysis of the hind quarters. Therefore, recommendations are usually given to treat as soon as practical after the first frost and to not treat after a certain date in the fall, depending on the region. These dates vary, but generally treatment should not be done after November 1.

Horn Flies

Horn flies resemble houseflies but are about half the size. Both sexes require blood meals and create considerable irritation and pain in their feeding activity, which peaks in the warmest months. During feeding, they hold their wings at about a 45° angle, with their heads downward. Horn flies are very common across the United States and are responsible for considerable economic loss resulting from reduced meat and milk production and hide damage. Several research trials have generated data that place the economic threshold for loss at about 100–200 flies per animal (cows and yearlings). The female fly will lay her eggs only in very fresh manure, and these eggs hatch to larvae in about 24 hours. Depending on the weather, the larva and pupa stages last from about 10 days to 4 weeks, after which the new flies emerge. The fly overwinters in the pupa stage (17).

Control strategies are usually aimed at the adult and the larval stages. Insecticides administered in dusts, sprays, backrubbers, and pour-on formulations are used to control adult horn flies, and the residual activity depends upon the compound used. Most need frequent reapplication. Within the past 20 years, insecticide-impregnated strips or ear tags have been used to provide long-term control. A significant amount of resistance to the pyrethroid class of insecticides, commonly used in these tags, has developed in horn fly populations in many regions across the country. Newer compounds in this class have improved activity, but resistance is still a problem. The amount of chemical available in these tags is limited to a finite time span. Putting the tags in too early in the season allows flies at the end of the season to be exposed to sublethal doses of the insecticide. This encourages resistance development in the flies. For this reason, tags should not be applied until economically significant numbers of flies are present on the cows, and they should be removed as soon as their effective life is over. The manufacturers of these tags indicate what that time period is for their product. Unfortunately, convenience in working the cattle is often the determinant for application and removal of the tags, and suboptimal performance and resistance development frequently result. Ear tag brand names and ingredients change rather rapidly. State extension entomologists can usually provide a current listing of available tags.

Plans for a walk-through fly trap were first released in 1930, and the development of resistance to chemicals and a desire to reduce dependency on them has led to some renewed interest in using mechanical removal. Details on the construction of a horn fly trap are available from several sources (including the University of Missouri Extension, HTTP://MUEXTENSION.MISSOURI.EDU/EXPLORE/MISCPUBS/MX1904.HTM, Plan MX1904C6 "Fly Trap") (8). The trap is placed in a "forced use" situation where cattle have to walk through it on a regular basis. Strips of canvas, carpet, or plastic brush off the flies as the animal walks through. Flies attempting to leave the trap are caught in screened trapping elements as they travel toward light. The trapped flies are collected within the screens and can be removed through clean-out doors. Field studies have suggested that the trap may reduce horn fly populations by 50% or more. Although chemical use may reduce populations to a greater degree, the trap can reduce fly populations below the economic threshold. The trap does not provide good control of face flies.

Feed additive larvacides or growth regulators can be used to control horn fly populations. These chemicals work by preventing the development of flies in the manure. They are administered in feed, loose mineral, or block form, and all animals in the group must consume the recommended dosage for effective control. Because flies can move from herd to herd over several miles, oral larvacides must be used rather extensively across an area or region to be effective. They are not effective in controlling fly populations that lay eggs in sites other than manure. Certain topically applied dewormers, now commonly used, also provide some measure of horn fly control. Some controversy exists surrounding the use of chemicals that provide control of fly larvae in the manure. Research has indicated that populations of some beneficial insects that use the manure for some part of their life cycle, such as the dung beetle, may be harmed or reduced. Dung beetles reportedly can aid in the control of horn flies by removing and burying manure before the life cycle of the horn fly is completed.

Face Flies

The face fly, which lays its eggs in fresh manure, is a relatively recent pest in North America. The larvae hatch and pupate in dried manure; the adults overwinter in buildings. The adults annoy horses and cattle at pasture, especially on warm sunny days, but are not inclined to go into barns. They feed on the secretions around the mouth, nose, and eyes. Possibly the most serious problem associated with the face fly is pinkeye infections in cattle. The bacteria that cause this disease can live on the legs of the fly for up to 3 days, and the fly serves to mechanically transmit the bacteria from animal to animal.

Face flies are usually controlled by the application of insecticidal sprays and dusts and by application of insecticide-impregnated ear tags. Drug resistance does not appear to be as significant a problem as that seen for horn flies. Oral larvacides appear to give variable to poor control.

Miscellaneous Flies

The nasal bot fly of sheep deposits its larvae on the sheep's nose during periods of bright sun. The larvae crawl up the nasal passages and develop through several stages in the nasal passages and sinuses. They are then sneezed out, and they pupate in the soil. The parasite overwinters in the soil or in the sheep's nose. These parasites cause annoyance to the sheep on pasture, and large numbers of larvae in the nose may cause nasal discharge (sometimes tinged with blood) and blockage. Their economic importance in the United States is unknown. The only available drug labeled for treatment is ivermectin, and routine treatment or prevention may not always be indicated.

Fly strike is caused by a number of species of flies (blowflies) that are attracted to odors produced by urine- or feces-soaked wool or hair, bacterial growth in hair or wool, or blood and serum contamination of the hair or wool such as might occur around the navel. Sheep, calves, and dogs are often victims. The flies lay their eggs near the soiled area, and the developing larvae feed on the debris. Severe infestations sometimes result in significant damage to the skin caused by the feeding activity and excretions of the larvae. Sometimes death occurs. The only prevention is to be vigilant for conditions that may attract the flies and prevent or clean them. Dogs used for predation prevention in sheep and other livestock species typically have long, dense hair coats that may become matted and dampened by rain. Bacterial growth in the hair mat may attract flies that lay their eggs on the hairs. Severe infestations can kill the dog, and owners may be unaware of the problem until it is advanced. Clipping the hair in the spring or early summer often makes the dog more comfortable and prevents fly strike. Wounds caused by injuries, castration, dehorning, or docking may attract flies. Use of an insecticidal repellent can prevent fly strike in these situations, but it may fail after a few days. Planning castration and dehorning activities for times when there is no fly activity is the best prevention when feasible.

Houseflies and stable flies annoy livestock with their feeding activity, and houseflies may become a nuisance to both the animals and humans. Unlike the horn and face flies that prefer manure in which to lay their eggs, these pests lay their eggs in damp, decaying organic matter such as grass clippings, silage or grain spills, manure and bedding piles, and sawdust piles. Although insecticides and repellents are available to assist in reducing fly numbers and their activity on livestock, the best control is to prevent accumulations of organic matter that can act as sites for egg laying. Around dairies, the weekly removal of calf manure, and spreading or composting it, disrupts the life cycle and can result in fewer flies.

Biological control of houseflies, stable flies, and some other species can be accomplished using several species of parasitic wasps. These tiny wasps attack the fly pupae in manure and decaying organic debris and lay their eggs in them. They do not attack humans or livestock. However, they are probably best used in an integrated approach to pest management. Insecticidal premise sprays and oral larvacides must not be used, or be used with caution, to avoid killing the parasitic wasps. Biological control can reduce the need for chemicals, but it must be carefully planned to be successful (24).

Deerflies and horseflies are aggressive feeders, and females require a blood meal for egg maturation. They parasitize many species of mammals. They usually concentrate along watercourses and lay their eggs on vegetation that overhangs water. During feeding, the horsefly cuts the skin with its mouth parts and laps up the blood. The bite is painful and often leads to interrupted feeding before the fly is finished with its blood meal. The fly may land on another animal to begin another attempt at a blood meal. In so doing, the fly may transmit blood-borne diseases such as anaplasmosis. Control of this fly is extremely difficult. Very large populations may require stabling the livestock during the day when these files are most active.

Lice

Lice are of two main types, biting lice and sucking lice (blood feeders). These insects spend their entire life cycles on the animal, and are species specific. In other words, the lice that infest cattle do not infest sheep. Lice are transmitted by close contact and contaminated equipment; they do not live long off the host. Louse infestations may cause their host discomfort, and damage to hides and wool occurs because of the rubbing and licking caused by the itching. However, serious infestations that result in severe anemia or loss of condition are relatively uncommon. The large populations that result in severe losses are often consequences of other concerns such as undernutrition, crowding as sometimes seen in winter housing, and animal stress from other disease or weather conditions.

Numerous compounds are available to control lice, and control is usually strategically applied in the fall before housing and winter weather ensues. Insecticides with little residual activity may need to be given twice, 30–60 days apart, to control the lice that have hatched from eggs after the first application. More recently, compounds in the avermectin/milbemycin group (i.e., eprinomectin, ivermectin, moxidectin, and doramec-

tin) have become available that have a longer duration of activity. The pour-on formulations of some of these products have extended activity. One application may be sufficient for control of lice all winter provided that infested new animals are not introduced to the treated group after the drug is no longer active.

Mites

Although several mites may attack livestock, the ones we usually think about are the "mange" mites. Mange is a dermatitis that is usually accompanied by intense itching, hair or wool loss, thickening of the skin, and sometimes weeping and crusting of the skin. It is sometimes referred to as "scabies" and is to be distinguished from the brain disease of sheep called "scrapie." If mange is diagnosed, it is important to differentiate the kind because two of them, psoroptic and sarcoptic mange, are reportable to regulatory veterinarians in state and federal government. Microscopic examination of skin scrapings is the method for making that diagnosis. Chorioptic mange in cattle usually appears in winter and primarily involves the tail head, hind legs, and the area between the tail head and udder. It sometimes affects the scrotum, and in sheep, the lower legs and scrotum of rams are most often affected. Mites spend their entire life cycle on the animal, and transmission is by direct contact and contaminated equipment.

Control of mange mites is usually only necessary when the herd or flock becomes infected, usually by the addition of new animals. The advent of the macrolide class of dewormers and insecticides (avermectins) has made treatment of mange much simpler because these compounds are systemically active and very effective. Routine use of them for treatment of lice and internal parasites has probably reduced the prevalence of mange in many livestock operations.

Ticks

On our domestic livestock, all ticks feed by blood sucking. There are numerous types of ticks and each has unique aspects to its life cycle. The basic life cycle involves the eggs, larvae, nymphs, and adults. In some cases, only one stage feeds on livestock, and some species are relatively host specific. The most important feature of tick parasitism is the fact that ticks transmit several important diseases of man and animals. A few of these include anaplasmosis, Texas fever, Rocky Mountain spotted fever, Lyme disease, and tularemia (also called rabbit fever). Effects that are more direct include a tick paralysis produced by toxins in tick saliva, painful bite wounds by some species that tend to become infected or inflamed, and loss of blood, unthriftiness, and reduced grazing activity in severe infestations. In the United States, tick infestations of livestock tend to be a greater concern in the subtropical climates, but ticks are present across most of the country.

Control of ticks, like horseflies and deerflies, is difficult because much of the life cycle is spent off the livestock host and because many species use other animals as intermediate hosts in their life cycle. Strategies designed to attack tick stages on intermediate hosts, or the intermediate hosts themselves, have had very limited success. Control must be tailored specifically to the region and specific tick species of concern. Some of the sprays and dips used for lice and flies have activity against ticks. They do have to be repeated at regular intervals if season-long con-trol is expected. The currently available avermectins that are widely used in internal parasite and louse control are not labeled for control of ticks.

RESPONSIBLE DRUG USE

In domestic livestock production today, producers must be more careful than ever to properly use and store chemicals used in the treatment and control of livestock diseases. Quality assurance programs require record keeping of the products used. Residues of drugs in animals entering the human food chain are unacceptable. Aside from the food safety and perception of quality issues, responsible and judicious use of drugs is important to prolong their effective life expectancy in the face of the potential for development of resistance by bacteria and internal and external parasites.

Most of the important points in responsible drug use can be distilled down to a few words: **read the label** and **keep good records**. Drug labels today give producers all they need to know to use drugs judiciously and effectively. Use in any other way than exactly as the label states is illegal and, in some cases, potentially dangerous. Except perhaps for sheep and goats, we now have a better selection of drugs to effectively treat internal and external parasites than ever before. It is wise to develop a good relationship with a veterinarian who can help design a total herd health plan and who can help select products that are not only effective for the use intended, but also cost-effective.

Abbreviations

ADF – acid detergent fiber

ADG – average daily gain

AI – artificial insemination

BCS – body condition score

BUN – blood urea nitrogen

CLA – conjugated linoleic acid

CNCPS – Cornell Net Carbohydrate Protein System

CP – crude protein

cwt – hundred weight

DHIA – Dairy Herd Improvement Association

DM – dry matter

DMI – dry matter intake

EAA – essential amino acids

EPD – expected progeny differences

EPG – eggs per gram (of feces)

FDA – U.S. Food and Drug Administration

GE – gaseous products of digestion

HE – total heat production

Mcal – megacalorie=1,000 kilocalories or 1,000,000 calories

M:G – milk:grain price ratio

MUN – milk urea nitrogen

NAHMS – National Animal Health Monitoring System

NDF – neutral detergent fiber

NEg – net energy gain

NEl – net energy for lactation

NEm – net energy for maintenance

NFC – nonfiber carbohydrate

NFF – nonforage fiber

NPN – nonprotein nitrogen

NRC – National Research Council

NSC – nonstructural carbohydrate

pTMR – "partial" total mixed ration

RDP – rumen-degradable protein

RUP – rumen-undegradable protein

SIP – soluble intake protein

TDN – total digestible nutrients

TMR – total mixed ration

TP – total protein

Conversion Tables

Type of measurement	To convert:	Into:	Multiply by:
Length	centimeters (cm)	inches (in)	0.394
	feet (ft)	centimeters (cm)	30.48
	feet (ft)	inches (in)	12
	feet (ft)	yards (yd)	0.33
	inches (in)	feet (ft)	0.083
	inches (in)	millimeters (mm)	25.4
	inches (in)	centimeters (cm)	2.54
	meters (m)	inches (in)	39.37
	meters (m)	feet (ft)	3.281
	meters (m)	yards (yd)	1.094
	yards (yd)	feet (ft)	3
	yards (yd)	centimeters (cm)	91.44
	yards (yd)	meters (m)	0.9144
Area	acres	square feet (ft^2)	43,560
	acres	square yards (yd^2)	4,840
	acres	hectares (ha)	0.4047
	hectares (ha)	acres	2.471
	hectares (ha)	square meters (m^2)	10,000
	square inches (in^2)	square centimeters (cm^2)	6.452
	square centimeters (cm^2)	square inches (in^2)	0.155
	square feet (ft^2)	square centimeters (cm^2)	929.09
	square feet (ft^2)	square meters (m^2)	0.0929
	square meters (m^2)	square feet (ft^2)	10.76
	square meters (m^2)	square yards (yd^2)	1.196
Weight	grams (g)	ounces (oz)	0.0353
	kilograms (kg)	pounds (lb)	2.205
	metric tons (megagrams)	short tons	1.1023
	ounces (oz)	pounds (lb)	0.0625
	ounces (oz)	grams (g)	28.35
	pounds (lb)	ounces (oz)	16
	pounds (lb)	grams (g)	453.6
	short tons	metric tons (megagrams)	0.9078
Volume, solids	bushels (bu)	cubic feet (ft^3)	1.24
	bushels (bu)	cubic meters (m^3)	0.352
	bushels (bu)	liters (L)	35.24
	cubic feet (ft^3)	liters (L)	28.32
	cubic feet (ft^3)	U.S. gallons (gal)	7.48
	cubic feet (ft^3)	cubic inches (in^3)	1,728
	cubic feet (ft^3)	cubic yards (yd^3)	0.037
	cubic feet (ft^3)	bushels (bu)	0.804
	cubic inches (in^3)	milliliters (ml)	16.39
	cubic meters (m^3)	cubic yards (yd^3)	1.308

Conversion Tables *(continued)*

Type of measurement	To convert:	Into:	Multiply by:
Volume, solids (continued)	cubic meters (m^3)	U.S. gallons (gal)	264.2
	cubic meters (m^3)	cubic feet (ft^3)	35.3
	cubic yards (yd^3)	cubic feet (ft^3)	27
	cubic yards (yd^3)	liters (L)	764.6
	cubic yards (yd^3)	cubic meters (m^3)	0.765
	cubic yards (yd^3)	bushels (bu)	21.7
	gallons, U.S. dry (gal)	cubic inches (in^3)	269
	liters (L)	cubic inches (in^3)	61.02
	milliliters (mL)	cubic inches (in^3)	0.0610
	quarts, dry (qt)	cubic inches (in^3)	67.2
Volume, liquids	cubic centimeters (cm^3 or cc)	milliliters (mL)	1
	cups (c)	fluid ounces (fl oz)	8
	gallons, U.S. (gal)	cups (c)	16
	gallons, U.S. (gal)	cubic inches (in^3)	231
	gallons, U.S. (gal)	quarts (qt)	4
	gallons, U.S. (gal)	liters (L)	3.785
	gallons, U.S. (gal)	gallons, Imperial (gal)	0.833
	gallons, Imperial (gal)	cubic inches (in^3)	277.42
	gallons, Imperial (gal)	liters (L)	4.546
	gallons, Imperial (gal)	gallons, U.S. (gal)	1.20
	liters (L)	pints (pt)	2.113
	liters (L)	quarts (qt)	1.057
	liters (L)	gallons, U.S. (gal)	0.2642
	milliliters (mL)	fluid ounces (fl oz)	0.0338
	pints (pt)	fluid ounces (fl oz)	16
	pints (pt)	cups (c)	2
	pints (pt)	quarts (qt)	0.5
	pints (pt)	cubic inches (in^3)	28.87
	pints (pt)	liters (L)	0.4732
	fluid ounces (fl oz)	cubic inches (in^3)	1.805
	fluid ounces (fl oz)	tablespoons (Tbsp)	2
	fluid ounces (fl oz)	teaspoons (tsp)	6
	fluid ounces (fl oz)	milliliters (mL)	29.57
	quarts (qt)	fluid ounces (fl oz)	32
	quarts (qt)	cups (c)	4
	quarts (qt)	pints (pt)	2
	quarts (qt)	U.S. gallons, liquid (gal)	0.25
	quarts (qt)	cubic inches (in^3)	57.7
	quarts (qt)	liters (L)	0.9463
	tablespoons (Tbsp)	teaspoons (tsp)	3
	tablespoons (Tbsp)	milliliters (mL)	15
	teaspoons (tsp)	milliliters (mL)	5

Conversion Tables *(continued)*

Weight per volume	grams/cubic centimeter (g/cm^3)	pounds/cubic foot (lbs/ft^3)	62.3
	tablespoons/bushel (Tbsp/bu)	pounds/cubic yard (lbs/yd^3)	1 (approx.)
	pounds/cubic yard (lbs/yd^3)	ounces/cubic foot (oz/ft^3)	0.6
	ounces/cubic foot (oz/ft^3)	pounds/cubic yard (lbs/yd^3)	1.67
	pounds/cubic yard (lbs/yd^3)	grams/liter (g/L)	0.595
	kilograms/cubic meter (kg/m^3)	pounds/cubic yard (lbs/yd^3)	1.6821

Parts per million (ppm) conversions

- 1 milligram/liter = 1 ppm
- 1 ounce/gallon = 7,490 ppm
- 1 ounce/100 gallons = 75 ppm

percent fertilizer element x 75 = ppm of element in 100 gallons of water per ounce of fertilizer

For example, for a 9-45-15 fertilizer, the ppm nitrogen (N) in 100 gallons of water per ounce of fertilizer would be:
0.09 (percent N) x 75 = 6.75 ppm N in 100 gallons of water per ounce of 9-45-15

If you want 150 ppm N, and each ounce gives 6.75 ppm, then you need:
150 ÷ 6.75 = 22.22 ounces of 9-45-15 fertilizer in 100 gallons of water

Temperature Conversion Formulas

- To convert °C to °F: (°C x 9/5) + 32 = °F
- To convert °F to °C: (°F - 32) x 5/9 = °C

Glossary

Acid detergent fiber – A laboratory estimate of the less digestible cellulose and lignin or "woody" fiber in the plant.

Anthelmintic – A drug that kills parasitic worms.

Body condition score – A subjective assessment of energy reserves of a livestock animal. It involves assigning a numerical score to an animal based on its relative amount of body energy reserves, primarily fat.

Coccidiosis – A disease caused by single-celled protozoan parasites called coccidia that reside in the ruminant's intestines. Destruction of the cells lining the intestine by the developing immature forms of the parasite leads to diarrhea, often bloody, and dehydration.

Conjugated linoleic acid – A fatty acid found in dairy products; may be beneficial to human health.

Creep feeding – Providing supplemental feed to nursing calves.

Crude protein – Estimated by measuring the amount of N in the forage sample, both true protein and nonprotein N, and multiplying this value by 6.25. Crude protein is the source of N and amino acids in feeds.

Expected progeny differences – Expected progeny differences (EPDs) provide an estimate of the genetic value of an animal as a parent. Differences in EPDs between two individuals of the same breed predict differences in performance between their future offspring when each is mated to animals of the same average genetic merit.

Flushing – The practice of increasing ewes' energy intake, and therefore body condition, during the 10–14 days prior to breeding. Leads to increased ovulation rates and thereby increased lambing percentage.

Grass tetany – A nutritional condition in grazing ruminants in which the concentration of magnesium in the blood is too low for good health, resulting in paralysis and death of the animal.

Heterosis – The superiority in performance of the crossbred animal compared to the average of the straightbred parents. Heterosis is maximized when the breeds crossed are genetically diverse. Also known as hybrid vigor.

Ionophores – Feed additives for ruminants that generally improve feed efficiency by decreasing feed intake.

Leader-follower grazing – The practice of first grazing a pasture with a class of animals whose nutritional requirements are high, such as lactating cows, then following that group with animals whose nutritional requirements are lower, such as bred heifers.

Nematode – A tiny wormlike organism that may feed on or in plants, including roots; may be referred to as roundworms, threadworms, or eelworms.

Neutral detergent fiber – An estimate of a plant's cell wall content, including the ADF fraction and hemicellulose.

Pasture carrying capacity – A measure of the number of animals that can be placed on a pasture for a season to achieve a targeted level of animal performance or economic production without causing deterioration of the pasture resource.

Periparturient – Around the time of giving birth, including the periods before and after parturition.

Phenology – The study of the relationships between climate and biological processes.

Polled – Having no horns.

Prussic acid poisoning – Also known as hydrocyanic acid or HCN, prussic acid is a potentially lethal poison produced during digestion of plant species with high concentrations of cyanogenic glycosides. Species such as sorghum, sudangrass, and johnsongrass can accumulate cyanogenic glycosides, particularly during drought and especially immediately after a drought has broken.

Slug feeding – When dairy cows are fed and consume large amounts of concentrates in a short period of time.

Stocking density – The number of animals present per unit land area at a given point in time.

References

CHAPTER 1

1. Arnold, G. W., and J. L. Hill. 1972. Chemical factors affecting selection of food plants by ruminants. pp. 72–101, In: J. B. Harborne (ed.). *Phytochemical Ecology*. Academic Press, NY.

2. Emmick, D. L. (ed.). 2000. Prescribed grazing and feeding management for lactating dairy cows. NY State Grazing Lands Conservation Initiative, USDA-NRCS. Syracuse, NY.

3. Emmick, D. L., and D. G. Fox. 1993. Prescribed grazing to improve pasture productivity in New York. USDA-NRCS and Cornell University, Ithaca, NY.

4. Goatcher, W. D., and D. C. Church. 1970. Taste responses in ruminants. IV. Reactions of pygmy goats, normal goats, sheep and cattle to acetic acid and quinine hydrochloride. *J. Anim. Sci.* 31: 373–382.

5. Hanley, T. A. 1982. The nutritional basis for food selection by ungulates. *J. Range Mngt.* 35: 146–152.

6. Hoeck, H. N. 1975. Differential feeding behavior of the sympatric hyrax *Procavia johnstoni* and *Heterohyrax brucei. Oecologia* 22: 15–47.

7. Hofman, R. R. 1989. Evolutionary steps of ecophysical adaptation and diversification of ruminants: A comparative view of their digestive system. *Oecologia* 78: 443–457.

8. Hofman, R. R., and D. R. M. Stewart. 1972. Grazer or browser: A classification based on the stomach structure and feeding habits of East African ruminants. *Mammalia* 36: 226–240.

9. Holechek, J. L., R. D. Piper, and C. H. Herbel. 1989. *Range Management: Principles and Practices.* Regents/Prentice Hall, NJ.

10. Holmes, C. W. 1987. Pastures for dairy cows. pp. 133–143, In: A. M. Nicol (ed.). *Livestock Feeding on Pasture.* NZ Soc. Anim. Prod. Occ. Publ. No. 10. Ruakura Agric. Res. Ctr., NZ.

11. Howery, L. D., F. D. Provenza, and G. B. Ruyle. 1998. How do domestic herbivores select nutritious diets on rangelands? Arizona Cooperative Extension publication AZ1023, University of Arizona, Tucson, AZ.

12. Janis, C. M., and D. Ehrhardt. 1988. Correlation of relative muzzle width and relative incisor width with dietary preference in ungulates. *Zool. J. Linn. Soc.* 92: 267–284.

13. Krueger, W. C., W. A. Laycock, and D. A. Price. 1974. Relationships of taste, smell, sight and touch to forage selection. *J. Range Mngt.* 27: 258–262.

14. Launchbaugh, K. L., J. W. Walker, and C. A. Taylor. 1999. Foraging behavior: Experience or inheritance? pp. 28–35, In: K. L. Launchbaugh, J. C. Mosley, and K. D. Saunders (ed.). *Grazing Behavior of Livestock and Wildlife.* Idaho Forest, Wildlife, and Range Experiment Station. Moscow, ID.

15. Mayland, H. F., and G. E. Shewmaker. 1999. Plant attributes that affect livestock selection and intake. pp. 70–74, In: K. L. Launchbaugh, J. C. Mosley, and K. D. Saunders (ed.). *Grazing Behavior of Livestock and Wildlife.* Idaho Forest, Wildlife, and Range Experiment Station. Moscow, ID.

16. Pfister, J. A. 1999. Behavioral strategies for coping with poisonous plants. pp. 45–59, In: K. L. Launchbaugh, J. C. Mosley, and K. D. Saunders (ed.). *Grazing Behavior of Livestock and Wildlife.* Idaho Forest, Wildlife, and Range Experiment Station. Moscow, ID.

17. Provenza, F. D. 1995. Postingestive feedback as an elementary determinant of food preference and intake in ruminants. *J. Range Mngt.* 48: 2–17.

18. Provenza, F. D. 1995. Role of learning in food preferences of ruminants: Greenhalgh and Reid revisited. pp. 231–245, In: W. V. Engelhardt, S. Leonhard-Marek, G. Breves, and D. Giesecke (ed.). *Ruminant Physiology: Digestion, Metabolism, Growth, and Reproduction. Proc. Eighth Int. Symp. Ruminant Physiol.*, Ferdinand Enke Verlag, Stuttgart, Germany.

19. Provenza, F. D. 1996. Acquired aversions as the basis for varied diets of ruminants foraging on rangelands. *J. Anim. Sci.* 74: 2010–2020.

20. Provenza, F. D., and D. F. Balph. 1990. Applicability of five diet selection models to various foraging challenges ruminants encounter. pp. 423–459, In: R. N. Hughes (ed.). *Behavioral Mechanisms of Food Selection.* Vol. 20: NATO ASI Series G: Ecological Sciences. Springer-Verlag, Heidelberg, Germany.

21. Provenza, F. D., and K. L. Launchbaugh. 1999. Foraging on the edge of chaos. pp. 1–12, In: K. L. Launchbaugh, J. C. Mosley, and K. D. Saunders (ed.). *Grazing Behavior of Livestock and Wildlife.* Idaho Forest, Wildlife, and Range Experiment Station. Moscow, ID.

22. Provenza, F. D., J. A. Pfister, and C. D. Cheney. 1992. Mechanisms of learning in diet selection with reference to phytotoxicosis in herbivores. *J. Range Mngt.* 45: 36–45.

23. Provenza, F. D., J. J. Villalba, C. D. Cheney, and S. J. Werner 1998. Self-organization of foraging behavior: From simplicity to complexity without goals. *Nutr. Res. Rev.* 11: 1–24.

24. Rhodes, D. F. 1979. Evolution of plant chemical defense against herbivores. pp. 3–54, In: G. A. Rosenthal and D. H. Janzen (ed.). *Herbivores: Their Interaction with Secondary Plant Metabolites.* Academic Press, NY.

25. Robbins, C. T., D. E. Spalinger, and W. Van Hoven. 1995. Adaptation of ruminants to browse and grass diets: Are anatomical-based browser-grazer interpretations valid? *Oecologia* 103: 208–213.

26. Scehovic, J., 1985. Palatability and the organoleptic characteristics of the cultivars and hybrids of tall fescue (*Festuca arundinacea*). pp. 317–319, In: *Proc. XV Intl. Grasslands Congress*, Kyoto, Japan.

27. Scehovic, J., C. Poisson, and M. Gillet. 1985. Palatability and the organoleptic characteristics of grasses I. Comparison between ryegrass and tall fescue. *Agronomie* 5: 347–354.

28. Shipley, L. A. 1999. Grazers and browsers: How digestive morphology affects diet selection. pp. 20–27, In: K. L. Launchbaugh, J. C. Mosley, and K. D. Saunders (ed.). *Grazing Behavior of Livestock and Wildlife.* Idaho Forest, Wildlife, and Range Experiment Station. Moscow, ID.

29. Shoemaker, G. E., H. F. Mayland, and S. B. Hansen. 1997. Cattle grazing preference among eight endophyte-free tall fescue cultivars. *Agron. J.* 89: 695–701.

30. Smith, R. L. 1980. *Ecology and Field Biology*, 3rd ed. Harper and Row, NY.

31. Stoddart, L. A., A. D. Smith, and T. W. Box. 1975. *Range Management*, 3rd ed. McGraw Hill Book Co., NY.

32. Van Soest, P. J. 1982. *Nutritional Ecology of the Ruminant.* O&B Books, Inc. Corvalis, OR.

33. Villalba, J. J., and F. D. Provenza. 2000. Roles of novelty, generalization and postingestive feedback in the recognition of foods by lambs. *J. Anim. Sci.* 78: 3060–3069.

34. Weiner, J. 1994. *The Beak of the Finch*. Vintage Books, NY.

35. Provenza, F. D. and J. J. Villalba. 2006. Foraging in domestic vertebrates: Linking the internal and external milieu. pp. 210–240 in V. L. Bels (ed.), *Feeding in Domestic Vertebrates: From Structure to Function.* CABI Publ., Oxfordshire, U.K.

CHAPTER 2

1. Abdalla, H. O., D. G. Fox, and R. R. Seaney. 1988. Variation in protein and fiber fractions in pasture during the grazing season. *J. Anim. Sci.* 66: 2663–2667.

2. Ball, P. R, and J. C. Ryden. 1984. Nitrogen relationships in intensively managed temperate grasslands. *Plant Soil* 76: 23–33.

3. Beever, D. E. 1982. Protein utilization from pasture. p. 99. In: T. W. Griffiths and M. F. Maguire (ed.). *Forage Protein Conservation and Utilisation.* Commission of the European Communities, Dublin, Ireland.

4. Belesky, D. P., K. E. Turner, J. M. Fedders, and J. M. Ruckle. 2001. Mineral composition of swards containing forage chicory. *Agron. J.* 93: 468–475.

5. Broderick, G. A. 1995. Desirable characteristics of forage legumes for improving protein utilization in ruminants. *J. Anim. Sci.* 73: 2760–2773.

6. Broderick, G. A. 1996. Quantifying forage protein quality. pp. 200–228. In: G. C. Fahey, Jr. (ed.). *Forage Quality, Evaluation, and Utilization.* ASA, CSSA, and SSSA, Madison, WI.

7. Broderick, G. A., R. J. Wallace, and E. R. Ørskov. 1991. Control of rate and extent of protein degradation. pp. 541–592. In: T. Tsuda, Y. Sasaki, and R. Kawashima (ed.). *Physiological Aspects of Digestion and Metabolism in Ruminants.* Academic Press, NY.

8. Burns, J. C., and J. E. Standaert. 1985. Productivity and economics of legume-based vs. nitrogen-fertilized grass-based pastures in the United States. pp. 56–71. In: R. F Barnes et al. (ed.). *Proc. Trilateral Workshop,* Palmerston North, N.Z.USDA-ARS, Washington, DC.

9. Christian, K. R. 1987. Matching pasture production and animal requirements. In: J. L. Wheeler, C. J. Pearson, and G. E. Robbards (ed.). *Temperate Pastures, Their Production, Use, and Management.* CSIRO, Australia.

10. Hammond, A. C., W. E. Kunkle, P. C. Genho, S. A. Moore, C. E. Crosby, and K. H. Ramsay. 1994. Use of blood urea nitrogen concentration to determine time and level of protein supplementation in wintering cows. *Prof. Anim. Scientist* 10: 24–31.

11. Hodgson, J. 1985. The control of herbage intake in the grazing ruminant. *Proc. Nutr. Soc.* 44: 339–345.

12. Jung, G. A., J. A. Shaffer, G. A. Varga, and J. R. Everhart. 1996. Performance of 'Grasslands Puna' chicory at different management levels. *Agron. J.* 88: 104–111.

13. Karnezos, T. P., A. G. Matches, R. L. Preston, and C. P. Brown. 1994. Corn supplementation of lambs grazing alfalfa. *J. Anim. Sci.* 72: 783–789.

14. Kohn, R. A., and M. S. Allen. 1995. Effect of plant maturity and preservation method on in vitro protein degradation of forages. *J. Dairy Sci.* 78: 1544–1551.

15. Macrae, J. C., J. S. Smith, P. J. S. Dewey, A. C. Brewer, D. S. Brown, and A. Walker. 1985. The efficiency and utilization of metabolizable energy and apparent absorption of amino acids in sheep given spring- and autumn-harvested dried grass. *Br. J. Nutr.* 54: 197–209.

16. Marten, G. C., C. C. Sheaffer, and D. L. Wyse. 1987. Forage nutritive value and palatability of perennial weeds. *Agron. J.* 79: 980–986.

17. Matches, A. G. 1989. A survey of legume production and persistence in the United States. pp. 37–44. In: G. C. Marten et al. (ed.). *Persistence of Forage Legumes.* ASA, CSSA, and SSSA, Madison, WI.

18. Maynard, L. A., and Loosli, J. K. 1969. *Animal Nutrition*, 6th ed., McGraw-Hill Book Company, NY, pp. 229–231.

19. Minson, D. J. 1990. *Forage in Ruminant Nutrition.* Academic Press, San Diego.

20. Moore, J. E. 1980. Forage crops. pp. 61–91, In: C. S. Hoveland (ed.). *Crop Quality, Storage, and Utilization.* ASA, Madison, WI.

21. Moore, J. E., W. E. Kunkle, and W. F. Brown. 1991. Forage quality and the need for protein and energy supplements. In: *40th Annual Florida Beef Cattle Short Course Proceedings.* Anim. Sci. Dept., Univ. Fla., Gainsville.

22. National Research Council. 1976. *The Nutrient Requirements of Beef Cattle,* 5th ed., National Academy Press, Washington, D.C.

23. National Research Council, 1981. *Nutritional Energetics of Domestic Animals and Glossary of Energy Terms.* National Academy Press, Washington, D.C.

24. National Research Council. 1988. *The Nutrient Requirements of Dairy Cattle,* 6th ed., National Academy Press, Washington, D.C.

25. National Research Council. 1985. *Nutrient Requirements of Sheep*, 6th ed., National Academy Press, Washington. D.C.

26. Nelson, C. J., and L. E. Moser. 1994. Plant factors affecting forage quality. pp. 115–154. In: G. C. Fahey, Jr. (ed.). *Forage Quality, Evaluation, and Utilization*. ASA, CSSA, and SSSA, Madison, WI.

27. Newbold, C. J., R. J. Wallace, N. D. Watt, and A. J. Richardson. 1988. Effect of the novel ionophore tetronasin (ICI 139603) on ruminal microorganisms. *Appl. Environ. Microbiol.* 54: 544–547.

28. Nocek, J. E., and J. B. Russell. 1988. Protein and energy as an integrated system, relationship of ruminal protein and carbohydrate availability to microbial synthesis and milk production. *J. Dairy Sci.* 71: 2070–2107.

29. Pfander, W. H., S. E. Grebing, C. M. Price, O. Lewis, J. M. Asplund, and C. V. Ross. 1975. Use of plasma urea nitrogen to vary protein allowances of lambs. *J. Anim. Sci.* 41: 647–653.

30. Pitt, R. E. 1990. *Silage and Hay Preservation*. Natural Resource, Agriculture, and Engineering Service, NRAES–5, Ithaca, NY.

31. Preston, R. L., D. D. Schnakenberg, and W. H. Pfander. 1965. Protein utilization in ruminants. I. Blood urea nitrogen as affected by protein intake. *J. Nutr.* 86: 281–288.

32. Poppi, D. P., and S. R. McLennan. 1995. Protein and energy utilization by ruminants at pasture. *J. Anim. Sci.* 73: 278–290.

33. Russell, J. B., and H. J. Strobel. 1989. Effect of ionophores on ruminal fermentation. *Appl. Environ. Microbiol.* 55: 1–6.

34. Squires, V. R. 1988. Water and its functions, regulation, and comparative use by ruminant livestock. pp. 217–226. In: D. C. Church (ed.). *The Ruminant Animal—Digestive Physiology and Nutrition*. Prentice Hall, Englewood Cliffs, NJ.

35. Turner, K. E., D. P. Belesky, and J. M. Fedders. 1999. Chicory effects on lamb weight gain and rate of in vitro organic matter and fiber disappearance. *Agron. J.* 91: 445–450.

36. Turner, K. E., D. P. Belesky, J. M. Fedders, and E. B. Rayburn. 1996. Canopy management influences on cool-season grass quality and simulated livestock performance. *Agron. J.* 88: 199–205.

37. Turner, K. E., and J. G. Foster. 2000. Nutritive value of some common browse species. *Amer. Forage Grassl. Proc.* 9: 241–245.

38. Turner, K. E., K. E. McClure, W. P. Weiss, R. J. Borton, and J. G. Foster. 2002. Alpha-tocopherol concentrations in lamb muscle and case life as influenced by concentrate or pasture finishing. *J. Anim. Sci.* 80: 2513–2521.

39. Underwood, E. J. 1981. *The Mineral Nutrition of Livestock*. Commonwealth Agricultural Bureau, London.

40. Whitehead, D. C. 1970. The role of nitrogen in grassland productivity. Commonwealth Bureau Pastures and Field Crops Bull. 48, p. 202. Commonwealth Agric. Bureau, Farnham Royal.

CHAPTER 3

1. Adams, N. R. 1995. Detection of the effects of phytoestrogens on sheep and cattle. *J. Anim. Sci.* 73: 1509.

2. Baker, M. J., E. C. Prigge, and W. B. Bryan. 1988. Herbage production from hay fields grazed by cattle in fall and springs. *J. Prod. Agric.* 1: 275–279.

3. Ball, D., G. Lacefield, C. Hoveland, and W. C. Young. Tall fescue/endophyte relationships. Mimeograph. Oregon Tall Fescue Commission, Salem, OR.

4. Bellows, R. A., R. E. Short, and G. V. Richardson. 1982. Effects of sire, age of dam and gestation feed level on dystocia and postpartum reproduction. *J. Anim. Sci.* 55: 18–27.

5. Bircham, J. S. 1980. Herbage mass and height: Their relevance to management systems. In: *Proc. Workshop on Mixed Grazing,* Agric. Inst. Ireland and Agric. Res. Inst. Iceland. pp. 93–98.

6. Bishop, D. K., R. P. Wettemann, and L. J. Spicer. 1994. Body energy reserves influence the onset of luteal activity after early weaning of beef cows. *J. Anim. Sci.* 72: 2703–2708.

7. Bryan, W. B., E. C. Prigge, D. J. Flaherty, and G. E. D'Sonza. 1997. Buffer grazing for a twelve month cow-calf production system. D. 2995–2996. In: *Proc. 18th Int. Grassland Congr.,* Vol. 2. Canada Forage Council. Calgary, AB.

8. Bryan, W. B., E. C. Prigge, M. Lasat, T. Paska, D. J. Flaherty, and J. Lozier. 2000. Productivity of Kentucky bluegrass pastures grazed at three heights and two intensities. *Agron. J.* 92: 30–35.

9. Butler, W. R. 1997. Effect of protein nutrition on ovarian and uterine physiology. *J. Dairy Sci.* 80 (Suppl. 1): 139.

10. Cantrell, J. A., J. R. Kropp, S. L. Armbruster, K. S. Lusby, R. P. Wettemann, and R. L. Hintz. 1981. The influence of postpartum nutrition and weaning age of calves on cow body condition, estrus, conception rate and calf performance of fall-calving beef cows. *Anim. Sci. Res. Rpt.* pp. 53–58.

11. Carroll, D. J., B. A. Barton, and D. R. Thomas. 1997. Review of protein nutrition-reproduction studies. *J. Dairy Sci.* 80 (Suppl. 1): 139.

12. Collins, M., and V. A. Balasko. 1981. Effects of N fertilization and cutting schedules on stockpiled tall fescue. I. Forage yield. *Agron. J.* 73: 803–807.

13. Comerford, J. W., E. H. Cash, H. W. Harpster, and L. L. Wilson. 1995. Effects of early weaning and return to pasture on health and performance of beef calves. *J. Anim. Sci.* 73 (Suppl. 1): 238.

14. Comerford, J. W., H. W. Harpster, and V. H. Baumer. 1996. Effects of grazing and protein supplementation for Holstein steers. *J. Anim. Sci.* 74 (Suppl. 1): 254.

15. Comerford, J. W., J. B. Cooper, L. L. Benyshek, and J. K. Bertrand. 1991. Evaluation of feed conversion in steers from a diallel of Simmental, Limousin, Polled Hereford, and Brahman beef cattle. *J. Anim. Sci.* 69: 2770.

16. Croom, W. J., W. M. Hagler, M. A. Froetschel, and A. D. Johnson. 1995. The involvement of slaframine and swainsonine in slobbers syndrome: A review. *J. Anim. Sci.* 73: 1499.

17. Cundiff, L. V., K. E. Gregory, and R. M. Koch. 1984. Germplasm evaluation program report No. 11. Roman L. Hruska Meat Animal Evaluation Center, U.S. Department of Agriculture, Agricultural Research Service, Clay Center, NE.

18. DeRouen, S. M., D. E. Franke, D. G. Morrison, W. E. Wyatt, D. F. Coombs, T. W. White, P. E. Humes, and B. B. Greene. 1994. Prepartum body condition and weight influences on reproductive performance of first-calf beef cows. *J. Anim. Sci.* 72: 1119–1125.

19. Dubey, J. P. 1999. Neosporosis in cattle: Biology and economic impact. *J. Am. Vet. Med. Asoc.* 214 (No.8): 1160.

20. Encinias, A. M., and G. Lardy. 2000. Body condition scoring I: Managing your cow herd through body condition scoring. *Beef InfoBase,* Version 1.2. Adds Center, Inc.

21. Essig, H. W., C. E. Cantrell, F. T. Withers, D. J. Lang, D. H. Loughlin, and M. E. Boyd. 1989. Performance profitability of cow-calf systems grazing on EF and EL KY-31 fescue. *Proc. Tall Fescue Toxicosis Workshop*, Atlanta, GA.

22. Farrell, C. L., and T. G. Jenkins. 1982. Energy utilization by mature cows. Beef Research Program Report No. 1. Roman L. Hruska Meat Animal Evaluation Center, U.S. Department of Agriculture, Agricultural Research Service, Clay Center, NE.

23. Froetshel, M. A., H. E. Amos, D. Kumar, V. Pattarajinda, and C. A. McPeake. 2000. Determining the energetic value of whole cottonseed as compared to corn and cottonseed meal in a block supplement for growing cattle based on broiler litter and molasses. *J. Anim. Sci.* 79 (Suppl. 1): 283.

24. Gay, N., J. A. Boling, R. Dew, and D. E. Miksch. 1988. Effects of endophyte-infected tall fescue on beef cow-calf performance. *Appl. Agr. Res.* 3: 182.

25. Gillespie, J. R. 1997. *Animal Science.* Delmar Publishers, Albany, NY.

26. Hanson, D., and Rossiter, C. 1999. What do I need to know about johne's disease in beef cattle? Veterinary Services Info Sheet. USDA-APHIS.

27. Hermel, S. R. 1997. *Buyer beware. BEEF.* Intertec Publishing, Overland Park, KS.

28. Hodgson, J. 1985. The significance of sward height characteristics in the management of temperate sown pasture. pp. 63–67, In: *Proc. Int. Grassland Congr.,* Kyoto, Japan.

29. Houghton, P. L., R. P. Lemenager, L. A. Horstman, K. S. Hendrix, and G. E. Moss. 1990. Effects of body composition, pre- and postpartum energy level and early weaning on reproductive performance of beef cows and preweaning calf gain. *J. Anim. Sci.* 68: 1438–1446.

30. House, B. R. 1992. Effects of Forage and Protein Source on Production of Holstein Steers. M.S. Thesis. The Pennsylvania State University, University Park.

31. Hudson, D. 1982. Foot rot. In: *Beef Infobase,* Version 1.1. Adds Center, Inc., Madison, WI.

32. James, L. F., W. J. Hartley, and K. R. Van Kampen. 1981. Syndromes of *Astragalus* poisoning in livestock. *J. Amer. Vet. Med. Assoc.* 178: 146.

33. Kunkle, W. H., and T. M. Bates. 2000. Evaluating feed purchasing options: Energy, protein, and mineral supplements. *Beef InfoBase,* Adds Center, Inc., Madison, WI.

34. Lalman, D. L., D. H. Keisler, J. E. Williams, E. J. Scholljegerdes, and C. M. Mallet. 1997. Influence of postpartum weight and body condition change on duration of anestrus by undernourished suckled beef heifers. *J. Anim. Sci.* 75: 2003–2008.

35. Lesmeister, J. L., B. J. Burfening, and R. L. Blackwell. 1973. Date of first calving in beef cows and subsequent calf production. *J. Anim. Sci.* 36: 1–6.

36. Loper, G. M., C. H. Hanson, and J. H. Graham. 1967. Coumestrol content of alfalfa as affected by selection for resistance to foliar diseases. *Crop Sci.* 7: 189.

37. McCann, M. A. 1995. Creep feeding beef calves. Leaflet 403, University of Georgia, Athens.

38. McCollum, T. 1997. The latest methods to determine when to supplement. *Beef InfoBase,* Adds Center, Inc. Madison, WI.

39. McDonald, W. T. 1989. Performance of Cows and Calves Grazing Endophyte-Infected Pasture. M.S. Thesis. University of Tennessee, Knoxville.

40. Majak, W., J. W. Hall, and W. P. McCaughey. 1995. Pasture management strategies for reducing the risk of legume bloat in cattle. *J. Anim. Sci.* 73: 1493.

41. Makarechian, M., and P. F. Arthur. 1990. Effects of body condition and temporary calf removal on reproductive performance of range cows. *Theriogenology* 34: 435–443.

42. Marston, T. T., K. S. Lusby, R. P. Wettemann, and H. T. Purvis. 1995. Effects of feeding energy or protein supplements before or after calving on performance of spring-calving cows grazing native range. *J. Anim. Sci.* 73: 657–664.

43. Morrison, D. G., J. C. Spitzer, and J. L. Perkins. 1999. Influence of prepartum body condition score change on reproduction in multiparous beef cows calving in moderate body condition. *J. Anim. Sci.* 77: 1048–1054.

44. National Research Council (NRC). 1996. *Nutrient Requirement of Beef Cattle,* 7th ed. National Academy Press, Washington, D.C.

45. National Cattlemans Association. 1995. *IRM-SPA Database.* National Cattlemans Association, Denver, CO.

46. Perry, R. C., L. R. Corah, R. C. Cochran, W. E. Beal, J. S. Stevenson, J. E. Minton, D. D. Simms, and J. R. Brethour. 1991. Influence of dietary energy on follicular development, serum gonadotropins, and first postpartum ovulation in suckled beef cows. *J. Anim. Sci.* 69: 3762–3773.

47. Prigge, E. C., W. B. Bryan, and E. S. Goldman-Innis. 1999. Early and late-season grazing of orchardgrass hayfields overseeded with red clover. *Agron. J.* 91: 690–696.

48. Prigge, E. C., W. B. Bryan, and E. L. Nestor. 1997. Sward height on performance of cow-calf units and yearling steers grazing cool season pasture. In: *Proc. 18th Int. Grassland Congr.,* Vol. 2. Canada Forage Council. Calgary, AB.

49. Rasby, R. J., R. P. Wettemann, R. D. Geisert, L. E. Rice, and C. R. Wallace. 1990. Nutrition, body condition and reproduction in beef cows: Fetal and placental development, and estrogens and progesterone in plasma. *J. Anim. Sci.* 68: 4267–4276.

50. Rayburn, E. B., R. E. Blaser, and D. D. Wolfe. 1979. Winter tall fescue yield and quality with different accumulation periods and N rates. *Agron. J.* 71: 959–963.

51. Richards, M. W., J. C. Spitzer, and M. B. Warner. 1986. Effect of varying levels of postpartum nutrition and body condition at calving on subsequent reproductive performance in beef cattle. *J. Anim. Sci.* 62: 300–306.

52. Rutter, L. M., and R. D. Randel. 1984. Postpartum nutrient intake and body condition: Effect on pituitary function and onset of estrus in beef cattle. *J. Anim. Sci.* 58: 265–274.

53. Rook, A. J., C. A. Huckle, and R. J. Wilkins. 1994. The effects of sward height and concentrate supplementation on the performance of spring calving dairy cows grazing perennial ryegrass-white clover sward. *Anim. Prod.* 58: 167–172.

54. Roquette, F. M. 2000. Matching forage quality to beef cattle requirements. *Beef InfoBase,* Adds Center, Inc., Madison, WI.

55. Selk, G. E., R. P. Wettemann, K. S. Lusby, J. W. Oltjen, S. L. Mobley, R. J. Rasby, and J. C. Garmendia. 1988. Relationships among weight change, body condition and reproductive performance of range beef cows. *J. Anim. Sci.* 66: 3153–3159.

56. Short, E. R., and R. A. Bellows. 1971. Relationship among weight gains, age of puberty and reproductive performance in heifers. *J. Anim. Sci.* 32: R 27–131.

57. Spitzer, J. C., D. G. Morrison, R. P. Wettemann, and L. C. Faulkner. 1995. Reproductive responses and calf birth and weaning weights as affected by body condition at parturition and postpartum weight gain in primiparous beef cows. *J. Anim. Sci.* 73: 1251–1257.

58. Tucker, C. A., R. E. Morrow, J. R. Gerrish, C. J. Nelson, G. B. Garner, V. E. Jacobs, W. G. Hires, J. J. Shinkel, and J. R. Forwood. 1989. Forage systems for beef cattle: Effect of winter supplementation and forage system on reproductive performance of cows. *J. Prod. Agric.* 2: 217.

59. Van Soest, P. J. 1994. *Nutritional Ecology of the Ruminant*, 2nd ed. Cornell Univ. Press, Ithaca, NY.

60. Vicini, J. L., E. C. Prigge, W. B. Bryan, and G. A. Varga. 1982. Influence of forage species and creep grazing on a cow-calf system. II. Calf production. *J. Anim. Sci.* 55: 759–764.

61. Vough, L. R., and E. K. Cassel. 1988. Prussic acid poisoning in livestock. Fact Sheet No. 427. University of Maryland, College Park.

62. Wedin, W. F., I. T. Carlson, and R. L. Vetler. 1966. Studies on nutritive value of fall-saved forage, using rumen fermentation and chemical analysis. pp. 424–428. In: *Proc. 10th Int. Grassl Congr.,* Helsinki, Finland. Valtineuvoston Kirjapiano.

63. Wilson, L. L. 1990. Comparisons of beef cow-calf creep feeding methods. *1990 Penn State Dairy and Animal Science Report.* Pennsylvania State University, University Park.

64. Wright, I. A., and T. K. Whyte. 1989. Effects of sward surface height on the performance of continuously stocked spring-calving beef cows and their calves. *Grass and Forage Sci.* 44: 259–266.

CHAPTER 4

1. Bargo, F., L. D. Muller, E. S. Kolver, and J. E. Delahoy. 2003. Invited review: Production and digestion of supplemented dairy cows on pasture. *J. Dairy Sci.* 86: 1–42.

2. Bargo, F., L. D. Muller, J. E. Delahoy, and T. W. Cassidy. 2002. Milk response to concentrate supplementation of high-producing dairy cows grazing at two pasture allowances. *J. Dairy Sci.* 85: 1777–1792.

3. Bargo, F., L. D. Muller, J. E. Delahoy, and T. W. Cassidy. 2002. Performance of high-producing dairy cows with three different feeding systems combining pasture or total mixed rations. *J. Dairy Sci.* 85: 2959–2974.

4. Dickinson, F. N., and R. W. Touchberry. 1961. Livability of purebred vs. crossbred dairy cattle. *J. Dairy Sci.* 44: 879–887.

5. Dillon, P., and F. Buckley. 1998. Managing and feeding high genetic merit dairy cows at pasture. Technical Bulletin Issue No. 2. R & H Hall, Dublin, Ireland.

6. Hongerholt, D. D., and L. D. Muller. 1998. Supplementation of rumen undegradable protein to the diets of early lactation Holstein cow grazing grass pasture. *J. Dairy Sci.* 81: 2204–2214.

7. James, R. E. 1999. Controlling feed costs. Third Annu. Conf. Prof. Dairy Heifer Grower's Assoc., Bloomington, MN, March 1999.

8. Kelly, J. L., E. S. Kolver, D. E. Bauman, M. E. Van Ambugh, and L. D. Muller. 1998. Effects of intake of pasture on concentrations of CLA in milk of lactating cows. *J. Dairy Sci.* 81: 1630–1636.

9. Kolver, E. S., L. D. Muller, M. C. Barry, and J. W. Penno. 1998. Evaluation and application of the Cornell Net Carbohydrate and Protein System for dairy cows fed pasture-based diets. *J. Dairy Sci.* 81: 2029–2039.

10. Kolver, E. S., and L. D. Muller. 1998. Performance and nutrient intake of high-producing Holstein cows consuming pasture or a total mixed ration. *J. Dairy Sci.* 81: 1403–1411.

11. Loor, J. J., J. H. Herbein, and C. E. Polan. 2002. Trans18:1 and 18:2 isomers in blood plasma and milk fat of grazing cows fed a grain supplement containing solvent-extracted or mechanically extracted soybean meal. *J. Dairy Sci.* 85: 1197–1207.

12. McDaniel, B. T., J. S. Clay, and C. H. Brown. 1999. Variances and correlations among progeny tests for reproductive traits or cows sired by Holstein bulls. *J. Dairy Sci.* 82 (Suppl. 1): 29. (Abstr.)

13. McDowell, R. E. 1982. Crossbreeding as a system of mating for dairy production. Southern Cooperative Series Bulletin No. 259.

14. Muller, L. D., and S. L. Fales. 1998. Supplementation of cool season grass pastures for dairy cattle. p. 335, In: *Grass for Dairy Cattle*. J. H. Cherney and D. J. R. Cherney (ed.). CAB International, Oxon, UK.

15. National Research Council. 2001. *Nutrient Requirements of Dairy Cattle,* 7th rev. ed. National Academy Press, Washington, DC.

16. Novaes, L. P., C. E. Polan, M. L. McGilliard, C. N. Miller, and W. Wark. 1991. Intake of grazing Holstein heifers in response to lasalocid and supplemental protein-energy as compared to drylot diet. *J. Dairy Sci.* 74 (Suppl. 1): 151.

17. Novaes, L. P., C. E. Polan, M. L. McGilliard, and C. N. Miller. 1991. Holstein heifer growth on grass-legume pastures supplemented with rumen undegradable protein. *J. Dairy Sci.* 74 (Suppl. 1): 304.

18. Polan, C. E., and W. A. Wark. 1997. High moisture corn, dry ground corn and zero supplement for grazing cows compared to TMR for milk yield and composition. *J. Dairy Sci.* 80 (Suppl. 1): 159.

19. Schroeder, G. F., G. A. Gagliosko, F. Bargo, J. E. Delahoy, and L. D. Muller. 2004. Fat supplementation on milk production and composition by dairy cows on pasture: A review. *Livestock Prod. Sci.* 86: 1–18.

20. Soder, K. J., and L. A. Holden. 1999. Use of anionic salts with grazing pre-partum dairy cows. *Prof. Anim. Sci.* 15: 278–285.

21. Soder, K. J., and C. A. Rotz. 2003. Economic and environmental impact of utilizing a total mixed ration in Pennsylvania grazing dairy herds. *Prof. Anim. Sci.* 19: 304–311.

22. Washburn, S. P., W. J. Silva, C. H. Brown, B. T. McDaniel, and A. J. McAllister. 2002. Trends in reproductive performance in Southeastern Holstein and Jersey DHI herds. *J. Dairy Sci.* 85: 244–251.

23. Weigel, K. A., and R. Rekaya. 1999. Genetic analysis of male and female fertility in California and Minnesota dairy herds. *J. Dairy Sci.* 82: Suppl. 1: 30. (Abstr).

24. White, S. L. 2000. Investigation of pasture and confinement dairy feeding systems using Jersey and Holstein cattle. M.S. Thesis, North Carolina State University. HTTP://WWW. WORLDCATLIBRARIES.ORG/WCPA/TOP3MSET/ 7FD675425C45A25CA19AFEB4DA09E526.HTML.

CHAPTER 5

1. National Research Council. 1985. *Nutrient Requirements of Sheep*, 6th ed. Washington, D.C.

2. Whittier, W. D., A. Zajac, and S. H. Umberger. 2003. Control of internal parasites in sheep. Virginia Cooperative Extension Publication No. 410-027. HTTP://WWW.EXT.VT.EDU/PUBS/SHEEP/410-028/410-028.HTML.

3. Whittier, W. D., and S. H. Umberger. 1996. Control, treatment, and elimination of foot rot from sheep. Virginia Cooperative Extension Publication No. 410-028. HTTP://WWW.EXT.VT.EDU/PUBS/ SHEEP/410-028/410-028.HTML.

4. Greiner, S. P. 2003. Sheep management schedule. Virginia Cooperative Extension Publication No. 410-365. HTTP://WWW.EXT.VT.EDU/PUBS/SHEEP/410-365/410-365.HTML.

5. Sheep and feeder lamb budgets. Virginia Cooperative Extension. HTTP://WWW.EXT.VT.EDU/DEPART-MENTS/AGECON/SPREADSHEETS/LIVESTOCK/SHEEP. HTML.

CHAPTER 6

1. AFRC. 1998. Technical Committee on Responses to Nutrients, Report No. 10. *The Nutrition of Goats.* CAB International, Wallingford, United Kingdom.

2. Ball, D. M., C. S. Hoveland, and G. D. Lacefield. 2002. *Southern Forages: Modern Concepts for Forage Crop Management,* 3rd ed. Graphic Communications Corp., Lawrenceville, GA.

3. Craddock, F., R. Machen, and T. Craig. 1994. Management tips for internal parasite control in sheep and goats. L-5092. Texas Agricultural Extension Service. The Texas A&M University System.

4. Gipson, T. A. 1995. Meat goat breeds and breeding plans. In: *Meat Goat Production and Marketing Handbook.* Southern States.

5. Haenlein, G. F. W., and D. L. Ace. 1984. *Extension Goat Handbook.* University of Maryland, College Park.

6. Lamand, M. 1981. Métabolisme et besoins en oligo-éléments des chèvres. pp. 210–217, In: P. Morand-Fehr, A. Bourbouze, and M. de Siminae (ed.). *Nutrition and Systems of Goat Feeding.* International Symposium. Vol. 1. ITOVIC-INRA. Tours, France.

7. *National Goat Handbook.* 1997. University of Maryland. HTTP://OUTLANDS.TRIPOD.COM/FARM/ NATIONAL_GOAT_HANDBOOK.PDF.

8. Food Animal Residue Avoidance Databank. 2003. A National Food Safety Project of U.S. Department of Agriculture Cooperative State Research, Education, and Extension Service. North Carolina State University, University of California-Davis, University of Florida. HTTP://WWW.FARAD.ORG/.

9. Hetherington, L., and J. G. Matthews. 1994. *All About Goats.* Farming Press Books. Ipswich, United Kingdom.

10. Luginbuhl, J.-M. 2000. Winter management tips for internal parasites of meat goats. Meat Goat Notes - 22. North Carolina State University, Raleigh.

11. Luginbuhl, J.-M. 2000. Gastrointestinal parasite management of meat goats. Meat Goat Notes - 24. North Carolina State University, Raleigh.

12. Luginbuhl, J.-M. 2000. Basic meat goat facts. ANS 00-606MG. Animal Science Facts. North Carolina State University, Raleigh.

13. Luginbuhl, J.-M. 2000. Breeds and production traits of meat goats. ANS 00-603MG. Animal Science Facts. North Carolina State University, Raleigh.

14. Luginbuhl, J.-M. 2000. Heat detection and breeding in meat goats. ANS 00-607MG. Animal Science Facts. North Carolina State University, Raleigh.

15. Luginbuhl, J.-M. 2000. Preparing meat goats for the breeding season. ANS 00-602MG. Animal Science Facts. North Carolina State University, Raleigh.

16. Luginbuhl, J.-M., J. T. Green, Jr., J. P. Mueller, and M. H. Poore. 1995. Grazing habits and forage needs for meat goats and sheep. pp. 105–112, In: D. S. Chamblee (ed.). *Production and Utilization of Pastures and Forages in North Carolina.* North Carolina Agricultural Research Service Technical Bulletin No. 305.

17. Luginbuhl, J.-M., J. T. Green, Jr., J. P. Mueller, and M. H. Poore. 2000. Forage needs and grazing management for meat goats in the humid southeast. ANS 00-604MG. Animal Science Facts. North Carolina State University, Raleigh.

18. Luginbuhl, J.-M., J. T. Green, Jr., M. H. Poore, and A. P. Conrad. 2000. Use of goats to manage vegetation in cattle pastures in the Appalachian region of North Carolina. *Sheep & Goat Res. J.* 16: 124–135.

19. Luginbuhl, J.-M., T. E. Harvey, J. T. Green, Jr., M. H. Poore, and J. P. Mueller. 1999. Use of goats as biological agents for the renovation of pastures in the Appalachian region of the United States. *Agroforestry Systems* 44: 241–252.

20. Luginbuhl, J.-M., J. P. Mueller, and A. P. Conrad. 2003. Winter annual grasses for meat goats. *J. Anim. Sci.* 81 (Suppl 2): 26.

21. Luginbuhl, J.-M., and M. H. Poore. 2000. Monitoring the body condition of meat goats: A key to successful management. ANS 00-605MG.

22. Luginbuhl, J.-M., M. H. Poore, and A. P. Conrad. 2000. Effect of level of whole cottonseed on intake, digestibility and performance of growing male goats fed hay-based diets. *J. Anim. Sci.* 78: 1677–1683.

23. Luginbuhl, J.-M., M. H. Poore, J. W. Spears, and T. T. Brown. 2000. Effect of dietary copper level on performance and copper status of growing meat goats. *Sheep & Goat Res. J.* 16: 65–71.

24. Luginbuhl, J.-M., J. T. Green, Jr., M. H. Poore, and A. P. Conrad. 2000. Use of goats to manage vegetation in cattle pastures in the Appalachian region of North Carolina. *Sheep & Goat Res. J.* 16: 124–135.

25. Morand-Fehr, P. 1981. Nutrition and feeding of goats: Application to temperate climatic conditions. pp. 193–232, In: C. Gall (ed.). *Goat Production.* Academic Press, London.

26. National Research Council. 1981. *Nutrient Requirements of Goats: Angora, Dairy, and Meat Goats in Temperate and Tropical Countries.* Number 15. National Academy Press. Washington, D.C.

27. Pond, K. R., J.-M. Luginbuhl, and D. S. Fisher. 1995. Grazing animal behavior. pp. 19–21, In: D. S. Chamblee (ed.). *Production and Utilization of Pastures and Forages in North Carolina.* North Carolina Agricultural Research Service Technical Bulletin No. 305.

28. Rayburn, E. B., and S. B. Rayburn. 1976. Feeding your dairy goat. *Dairy Goat J.* 54: 9–10, 13.

29. Smith, M. C., and D. M. Sherman. 1994. *Goat Medicine.* Lea & Febiger, Philadelphia.

30. Spears, J. W. 1995. Minerals in forages. In: G. C. Fahey, Jr., L. E. Moser, D. R. Mertens, and M. Collins (ed.). *Forage Quality, Evaluation, and Utilization.* ASA, CSSA, and SSSA, Madison, WI.

31. Thedford, T. R. 1983. *Goat Health Handbook: A Field Guide for Producers with Limited Veterinary Services.* Winrock International.

32. Zajac, A. 1996. Goat parasites and their control. pp. 1–6, In: *Goat Expo 1996. Theme: Marketing & Economics.* Meat Goat Program Virginia State University.

33. Zajac, A. M., and G. A. Moore. 1993. Treatment and control of gastrointestinal nematodes of sheep. *The Compendium* 15: 999–1011.

CHAPTER 7

1. Aiello, S.E., and A. Mays, (ed.). 1998. *The Merck Veterinary Manual.* Merck & Co., Inc., Whitehouse Station, NJ.

2. Archer, M. 1973. The species preferences of grazing horses. *J. Br. Grassl. Soc.* 28: 123–128.

3. Archer, M. 1978. Studies on producing and maintaining balanced pastures for studs. *Equine Vet. J.* 10: 54–59.

4. Archer, M. 1980. Grassland management for horses. *Vet. Rec.* 107: 171–174.

5. Avery, A. 1996. *Pastures for Horses: A Winning Resource.* RIRCD, Victoria.

6. Barnett, D. T., S. G. Jackson, and J. P. Baker 1985. Endophyte-infected tall fescue effects on gravid mares. In: *Proc. 9th Equine Nutrition and Physiology Soc. Symp.*, Michigan State University, East Lansing, MI.

7. Dorsett, D. J., and D. D. Householder. Horse pastures for Texas. Texas A&M U. Ag. Ext. Serv. Anim. Sci. Unit. HTTP://ANIMALSCIENCE.TAMU.EDU/ SUB/ACADEMICS/EQUINE/HRG006_HPASTURES.PDF.

8. Emmick, D. L., and D. G. Fox. 1993. Prescribed grazing management to improve pasture productivity in NY. USDA Soil Cons. Serv. and Cornell Dept. Anim. Sci.

9. Ensminger, M. E. 1991. *Horses and Tack.* Houghton Mifflin Co., Boston.

10. Evans, J. L. 1995. Forages for horses. pp. 303–311, In: Barnes, R. F., D. A. Miller, and C. J. Nelson (ed.). *Forages, Vol. II: The Science of Grassland Agriculture.* Iowa State Univ. Press.

11. Gallagher, J. R. 1996. The potential of pasture to supply the nutritional requirements of grazing horses. *Austral. Vet. J.* 73: 67–68.

12. Greene, D. L. 1997. Maintaining permanent pastures for livestock. Maryland Coop. Ext. Fact Sheet 720.

13. Henneke, D. R. 1983. Relationship between condition score, physical measurements and body fat percentages in mares. *Equine Vet. J.* 15 (4): 371–372.

14. Henneke, D. R. 1985. A condition score system for horses. *Equine Practice* 7(8): 13–15.

15. Henning, J. C., and W. Loch. 1993. Horse pastures. U. of Missouri-Columbia Ag. Pub. G4695.

16. Henning, J. C. 1994. The establishment and management of horse pastures. In: *Proc. Kentucky Forage and Grassl. Council Ann. Mtg.*, Lexington, KY.

17. Huesner, G. 1995. Common questions and answers about horses. Bulletin 1113. Univ. of Georgia Extension Service. Athens, GA.

18. Johnson, K. D., and M. A. Russell. 1993. Maximizing the value of pasture for horses. Purdue Univ. Coop. Ext. Serv. Forage Information Series, ID-167.

19. Jordan, S. A., K. R. Pond, J. C. Burns, D. S. Fisher, D. T. Barnets, and P. A. Evans. 1995. Controlled grazing of horses with electric fences. In: *Proc. Am. Forage and Grassl. Council,* Vol. 4. Lexington, KY.

20. Kohnke, J. 1992. *Feeding and Nutrition, The Making of a Champion.* Birubi Pacific. Rouse Hill, NSW, Australia.

21. Lewis, L. D. 1995. *Feeding and Care of the Horse.* Williams & Wilkins.

22. McCall, C. A. 1994. Decreasing the costs of feeding horses. Circular ANR-849. Alabama Coop. Extension Service, Auburn Univ.

23. Medica, D. L., M. J. Hanaway, S. L. Ralston, and M. V. K. Sukhdeo. 1996. Grazing behavior of horses on pasture: Predisposition to strongylid infection? *J. Equine Vet. Sci.* 16: 421–427.

24. Moffitt, D. L., T. N. Meacham, J. P. Fontenot, and V. G. Allen. 1987. Seasonal differences in apparent digestibility of fescue and orchardgrass/clover pastures in horses. Equine Nutrition and Physiology Soc. Symp. Colorado State University.

25. National Research Council. 1989. *Nutritional Requirements of Horses*, 5th ed. National Academy Press, Washington, D.C.

26. Nielsen, D. B. 1997. Observations on pasture management and grazing. Utah State Univ. Ext. AG 502.

27. Odberg, F. O., and K. Francis-Smith. 1976. A study on eliminative and grazing behaviour—The use of the field by captive horses. *Equine Vet. J.* 8: 147–149.

28. Sandage, L. J. and B. J. Hankins. 1994. Comparing various grazing management systems. U. of Arkansas Coop. Ext. Serv. FSA 2129.

29. Singer, J. W., N. Bobsin, W. J. Bamka, and D. Kluchinski. 1999. Horse pasture management. *J. Equine Vet. Sci.* 19: 540–545, 585–592.

30. Singer, J. W., W. J. Bamka, D. Kluchinski, and R. Govindasamy. Using the recommended stocking density to predict equine pasture management. *J. Equine Vet. Sci.* 22 (2): 73–76.

31. Singer, J. W., N. Bobsin, D. Kluchinski, and W. J. Bamka. 2001. Equine stocking density effect on botanical composition, species density and soil phosphorus. *Commun. Soil Sci. Plant Analysis* 32 (15/16): 2549–2559.

32. Sukhdeo, M. V. K. 2000. Inside the vertebrate host: Ecological strategies by parasites living in the third environment. pp. 43–62, In: R. Poulin, S. Morand, and A. Skorping (ed.). *Evolutionary Biology of Host-Parasite Relationships.* Developments in Animal and Veterinary Sciences 32. Elsevier.

33. Undersander, D., B. Albert, P. Porter, A. Crossley, and N. Martin. 1993. Pastures for profit: A guide to rotational grazing. U. of Wisconsin-Ext. and Minnesota Ext. Serv. A3529.

34. U.S. Department of Agriculture-Animal Plant Health Inspection Service. 2000. Endophytes in US horse pastures info sheet. HTTP://WWW.APHIS.USDA.GOV/VS/CEAH/NCAHS/NAHMS/EQUINE/EQUINE98/EQ98ENDOPH.PDF. USDA-APHIS, Washington, D.C.

35. U.S. Department of Agriculture-National Agriculture Statistics Service. 1999. 1999 US Equine Inventory Report. HTTP://USDA.MANNLIB.CORNELL.EDU/USDA/NASS/EQUINE/EQUI1999.TXT. USDA-NASS, Washington, DC.

36. Wagoner, D. M. (ed.) 1977. *Veterinary Treatments and Medications for Horsemen.* Equine Research Publications, Dallas, TX.

37. Webb, G. W., B. E. Conrad, M. A. Hussey, and G. D. Potter. 1989. Growth of yearling horses managed in continuous or rotational grazing systems at three levels of forage on offer. *J. Equine Vet. Sci.* 9: 258–261.

38. Wood, C. H. 1994. Managing horses and cattle on horse farms. In: *Proc. Kentucky Forage and Grassl. Council Ann. Mtg.,* Lexington, KY.

CHAPTER 8

1. Bowman, D. D., R. C. Lynn, and M. L. Eberhard. 2003. *Georgi's Parasitology for Veterinarians*, Eighth Edition. WB Saunders Co., Philadelphia.

2. Bairden, K., J. Armour, and J. L. Duncan. 1995. A 4-year study on the effectiveness of alternated grazing of cattle and sheep in the control of bovine parasitic gastroenteritis. *Vet. Parasit.* 60: 119–132.

3. Barger, I. 1997. Control by management. *Vet. Parasit.* 72: 493–506.

4. Coles, G. C., C. Bower, F. H. M. Borgsteede, et al. 1992. World Association for the Advancement of Veterinary Parasitology (WAAVP) guidelines for the detection of anthelmintic resistance in nematodes of veterinary importance. *Vet. Parasit.* 44: 35–44.

5. Courtney, C. H., C. F. Parker , K. E. McClure, et al. 1983. Population dynamics of *Haemonchus contortus* and *Trichostrongylus* spp. in sheep. *Int. J. Parasitology* 6: 557–560.

6. Craig, T.M. and S. E. Wikse. 1995. Control programs for internal parasites of beef cattle. *Comp. Cont. Ed. Pract. Vet.* 17: 579–587.

7. Fleming, S. A., T. M. Craig, R. M. Kaplan, et al. 2006. Anthelmintic resistance of gastrointestinal parasites in small ruminants. *J. Vet. Intern. Med.* 20: 435–444.

8. Hall, Robert D. 1996. "Walk-through trap to control horn flies on cattle," Missouri Agricultural publication G1195. University Extension, University of Missouri. HTTP://MUEXTENSION.MISSOURI.EDU/XPLOR/AGGUIDES/AGENGIN/G01195.HTM

9. Herd, R. P. and L. E. Heider. 1980. Control of internal parasites in dairy replacement heifers by two treatments in the spring. *J. Am. Vet. Med. Assoc.* 17:51–54.

10. Herd R. P. 1983. A practical approach to parasite control in dairy cows and heifers. *Comp. Cont. Ed. Pract. Vet.* 5: S73–S80.

11. Herd, R. P. 1991. Cattle practitioner: Vital role in worm control. *Comp. Cont. Ed. Pract. Vet.* 13: 879–885.

12. Kaplan, R. M. 2004. Drug resistance in nematodes of veterinary importance: a status report. *Trends. Parasitol.* 20: 477–481.

13. Kaplan, R. M., J. M. Burke, T. H. Terrill, et al. 2004. Validation of the FAMACHA eye color chart for detecting clinical anemia in sheep and goats on farms in the southern United States. *Vet. Parasitol.* 123:105–120.

14. Kaplan, R. M., A. M. Vidyashankar, and S. B. Howell. 2007. A novel approach for combining the use of in vitro and in vivo data to measure and detect emerging moxidectin resistance in gastrointestinal nematodes of goats. *Int. J. Parasitol.* 37:795–804.

15. Molento, M. B., J. A. van Wyk, G. C. Coles. 2004. Sustainable worm management. *Vet. Rec.* 155:95–96.

16. Nodtvedt, A., I. Dohoo, J. Sanchez, et al. 2002. Increase in milk yield following eprinomectin treatment at calving in pastured dairy cattle. *Vet. Parasitol.* 105:191–206.

17. Powell, P. K. 1995. "Horn fly biology and management." Integrated Pest Management. West Virginia University Extension Service. HTTP://WWW.IPMCENTERS.ORG/CROPPROFILES/DOCS/WVBEEFCATTLE.HTML

18. Powell, P. K. 1995. "Cattle grub biology and management." Integrated Pest Management. West Virginia University Extension Service. HTTP://WWW.IPMCENTERS.ORG/CROPPROFILES/DOCS/WVBEEFCATTLE.HTML

19. Reinemeyer, C.R. 1990. Prevention of parasitic gastroenteritis in dairy replacement heifers. *Comp. Cont. Ed. Pract. Vet.* 12:761–766.

20. Reinemeyer, C.R. 1992. The effects of anthelmintic treatment of beef cows on parasitologic and performance parameters. *Comp. Cont. Ed. Pract. Vet.* 14:678–687.

21. Reinemeyer, C. R. 1995. Should you deworm your clients' dairy cattle? *Vet. Med.* 90:496–502.

22. Sanchez, J., A. Nodtvedt, I. Dohoo, et al. 2002. The effect of eprinomectin treatment at calving on reproduction parameters in adult dairy cows in Canada. *Prev. Vet. Med.* 56:165–177.

23. Stromberg, B.E. and G. A. Averbeck. 1999. The role of parasite epidemiology in the management of grazing cattle. *Int. J. Parasit.* 29:33–39.

24. Townsend, L. "Biological control of flies." University of Kentucky Entomology. Cooperative Extension Service, University of Kentucky.